FUTURE CROPS

FUTURE CROPS

— Volume 1 —

Editor

Professor K.V. Peter

2011

DAYA PUBLISHING HOUSE®

Delhi - 110 035

ISBN 978-81-7035-753-7 (Series)
ISBN 978-81-7035-754-4 (Volume 1)

Published by : **Daya Publishing House®**
4760-61/23, Ansari Road, Darya Ganj,
New Delhi - 110 002
Phone: 23245578, 23244987
Fax: (011) 23260116
E-mail: dayabooks@vsnl.com
website: www.dayabooks.com

Laser Typesetting : **Classic Computer Services**
Delhi - 110 035

Printed at : **Chawla Offset Printers**
Delhi - 110 052

PRINTED IN INDIA

Acknowledgements

I acknowledge Prof. P.I. Peter, Chairman, NoniBiotech, Chennai for all the physical and spiritual support for editing this volume. I acknowledge Dr. Kirti Singh, Chairperson, World Noni Research Foundation, Chennai for the continued guidance. I also thank my wife Vimala, sons Anvar K. Peter and Ajay K. Peter and Daughters in law Reeba and Cynara for the patience. The help received from Mr. Thanigai Kumar, DTP Section, NoniBiotech is duly appreciated. My thanks are due to all the contributors to the 14 chapters which made the volume rich in content. I thank Dr. H.S. Gupta, Director, IARI New Delhi for the PREFACE and Dr. R.S. Paroda, Former Director General, ICAR, New Delhi for the FOREWORD.

Devotion

I devote the First volume-FUTURE CROPS to my eldest brother Shri. Kuruppacharil Varkey Joseph and sister-in-law Mrs. Mary Joseph, both teachers who showed the way and road to academic enlightenment to me. I chose the solemn Profession of Agriculture as a way of life at the guidance of my eldest brother.

Trust for Advancement of Agricultural Sciences

taas
Progress Through Science

Avenue II, Indian Agricultural Research Institute, New Delhi - 110 012
Phone : 011-65437870; Fax : 011-25843243
E-mail : taasiari@yahoo.co.in; Website : www.taas.in

Foreword

Science and Technology led Horticulture would lead India to meet the basic three Millennium Development Goals (MDG) set to achieve by 2015. Reducing the number of people below poverty line to half (28 to 14%), reducing infant mortality and malnutrition of pregnant women and reducing new lifestyle diseases like diabetes, obesity, cardiovascular diseases, cancer and spread of HIV/AIDS. Shrinking area under agriculture/horticulture, reduced availability of water for irrigation, non availability of farm labour due to migration for high income jobs compounded by general apathy to agriculture/horticulture put policy makers to tight corner the possibility of achieving the set goals. The inter-state and inter country migration throws newer opportunities for type and nature of food. Rice consumption is going down in developed countries and getting reduced in developing countries. The same is true for wheat and barley. Hitherto underexploited and underutilized horticultural crops are getting attention because of their lesser water requirement, higher productivity/unit of time and space and richness in nutrients-micro, macro, antioxidants. These are time tested, region specific, consumer centered and technology backed crops.

At a point of time they were major crops, got relegated to underutilized ones and now projected as FUTURE CROPS. Horticulture of the future is challenging with a lot of opportunities and avenues for enabling and empowering farmers and consumers with increased purchasing power. I envisage 10 cardinal characteristics for a crop to be called FUTURE CROPS. Future of horticulture is riddled with problems of climate change, threats on biodiversity, widening gap between haves and have notes, uneven benefit sharing as a result of Globalization, significant harvest and postharvest losses in perishable crops, lesser roles to farmers in marketing, patented technology(s) for value added final products and a very weak scientists-extension- farmers-consumers linkage. A FUTURE CROP is a continuum crop known for its lesser water requirement, higher productivity/unit of space, can be consumed directly

taas
Progress Through Science

Trust for Advancement of Agricultural Sciences

Avenue II, Indian Agricultural Research Institute, New Delhi - 110 012
Phone : 011-65437870; Fax : 011-25843243
E-mail : taasiari@yahoo.co.in; Website : www.taas.in

or along with other crops after processing. Nutrient recovery and absorption are better and wastes after processing can go as manure to the crop as well. The crop is amenable to processing to value added products.

I congratulate Prof. K V Peter, Former Vice-Chancellor, Kerala Agricultural University and currently the Director, World Noni Research Foundation, Chennai for conceiving a series on FUTURE CROPS. The First volume has 15 chapters dealing with 13 crops and 2 chapters dealing with vegetables of Kashmir and Medicinal and aromatic plants of Mizoram, two hot spots of biodiversity in India. Thirty one scientists from 15 Institutes have contributed scientific chapters to this volume. I congratulate all of them for this useful initiative. I also appreciate the efforts of Daya Publishing House, New Delhi for publishing this volume.

R.S. Paroda

भारतीय कृषि अनुसंधान संस्थान, नई दिल्ली-110012 (भारत)
INDIAN AGRICULTURAL RESEARCH INSTITUTE
(A University Under Section 3 of UGC Act. 1956)
NEW DELHI-110012 (INDIA)

भाकृ अनुप.
ICAR

हरि शंकर गुप्त, पीएचडी (आई.आई.टी), एफ.एन.ए.ए.एस.
निदेशक

H.S. Gupta, Ph.D. (IIT KGP), FNAAS
Director

Phones : (Off.) 011-2573 3367, 2584 3375
(Res.) 011-2584 6774, 2573 3973
Fax : +91-11-2584 6420
E-mail (off.): director@iari.res.in
personal : hsgupta@lycos.com

Preface

Horticultural crops comprise of fruits, vegetables, ornamentals, spices, medicinal plants, aromatic plants, tubers and mushrooms. They are basics to nutritional security and are prerequisites to balanced human diet. Medicinal values of horticultural crops are well known and National Institute of Nutrition has made recommendation on minimum requirement of vegetables (250g/capita/day) and fruits (150g/capita/day). India is known world over as the land of spices. Plantation crops like arecanut, cardamom, coconut, coffee, cocoa, cashew, tea, rubber, palmyrah, oilpalm and cinchona bring much needed foreign exchange to India. India grows an array of fruits like apple, pear, plum, mango, pineapple, banana, sapota, grapes, citrus, papaya, arid fruits like ber, bael and dates. India is also abode to an array of vegetables-roots, fruits, leaf, stem. Despite the favourable soil and climatic conditions of growing the above crops, our intake is one of the lowest in the world leading to acute and endemic malnutrition among women, children and aged. The figures on malnutrition in India is mind bogling and bring no credit to the Nation and its people. Booming population, widening disparity in income, climate change, dwindelling water and space and reduced availability of labour are aggravating already graver issues further fuelled by inflation and economic deprivation. There is need to think over emerging future crops in horticulture meeting the expected future scenarios based on climate change, globalization, urbanization, high technology driven processing methods, astronomical price difference between raw materials and finer processed products. India has made success in processing of spices, rubber, tea, coffee and cashew where the higher values realized at consumer level are transferred to farming level to a significant way. The provisions of Free Trade Agreement under WTO have been used wisely to protect the interest of farmers of rubber, tea, coffee and spices. But in perishable vegetables and fruits, the farmers have lost a lot.

The FUTURE CROPS should meet future challenges in climate, policies, depleted biodiversity, newer life styles and food habits, urbanization and above all economic affordability to majority of people. The First volume of FUTURE CROPS covers 15 chapters dealing 13 crops and 2 chapters dealing vegetables of Kashmir and Medicinal and Aromatic Plants of Mizoram, two hot spots of biodiversity in India.

I congratulate Prof. K.V. Peter and his colleagues for bringing out this volume which have academic excellence.

07/2/2010

H.S. Gupta

Contents

	Acknowledgements	*v*
	Devotion	*vii*
	Foreword	*ix*
	Preface	*xi*
	List of Contributors	*xv*
	Introduction	*xix*
1.	Chayote [*Sechium edule* (Jacq.) Swartz.] *Amish K. Sureja*	1
2.	Cluster Bean (*Cyamopsis tetragonoloba* L.) *H.R. Mahla, B.R. Choudhary and Dheeraj Singh*	12
3.	Date Palm (*Phoenix dactylifera* L.) Under Hot Arid Region *R.S. Singh, R. Bhargava and D.G. Dhandar*	22
4.	Grain Amaranth (*Amaranthus* spp.) *K. Krishnakumary*	40
5.	Kair [*Capparis decidua* (Forsk.) Edgew.] *Dheeraj Singh, H.R. Mahla, M.K. Choudhary, H. Dayal, M.L. Meena and H.S. Jat*	55

6. **Lettuce (*Lactuca sativa* L.)** **79**
Anant Bahadur and K.P. Singh

7. **Lichens and Future Food** **92**
Trupti Mohaptra, R. Valarmathi and G.N. Hariharan

8. **Longan (*Dimocarpous longan* Lour)** **109**
Rajesh Kumar

9. **Noni (*Morinda citrifolia* L.)** **143**
D.R. Singh, Shrawan Singh and R.C. Srivastava

10. **Rambutan (*Nephelium lappaceum* L.)** **172**
Rajesh Kumar

11. **Saffron (*Crocus sativus* L.)** **203**
Mohammad Ali Behdani

12. **Spearmint (*Mentha spicata* L.)** **209**
A. Mishra

13. **Threatened Medicinal and Aromatic Plants of Mizoram** **218**
T.K. Hazarika and B.P. Nautiyal

14. **Underutilized Vegetables of Kashmir** **243**
N. Jabeen, S. Mufti and S.H. Khan

15. **Vegetable Pigeonpea [*Cajanus cajan* (L) Millspaugh]** **260**
C.L.L. Gowda, R.K. Srivastava, K.B. Saxena and R.V. Kumar

Index **277**

List of Contributors

Bahadur, Anant
Indian Institute of Vegetable Research, Varanasi – 221 305, U.P.

Behdani, Mohammad Ali
Saffron Research Group, Faculty of Agriculture, The University of Birjand, P.O. Box: 97175/331, Birjand, Iran
E-mail: mabehdani@yahoo.com

Bhargava, R.
Central Institute for Arid Horticulture, Bikaner – 334 006 (Rajasthan)
E-mail: ciah@nic.in

Choudhary, B.R.
Agricultural Research Station, Rajasthan Agriculture University, Mandor, Jodhpur

Choudhary, M.K.
Krishi Vigyan Kendra, CAZRI, Pali-Marwar, Rajasthan

Dayal, H.
Krishi Vigyan Kendra, CAZRI, Pali-Marwar, Rajasthan

Dhandar, D.G.
Central Institute for Arid Horticulture, Bikaner – 334 006 (Rajasthan)
E-mail: ciah@nic.in

Gowda, C.L.L.
International Crops Research Institute for Semi Arid Tropics (ICRISAT), Patancheru – 502 324, A.P.
E-mail: c.gowda@cgiar.org

Hariharan, G.N.
Lichen Ecology and Bioprospecting Laboratory, M.S. Swaminathan Research Foundation, 3rd Cross Street, Taramani Institutional Area, Taramani, Chennai – 600 113
E-mail: gnhariharan@mssrf.res.in

Hazarika, T.K.
Department of Horticulture, Aromatic and Medicinal Plants, Mizoram University, Aizawl – 796 009
E-mail: tridip28@gmail.com

Jabeen, N.
Division of Olericulture, Sher-e-Kashmir University of Agricultural Sciences and Technology of Kashmir, Shalimar, Srinagar – 191 121, J & K (India)
E-mail: nayeem_jabeen@yahoo.co.in

Jat, H.S.
Senior Scientist, CSSRI, Karnal, Haryana

Khan, S.H.
Division of Olericulture, Sher-e-Kashmir University of Agricultural Sciences and Technology of Kashmir, Shalimar, Srinagar – 191 121, J & K (India)

Krishnakumary, K.
Associate Professor, Kerala Agricultural University, Vellanikkara – 680 656
E-mail: krishnakothottil@rediffmail.com

Kumar, R.V.
International Crops Research Institute for Semi Arid Tropics (ICRISAT), Patancheru – 502 324, A.P.

Kumar, Rajesh
National Research Centre for Litchi, Muzaffarpur – 842 002, Bihar
E-mail: rajeshkr_5@yahoo.com

Mahla, H.R.
Senior Scientist, Central Arid Zone Research Institute, Regional Research Station, Jaisalmer – 345 001, Rajasthan

Meena, M.L.
Krishi Vigyan Kendra, CAZRI, Pali-Marwar, Rajasthan

Mishra, A.
Plant Physiology and Biochemistry, Central Institute of Medicinal and Aromatic Plants, P.O. CIMAP, Lucknow – 226 015
E-mail: amisracimap@yahoo.co.in

Mohaptra, Trupti
Lichen Ecology and Bioprospecting Laboratory, M.S. Swaminathan Research Foundation, 3rd Cross Street, Taramani Institutional Area, Taramani, Chennai – 600 113

Mufti, S.
Division of Olericulture, Sher-e-Kashmir University of Agricultural Sciences and Technology of Kashmir, Shalimar, Srinagar – 191 121, J & K (India)

Nautiyal, B.P.
Department of Horticulture, Aromatic and Medicinal Plants, Mizoram University, Aizawl – 796 009

Saxena, K.B.
International Crops Research Institute for Semi Arid Tropics (ICRISAT), Patancheru – 502 324, A.P.

Singh, D.R.
Central Agricultural Research Institute, Port Blair – 744 101, Andaman and Nicobar Islands
E-mail: drsingh1966@yahoo.com

Singh, Dheeraj
Programme Coordinator, Krishi Vigyan Kendra, CAZRI, Pali-Marwar, Rajasthan
E-mail: dheerajthakurala@yahoo.com

Singh, K.P.
Indian Institute of Vegetable Research, Varanasi – 221 305, U.P.
E-mail: singhab98@gmail.com

Singh, R.S.
Central Institute for Arid Horticulture, Bikaner – 334 006 (Rajasthan)
E-mail: ciah@nic.in

Singh, Shrawan
Central Agricultural Research Institute, Port Blair – 744 101, Andaman and Nicobar Islands

Srivastava, R.C.
Central Agricultural Research Institute, Port Blair – 744 101, Andaman and Nicobar Islands

Srivastava, R.K.
International Crops Research Institute for Semi Arid Tropics (ICRISAT), Patancheru – 502 324, A.P.

Sureja, Amish K.
Senior Scientist, Division of Vegetable Science, Indian Agricultural Research Institute, Pusa, New Delhi – 110 012
E-mail: aksureja@rediffmail.com

Valarmathi, R.
Lichen Ecology and Bioprospecting Laboratory, M.S. Swaminathan Research Foundation, 3rd Cross Street, Taramani Institutional Area, Taramani, Chennai – 600 113

Introduction

Nature bestowed man with about 2. 5 lakh plant species and perhaps the same number of unknown species. Out of this plant wealth, 80000 plant species are edible but man exploited only 3000 of these species for food, shelter and clothing. Commercial scale cultivation is limited to 150 species and consumption for food is limited to 20 different kinds of crops, the world over. Four crops-rice, wheat, maize and potato alone supply about 60 per cent of total calories intake. The food basket is shrinking consequent to monocrop culture and reduced area under cropping. There is reduction in the consumption of organic compounds making ourselves vulnerable to stresses and blockage of immuno regulators in the body system. Awareness to enlarge food basket incorporating edible but so far underutilized and underexploited crops has come up as a positive development in food and nutrition. Traditional and ethnic foods are much valued for its presence of lesser known vegetables, fruits, spices, aromatic plants, tuber crops and emerging edible gymnosperms. A few of the underutilized vegetables contain antinutrient factors like free nitrates and oxalates. They supply micronutrients, essential amino acids, flavourants and pigments. The Indian floristic diversity illustrates 17, 500 angiospermic species belonging to 4000 genera of 315 families which include over 5000 endemic taxa. It has domesticated about 152 species including vegetables, fruits, ornamentals, plantation crops, spices, medicinal plants, aromatic plants and industrial crops. By 2020 India alone requires 200 million tones of vegetables, 180 million tones of fruits to meet the nutrient requirements of 1250 million people. It is foretold that by 2020 availability of land for agriculture/horticulture will get reduced to 50 per cent of the present, irrigation water by 40 per cent and labour availability by 40 per cent. Climate change by 2 degree centigrade will induce early flowering and lesser yield compounded by drought and flood. The urban and peri-urban agriculture/horticulture will be under severe stress of shrinking space, pollution and garbage accumulation. Heavy metal pollution of soil and depletion of soil micro flora would be taking its own toll. New life style diseases like obeisity, diabetes, vardiovascular diseases, HIV/AIDS and cancer would tell up on the quality of resources–soil, water, microflora, biodiversity. Use of plant based renewable sources of energy, feed for milch animals, piggery, poultry, pet animals and birds are putting heavy demand on the currently available food basket.

Future horticultural crops-vegetables, fruits, plantation crops, spices, ornamentals, medicinal and aromatic plants-would extent the food basket to meet the emerging requirements in consonance with changes in climate, human migration, urbanization, limited resources and maximization of efficiency. The future crops are grouped as Energy Crops, Protein crops, Food fibre and mineral crops, Alkaline food crops, Antioxidants, aromatic and medicinal crops, Industrial crops, Novel crops and GM plants at the end of the series. Volume 1 deals with Chayote [*Sechium edule* (Jacq.) Swartz], Cluster bean (*Cyamopsis tetragonoloba* L.), Date palm (*Phoenix dactylifera* L.), Grain amaranth (*Amaranthus spp.*), Kair (*Capparis decidua* (Forsk.) Edgew), Lettuce (*Lactuca sativa* L.), Litchens, Longan (*Dimocarpous longan* Lour.), Noni (*Morinda citrifolia* L.), Ramputan (*Nephelium lappaceum* L.), Saffron (*Crocus sativus* L), Spear mint (*Mentha spicata* L.) and Vegetable Pigeonpea (*Cajanus cajan* L). Two general chapters 'Potential vegetables of Kashmir' and 'Threatened medicinal and aromatic plants of Mizoram' are also included in volume 1.

Chayote, a lesser known member of Cucurbits is gaining popularity and importance as a food crop world wide. Chayote originated in humid tropical regions of Mexico, Central America and the Carribeans. Chayote is basically used for human consumption. Besides fruits, tender shoots are used as green vegetables and edible subterranean tuberous storage roots are also eaten. Chayote has high caloric value resultant of high carbohydrate content in young stems, roots and seed. Chayote may not replace other cucurbits like cucumbers, melons and gourds, it will add to the cucurbit vegetable basket especially in times of cucurbits scarcity.

Cluster bean is a protein rich pod vegetable grown industrially for its gum, the endosperm. An arid crop cultivated under rainfed conditions, it enriches soil with nitrogen in association with rhyzobium bacteria. India is the largest producer of gum in the world, Rajasthan having more than 80 per cent of area under cluster bean in India. The vegetable types with tender pods are rich source of protein, fibre and minerals. Legume vegetables like cowpeas, peas and beans are irrigated crops and as such grown only under copious irrigation. Cluster beans being drought tolerant and a limited water requiring crop has place in protein food basket, so vital in human and animal nutrition. Being grown for gum production, the vegetable version has all the characteristics of taste, flavour, edible fibre content and nutritional value.

Date palm is emerging as an arid fruit of wider adaptability. The commercial date palm growing countries are Iraq, Saudi Arabia, Algeria, Iran, Egypt, Morocco, Tunisia, Pakistan and Sudan besides many other Near East and North African Countries. In India date palms are grown under extremely hot climates in Jaisalmer, Barmer, Bikaner and Jodhpur in Rajastan;Bhatinda, Fazilka, Abihar in Punjab;Sirsa, Hisar, Mahendergarh, Bhiwani in Haryana and Kachchh in Gujarat. India imports date fruits from Gulf countries (0.25 million tones valued US$ 72.36 million during 2007. Per capita consumption of dates in India is negligible as compared to 28.9 kg/capita/year in Saudi Arabia and 28.0 kg/capita/year in Libya. Fresh dates (100g) contain 1.2 g protein, 0.4 g fat, 1.7 g minerals, 3.7 g fibre, 33.8 g carbohydrates, 22.0 mg calcium and 38.0 mg phosphorus. Dates have place in the future mineral and vitamin food basket of developing countries. The Central Institute of Arid Horticulture (CIAH) under ICAR has ambitious programmes on introduction of date palm lines, date palm improvement and area expansion. Water being a limiting factor in future horticulture and salinity looming large, date palm will usher in nutritional security added with purchasing power to growers and processors.

In developing countries around 192 million children and 2000 million others experience micronutrient deficiencies leading to infant mortality, stunted growth and consequent retarded and impeded growth. The National Academy of Sciences, USA selected grain amaranth as one of the four prospective crops from among 400 plants to combat malnutrition. A survey conducted in 1980 identified grain amaranth as one of the five crops which should receive immediate attention as a food crop better adapted to environmental stresses like drought, salinity and heavy metal pollution. It is a multipurpose crop where leaves, stem and grains are utilized. Being a C_4 plant, it is one of the a few dicots in which the first product of photosynthesis is a four carbon compound. The combination of unique anatomical features and C_4 metabolism results in increased efficiency to use CO_2 under a wide range of both temperature and moisture stress environmental conditions. Grain amaranth grows well in hot dry regions since they use lesser amount of water than conventional cereals and millets. In fact they grow the best under dry warm conditions. Amaranth is quite nutritious and the interest in amaranth has recently been aroused due to it's high nutritive value. The grain (seed) is rich in protein (15-18 per cent) and contains lysine (5 per cent) and methionine, two essential amino acids frequently not found in cereals. Amaranth protein has twice the lysine content of wheat protein, three times that of maize and the same as in milk. Supplementation of present day diet with amaranth grains is the economical way of combating protein energy malnutrition. The protein is also relatively rich in the sulphur containing amino acids which are normally limiting in pulse crops. The amaranth leaves are edible with biological value comparable or even superior to common vegetables of popular consumption consisting of less oxalates and nitrates. Amaranth has a wide variety of applications in the food industry. It can be puffed or ground into whole grain flour. As a future crop to be grown under limited water, space and light energy, amaranth is a prospective candidate.

Kair is an indigenous multipurpose shrub which provides food (pickle and vegetable), fodder and fuel and biofence around farms and homesteads. The wood is resistant to termite attack and used for making handles and cartwheels. The mature fruits serve as valuable source of nutrition for villagers of arid and semi arid tracts in Rajasthan. It has fair tolerance to salinity and alkalinity and can help to improve fertility of sand dunes and reduce alkalinity. Uses vary from food, fodder, fuel, timber, oil, medicinal purposes (stomachache, diabetes, constipation, toothache, rheumatism, eczema). As a future crop for the arid and semi-arid tracts, kair has important place. Kair possesses genetic traits useful for developing other horticultural crops for water deficit areas and denuded lands.

Lettuce is a leaf vegetable eaten raw in salads, sandwiches, hamburgers, tacos and in many other dishes. In China the largest producer and Egypt, stems instead of leaves of lettuce are consumed either cooked, raw, pickled, dried or as a sauce. Lettuce contains organic acids, especially citric acid which soothes the nervous system and improves the sleep. Consumption of lettuce reduces blood cholesterol, obesity and diabetes and improves blood structure. Fresh lettuce juice can be taken in chronic ulcers. They also contain vitamins C, B_1, B_2, K and A. The high nutritive values of lettuce make it a potential future crop for micro-nutrient deficient dietary regions. Lettuce is also an ideal crop to be grown under protected cultivation.

Declining land and water resources, increasing pests and pathogens and exploding population necessitate greater demand for food, fibre, medicine and fuel energy. The current agricultural practices emphasize production of three major cereals (rice, wheat and maize) and fail to pay attention to the large scale production of new sources of food (*e.g.* mushrooms, tubers, vascular and non-vascular epiphytes) as future food stuff. Lichens are potential food stuffs with considerable energy, nutrient, medicinal and phytochemical values. Lichens are edible fungi (between fungi and algae) eaten by people in North America, Asia and Africa and perhaps elsewhere. Manna lichen is described in Bible sent miraculously to the Israelites during their forty years of stay in the desserts. Lichens being rich in carbohydrates are used indirectly to make alcohol and also to feed livestock. As a biological system, lichens contain proteins, fats, fibers, vitamins and micro-nutrients. Litchens are used as folk medicines for centuries. The 50 per cent of all lichen species have antibiotic properties. They grow anywhere where a stable and reasonably well–lit surface occurs. Ecology of litchens refers to the place they hold in ecosystems. Nearly 80 per cent of the earths terrestrial surface is covered by litchens. Different lichen species range 15,000 to 20,000 worldwide. In India 2000-2200 species are reported. Being the richest source of carbohydrates, litchen emerges as a very potential biological system for future.

Longan known locally as *kathlitchi* in India is referred as the little brother of litchi. In China, the longan cultivation is running parellel to litchi. Its role in the rural economy is increasing, especially for employment and income generation. Longan is a subtropical tree which grows well in the tropics but requires a prominent change of seasons for flowering. Longan though the least exploited is becoming popular among Asians particularly Chinese. A few in China and Thailand claim the taste of longan equals or is superior to that of litchi. Processing/packing industries can be engaged for longer period even after end of litchi harvesting period for longan processing and value addition. All these factors render longan fruits highly nutritive potential crop for forthcoming period in the Indian subcontinent too.

Noni is called a wonder plant for its nutritional, medicinal, nutraceutical and wellness values. Polynesians used the fruits and leaves for centuries as maintenance food. A lot of divinity was attributed to the crop for its manifold uses against diseases and disorders. Research on this crop is being conducted in over 48 Universities and Institutes. Over 160 nutraceuticals including antioxidants, alkaloids, enzymes and anti-cancer substances are identified. The Chennai based NoniBiotech under the leadership of Prof. P.I. Peter (pipeter@nonifamily.net) has developed several products out of which the Divine Indian Noni is the most sought after. Prof. P I Peter has authored an all inclusive 'monograph on Noni', 'A compendium on Noni research' and a much read book 'All About Noni'. 'A text book on Noni' is also authored by him. Noni cultivation is promoted by Indian Noni Cultivation Council. Indian Horticulture Data Base-2009 published by National Horticultural Board reports that Noni is grown by 348 cultivators in 9 States and Union Territories of India over 250 ha.

Rambutan is yet another evergreen tree with its abundant coloured fruits rich in sugars, vitamins C, B_2 and Calcium. It is an emerging tropical fruit crop in South East Asian countries, Sri Lanka, Vietnam, Malaysia, Zanzibar, Philippines, Indonesia, Taiwan and India. The Rambutan prefers a tropical climate (within 15 degree of the equator) on land below 500 m elevation. The food values of rambutan show the aril as a valuable contributor to our daily nutritional needs. The fruit is taken to relieve diarrohea and dysentery. In Java, the toxic saponin from the seeds is dried and used as a medicine. In Malaysia, roots are used in a decoction for treating fever and the bark as an astringent for tongue diseases. Rambutan roots, bark and leaves have various uses in medicine and in the production of dyes. Rambutan has good prospects as the yield can be increased significantly with improved cultural techniques, better cultivars and postharvest treatment. The fruits have immense potential but require concerted efforts towards R and D on development of suitable cultivars, production techniques and improved management practices at pre and postharvest level in humid and sub-humid tropics.

Saffron is the costliest of all spices because of its high demand and short supply. It is predominantly used to give colour, flavor and aroma to food. Increasing world interest in the use of fragrance, colour, flavor and medicinal sources of plant origin have broadened horizon for production and consumption of saffron. It is grown on commercial scale in Iran, Spain, India, Greece and on a lesser scale in Italy, Morocco and France. Use of saffron to treat different human diseases is emerging. In view of its wide range of medicinal uses, the saffron has undergone extensive phytochemical and biochemical studies and a variety of biologically active ingredients are isolated. In view of its spices values and medicinal properties saffron is emerging as a prospective crop of 21st century.

Mints are a group of plants grown for essential oils. Spearmint (*Mentha spicata* L.) indigenous to Northern England (U.K.) is cultivated in USA in the States of Indiana, Michigan and Oregon. Spearmint is grown in Utrecht in the Netherlands since 1770 and soon spread to Germany, France and other European countries. In India, the cultivation is taken up in Badann and Barabanki districts of U.P. Due to pleasant fragrance of the spearmint oil, it is highly used in luxury soap manufacture in the USA. It is also used as a flavouring agent in confectionaries, tooth pastes, pickles and spices. Spearmint oil is also used as carminative and stimulant in pharmaceutical preparations as a non-bitter intensely sweet flavouring agent. The oil showed antimicrobial activity against pathogenic bacteria. The current and next century will witness growth in fast food industry and ready to eat cookery. Spearmint and mint oil will have heavy demand and more over spearmint is one of the most suited crops for protected cultivation.

Jammu and Kashmir represent temperate agro-eco system so favourable to a set of potential vegetables for earliness, more yield for a longer period, tolerance to biotic and abiotic stresses, sources of vitamins, minerals, proteins and dietary fibre and additional source of income to farmers. Dandelion, mallow, potulaca (common purslane), spiny amaranth, rhubarb rumex, top onion, lotus stem and centaurea are a few such underexploited leaf vegetables of potential for better nutrition. Mizoram in the North-Eastern region (NER) sustains a number of threatened medicinal and aromatic plants. Out of 9 important vegetation types of India, 6 are in NER and according to cumulative estimate the region harbours 8000 flowering plant species including 825 species of orchids, 80 species of rhododendrons, 17 species of citrus with 52 variants, 16 species of banana (*Musa*), 60 species of bamboo and 25 species of canes. Besides 40 gymnosperms, 500 pteridophytes and rich genetic diversity of many other species are also reported from the region. Owing to rich biodiversity, all eight states in NER are included under Indo-Burma hotspot for mega biodiversity. According to Indian Red Data Book, nearly 800 species of the region are already facing severe threats. The threatened Medicinal and Aromatic Plants of NER especially Mizoram are highlighted for scientific attention as a few plants as possible crops for human wellness and welfare.

Protein deficiency is endemic and manifested as stunted growth in developing countries. Majority of food proteins of plant origin are derived from legumes and pulses. Among the legumes, pigeon pea or red gram occupies an important role in rainfed agriculture. Although is grown in 22 countries, major producers are Myanmar, China and Nepal in Asia. Kenya, Malawi, Uganda, Mozambique and Tanzania also grow considerable amount of pigeonpea. In India dehulled split cotyledons of dry pigeonpea seed are cooked to make dal or sambar for eating with chapatti or rice; while in Southern and Eastern Africa and South America, whole dry seeds are used for porridge and its green seeds are used as fresh, frozen or canned vegetable. Broken seeds, husks and pod shells are fed to cattle; while the dry stems are used as household fuel wood. Pigeon pea is credited to be the most suitable crop for subsistence agriculture that needs minimum external inputs. The vegetable bush type is projected as the most promising supplier of protein to the imbalanced diet in developing countries.

The Volume 1 of FUTURE CROPS covers 15 chapters contributed by the working scientists in the field.

K.V. Peter

2011, Future Crops Vol. 1 ***Pages 1–11***
***Editor:* Professor K.V. Peter**
***Published by:* DAYA PUBLISHING HOUSE, NEW DELHI**

Chapter 1
CHAYOTE
[*Sechium edule* (Jacq.) Swartz.]

Amish K. Sureja
Senior Scientist, Division of Vegetable Science,
Indian Agricultural Research Institute, Pusa, New Delhi – 110 012
E-mail: aksureja@rediffmail.com

Chayote, a lesser-known member of the gourd family, is gaining in popularity and importance as a food crop worldwide. Scientific name for chayote is *Sechium edule* (Jacq.) Swartz. Currently, the major chayote growing countries are Mexico, Brazil, Costa Rica, Guatemala, Peru, Dominican Republic, Jamaica, New Zealand, India, Nepal, China, Papua New Guinea, Indonesia, Malaysia, Vietnam, Thailand, Cambodia, Taiwan and some European countries like Italy and Yugoslavia. In India, chayote is cultivated in North-Eastern states (high hills of Meghalaya, Manipur, Mizoram, Nagaland and Sikkim), West Bengal (Darjeeling), Tamil Nadu (Madurai, Dindigul and Nilgiri districts) and coastal and low hills (1000-2000 m) of Maharashtra and Karnataka.

Costa Rica is the leading trader and exporter of chayote and exports mainly to the United States, Canada, Colombia, Guatemala, Netherlands, Antilles, Nicaragua, Honduras, France, Germany, Great Britain, the Netherlands, Belgium, Luxembourg and Japan. Mexico is in second place. Exports from other countries like Peru, Dominican Republic and Brazil are in far smaller quantities.

Synonyms

Chayote is widespread throughout many regions of the world, hence known by a wide variety of names in different languages and countries given in Table 1.1

Uses and Properties

Chayote is eaten in many ways and can be casseroled, creamed, buttered, curried, fried, scalloped, stuffed, baked, frittered, boiled, mashed, used in salads, or in pies, salted or pickled to suit different palates (Aung *et al.*, 1990). Commercially, the biggest use appears to be for pickling.

Table 1.1

Country	Common Names
Mexico	'apupo', 'apopu', 'niktin', 'naña' 'itú-tse', 'jit-jiap', 'yape', 'aj-shá', 'rign', 'nñ', 'mishi', 'cal-mishi', 'tzihu', 'tzihub', 'mú-u', 'shamú', 'xamú', 'chayoj', 'chayojtli', 'macltucún', 'multucún', 'huisquilitl'
Guatemala and El Salvador	'güisquil', 'bisquil', 'huisquil', 'chuma', 'chima', 'chimaa', 'huisayote', 'güisayote', 'perulero'.
Honduras	'huisquil', 'patasté', 'patastilla'
Nicaragua	'chaya'
Costa Rica	'pís', 'pog-pog-iku', 'seuak', 'surú', 'tsua-uá'.
Cuba, Puerto Rico, the Dominican Republic, Jamaica	'tayote', 'tayón', 'chocho', 'chiote'
Colombia	'cidrayota', 'chayota', 'papa de pobre' (poor man's potato)
Venezuela	'chayoto'
Brazil	'cho cho', 'xuxu', 'chuchu'
Argentina	'cayota', 'papa del aire' (air potato)
Peru	'gayota'
USA	'christophine', 'mirliton'
Bolivia	'zapallo'
Ecuador	'achocha', or 'achojcha'
Portugal	'chahiota', 'cahiota', 'caiota', 'pepinella', 'pipinella
England, Australia, Madagascar, Reunion and Mauritius	'chocho', 'choko', 'chocho', 'chowchow', 'chouchou', 'chouchoute'
Russia	'cajot'
Italy	'zucca', 'zucca centenaria'
India	'vilaiti vanga'
Indonesia	'leong-siam'
Malaysia	'labooh selyem'
Java	'labooh tjena'
Cambodia and Vietnam	'su-suu'
Laos	'savëëx', 'nooy th'ai'
Thailand	'ma-kheua-kreua', 'aeng-kariang'

Chayote is basically used for human consumption. Besides the fruits, tender shoots are used as green vegetables and edible subterranean tuberous storage roots are also eaten. They are eaten as a vegetable and are either boiled or used in stews and desserts. The softness of the fruit flesh makes it suitable for giving consistency to baby foods, juices, sauces and pastes. Because of the flexibility and strength of the stems, they are used in some places in handicrafts to make baskets and hats. In India, as in the Americas, the fruits and roots are not only used as food but also as fodder for cattle (Chakravarty, 1990). Fruits can be dehydrated to increase the shelf-life. Jam is also prepared from fruits. Vines have ornamental or fodder value.

Medicinal use of chayote is documented in literature. The decoctions from the leaves or fruits are used to relieve urine retention and burning during urination or to dissolve kidney stones, and as a treatment for arteriosclerosis and hypertension (Lira, 1988; Flores, 1989; Yang and Walters 1992). The leaves and seeds have diuretic properties. The leaves and fruits have cardiovascular and anti-inflammatory properties.

Leaf and seed extracts of chayote are exploited as biopreservatives in food applications as well as for health supplements or functional food, to alleviate oxidative stress (Ordonez *et al.*, 2006). Peroxidases from chayote have a great potential for decontamination of waste waters containing phenolics while immobilized into an insoluble organic support (Villegas-Rosas *et al.*, 2003). A ribosome-inactivating protein (RIP), sechiumin from chayote has use for the preparation of immunotoxin as a potential cancer chemotherapeutic agent (Wu *et al.*, 1998).

Nutritional Value

The edible parts of *S. edule* are relatively low in fibre, protein and vitamins compared with other vegetables. Nevertheless, they have a high caloric value and carbohydrate content, especially in young stems, root and seed, and the micro and macro-nutrient content of the fruit are adequate. The fruits and the seed are rich in several important amino acids like aspartic acid, glutamic acid, alanine, arginine, cysteine, phenylalanine, glycine, histidine, isoleucine, leucine, methionine (only in the fruit), proline, serine, tyrosine, threonine and valine (Flores, 1989). Edible tubers are rich in starch. The fruit has low edible sugar content.

Siciliano *et al.* (2004) reported that the highest total amount of flavonoids in chayote is in the leaves (35.0 mg/10 g of dried part), followed by roots (30.5 mg/10 g), and finally by stems (19.3 mg/10 g). Dutta (1994) reported chayote fruit is rich in calcium (17-140 mg/100g), and is a good supplement in tropical diets.

Table 1.2: Chemical composition (per cent or mg/100 g) of fruit, young stems and roots of *Sechium edule*

Components	*Fruit*	*Stem*	*Root*
Calories	26.0–31.0	60.0	79.0
Humidity (per cent)	89.0–93.4	89.7	79.7
Soluble sugar (per cent)	3.3	0.3	0.6
Starch (per cent)	0.2	0.7	13.6
Proteins (per cent)	0.9–1.1	4.0	2.0
Fats (per cent)	0.1–0.3	0.4	0.2
Carbohydrates (per cent)	3.5–7.7	4.7	17.8
Fibre (per cent)	0.4–1.0	1.2	0.4
Ashes (per cent)	0.4–0.6	1.2	1.0
Ca (mg)	12.0–19.0	58.0	7.0
P (mg)	4.0–30.0	108.0	34.0
Fe (mg)	0.2–0.6	2.5	0.8
Thiamin (mg)	0.03	0.08	0.05
Riboflavin (mg)	0.04	0.18	0.03
Niacin (mg)	0.4–0.5	1.1	0.9
Ascorbic acid (mg)	11.0–20.0	16.0	19.0
Vitamin A (mg)	5.0	615.0	–

Sources: Engels, 1983; Aung *et al.*, 1990.

Origin and Distribution

Chayote originated in humid tropical regions of Mexico, Central America and the Carribean. Newstrom (1990) proposed Chayote was domesticated in Mexico and Guatemala in pre-Columbian times either from wild forms of the species or from its closest relative *S. compositum*. The widest variety of cultivated chayote is found in Southern Mexico, Guatemala and Costa Rica, at altitudes of 500-1500m.

The Spanish introduced chayote into Costa Rica (Bukasov, 1981). The chayote is introduced into South America during the 18th and 19th centuries (Bukasov, 1981; Newstrom, 1991). It was introduced into the Caribbean prior to this. During this same period, chayote was introduced into Europe and, from there, it was taken to Africa, Asia and Australia. However, it was not introduced into the United States until the end of 19th century (Cook, 1901; Flores, 1989; Newstrom, 1991).

Climate and Soil

In India, chayote is traditionally cultivated in empty patches, backyards and market gardens. It is a medium- to high-altitude crop (800-1800 m amsl) and requires a high relative humidity (80-85 per cent), well-distributed annual precipitation of at least 1500-2000 mm and 12 hours daylight to initiate flowering. An elevation of 1000-1200 m amsl is the best for the crop. The most suitable average temperature for crop growth is 13-21°C. Temperature less than 13°C damages young fruits while temperature above 28°C favours excessive growth.

It is highly sensitive to prolonged hard frost. In temperate regions, it is suggested that young plants are raised from seed, under protected cultivation with transplanting outdoors when the danger of frost has passed. After 6-8 weeks of growth, vines should be shaded to give an 8-h photoperiod for 4-6 weeks. Following flower induction, the vines can be grown under natural daylengths to produce a reasonable fruit yield (Aung *et al.*, 1990)

Chayote yields greater in deep soil with plenty of organic matter. It is highly susceptible to frosts, droughts and excessive humidity. The edible starchy tubers are distributed irregularly in the soil. These act as nutrient reserves when the plant sprouts at the start of the rainy season, as, during the dry season, the above-ground parts mostly die away.

Taxonomy

Jeffrey (1990) placed the genus *Sechium*, to which chayote belongs, in the subtribe Sicyinae of the tribe Sicyeae. The members of this subtribe are characterized by having spiny pollen, a single pendulous ovule and single-seeded fruits. This species was originally discovered by Browne (1756) in Jamaica. Swartz (1800) became the first to include this species in *Sechium*, when he proposed the combination by which it is still known, *S. edule* (Jacq.) Swartz.

Sechium is a well-defined genus composed of 11 species. Of these, nine are wild species distributed throughout central and southern Mexico, up to Panama. Of the two remaining species, *S. tacaco* is only cultivated in Costa Rica, and the other, *S. edule*, is cultivated worldwide in tropical and subtropical climates, with wild populations in Southern Mexico.

Taxonomic Classification of the Genus *Sechium* P.Br.

Section Sechium

1. *S. compositum* (J.D. Smith) C. Jeffrey
2. *S. chinantlense* Lira & Chiang

3. *S. edule* (Jacq.) Swartz (wild and cultivated)
4. *S. hintonii* (P.G. Wilson) C. Jeffrey
5. *S. tacaco* (Pittier) C. Jeffrey
6. *S. talamancense* (Wunderlin) C. Jeffrey

Section Frantzia

1. *S. panamense* (Wunderlin) Lira & Chiang
2. *S. pittieri* (Cogn.) C. Jeffrey
3. *S. venosum* (L.D. Gómez) Lira & Chiang
4. *S. villosum* (Wunderlin) C. Jeffrey
5. *Sechium* sp.

Botany and Reproductive Biology

Plant Habit

The chayote is a robust perennial, monoecious, vigorous herbaceous creeper. It grows from a single thick root which produces adventitious tuberous roots. The stems are angular-grooved, glabrous, thick woody towards the base and thin, firm, herbaceous towards the apex. The leaves are strongly three- angled, rough textured, and deep green with white veins. They are glabrous and have grooved petioles. Leaf margin is totally to slightly dentate. Plants have sturdy, branched, furrowed and glabrous tendrils which provide support.

Flower

They are unisexual. Inflorescence is 10-30 cm long erect raceme. Male/staminate flowers are arranged in fascicular or subracemose clusters. The pedicels are 1-2 mm long and are puberulent. Sepals are five and narrow triangular. They are 4 mm long and almost 1 mm wide. Petals are five, green to greenish-white, widely triangular, obtuse to acute, 6-7 mm long and 2-3 mm wide. Stamens are five. Filaments are fused along almost all of the length, forming a thick column, which normally separates into five short branches. The anthers are oblong and develop at the apex of the short branches of the filaments. About 10 pore-like uncovered nectaries are found at the base of the receptacle surrounding the staminal column.

The pistillate flowers are light to cream-yellow and are usually solitary. They develop in the same axilla as the staminate ones. Occasionally pistillate flowers might grow in pairs or three on rare occasions. The pedicel is thin, grooved, glabrous and is 1-3.5 cm long, growing up to 8-9 cm in the fruit. The styles are joined together in a thin column. The stigma is subglobose. The nectaries of the receptacle base are similar to those of the staminate flowers.

Among cultivated types, variations are found in the rate of production of staminate and pistillate flowers. This may be due to genetic, environmental, seasonal factors and plant age.

Pollination

Pollination is carried out by several insect species. Bees of the genus *Trigona* and *Apis mellifera* are the most common pollinators. Secondary pollinators of chayote include wasps from the genera *Polybia, Synoeca* and *Parachrataegus*.

Eswarappa *et al.* (2005) reported that the honey bees *Apis dorsata, A. cerana, A. florea* and *Trigona iridipennis* account for more than 82 per cent of the total insect pollinators visiting chow-chow. The activity of all insect pollinators is observed during mid-morning (10.00-11.00 h).

Fruit and Seed

The fruits are fleshy or fleshy-fibrous, single-seeded and grow either individually or in pairs. They may have 5-7 longitudinal ridges or furrows. They may have woody ridges or lenticels on the surface, especially when ripe. Fruit shape may be globose, ovoid, subovoid, pyriform, or elongated pyriform with sizes 4.3-26.5 cm long and 3-11 cm wide. Fruit colour ranges from white, light green, dark green to pale yellow. They may be unarmed and smooth, or with varied indumentum or armature/ prickles. The pulp is pale green or whitish and tastes sweet or insipid in cultivated plants.

The average total fruit yield is 50-60 tons/ha. A single plant bears 70-100 fruits (30-50 kg) during the 1st year, increasing to 500-600 fruits in the 2nd and 3rd years (Dutta, 1994). Fruits for human consumption are harvested 28-36 days after fertilization, before viviparous germination starts.

The fruit has deep fissure at the blossom end, which is the site for seed. The seed is large, 3-5 cm long, white, ovoid, compressed/flat and smooth. It is viviparous (germinates within the fruit when the fruit is still on the plant).

Cytogenetics

The cultivated chayote is diploid with 24 chromosomes (2n=24) (Sobti and Singh, 1961; Newstrom, 1990). The chromosome size ranges from 0.7 to 1.9 µm (Donato *et al.*, 1994).

Pests and Diseases

Tetranychus mite (*Tetranychus urticae*) is one of the important pests which appear during the dry season and causes yellowing of the leaves and clear scabs on the fruit. This pest transmits the fruit disease known as 'sarna' or 'roña' (scabies/mange) since the scabs increase as the plants become more infested (Vargas, 1991). Secondary pests include *Diaphania hyalinata* and *D. nitidalis* (pickle worm), which perforate unripe fruit, opening up ways of access for fungal and/or bacterial diseases. *Aphis* spp. and *Bemisia tabaci* suck the leaves, fruit peduncles and occasionally the fruit, causing dark blemishes which make the fruit unmarketable. Spotted cucumber beetle(*Diabrotica undecimpunctata*) mainly attacks the fruit and also the seedling. The most important nematode is *Meloidogyne incognita* which causes rotting of the roots.

The most important pathogens include *Ascochyta phaseolorum* and various species of *Fusarium, Macrophomina* and *Colletotrichum*. Mandal *et al.* (2004) reported occurrence of a new yellow mosaic disease of chayote in West Bengal, associated with a begomovirus. The leaves of affected plants develop yellow spots, mosaic and upward curling with occasional enations. Powdery mildew was also reported from Manipur (Kabitarani and Bhagirath, 1991). The insect, *Empoasca solana* is the vector of chayote curl disease.

Since the fruits are made unmarketable by many of these and other diseases, it is of vital importance for the commercial development of the crop where more efforts are made to control such diseases.

Breeding Objectives

The above information makes clear that the commercial production of chayote is affected by pests and diseases. Accordingly, it is necessary to develop a breeding programme that takes into consideration these aspects.

The other main objectives of chayote breeding are:

1. To systematically collect germplasm for the entire genus *Sechium*.
2. To evaluate germplasm collected from various countries, to select plants demonstrating adaptability to annual cultivation (early maturity, compact habit), marketable qualities (excellent flavour, appropriate size and texture), root qualities (length, starch content) and desirable agronomic characters (high yield, resistance to drought and heat).
3. To develop varieties suitable for processing.
4. Study of seed germination of wild and cultivated species in the genus and the vegetative propagation of wild species and of the other cultivated species of *Sechium*.
5. To develop cultural practices to optimize fruit, tuber or shoot production in tropical, subtropical and temperate regions.
6. Creation of a database consisting of information related to characterization, conservation, improvement and use.
7. To evaluate wild species for some of the objectives mentioned above, particularly disease resistance.
8. To disseminate the improved lines and agronomic information to subsistence and commercial growers, research scientists and government agencies.
9. *In situ* conservation of wild species and development of efficient methods for the phytosanitary handling of collections.

Breeding Methods

Little or nothing is known about the heredity of many morphological features of chayote.

Rai *et al.* (2005) studied genetic variation among 8 genotypes of *S. edule* for 8 traits (leaf length, fruit length, fruit diameter, pulp thickness, TSS [total soluble solids], acidity, juice (per cent) and yield) in Meghalaya. The phenotypic coefficient of variation (PCV) was higher than the genotypic coefficient of variation (GCV). Leaf length, fruit pulp thickness, TSS, juice (per cent) and yield were the least affected by the environment, whereas fruit length, fruit diameter and acidity were highly affected by the environment. The highest GCV and PCV were recorded for fruit yield, suggesting that this trait may be improved by simple selection. All traits were positively correlated with yield at the phenotypic and genotypic levels. Leaf length was positively correlated with pulp thickness at the genotypic level. Fruit length showed a highly significant positive correlation with fruit diameter, acidity and fruit yield at the genotypic level. Leaf length, fruit diameter, juice (per cent) and TSS (Total soluble solids) exhibited positive and high direct effects on fruit yield.

Singh *et al.* (2002) studied the nature and magnitude of genetic variability for 11 traits of *S. edule* (number of days to sprouting, vine length, leaf width, leaf length, fruit diameter, fruit length, petiole length, fruits/plant, days to 50 per cent flowering, days to 50 per cent fruit set and fruit weight) using 11 genotypes in Nainital(Uttarkhand). The PCV was higher than the GCV. The highest PCV and GCV were recorded for fruits/plant. Maximum heritability and genetic advance were recorded for fruit diameter. The days to sprouting and fruit weight were also characterized by high heritability. Fruit diameter, fruits/plant, days to 50 per cent flowering, days to 50 per cent fruit set and fruit weight comprised the major yield components.

Yadav *et al.* (2005) evaluated 26 diverse genotypes of chow-chow from Arunachal Pradesh and West Bengal for fruit shape, size, colour and presence of spines. Most of the genotypes collected from

Meghalaya had high fruit polar diameter. Fruit weight was the lowest in the genotype Sikkim collection-3 (79.3 g) and the highest in Meghalaya collection-5 (734.3 g). Genetic and phenotypic coefficients of variation were high for fruit equatorial diameter (top) and fruit weight, indicating possibility of improvement of these characters through simple selection. Broad-sense heritability was high for all characters, indicating the major role of genotype in their expression. Genetic advance was low in most of the characters, but it was high in fruit weight, indicating the involvement of additive gene effect in the expression of this character. Correlation studies showed that all the characters showed positive and significant correlation with fruit weight.

Genetic Resources and Conservation

Descriptor

The descriptor for characterization and evaluation of chayote germplasm includes the following traits:

Length (cm), Width (cm), Thickness (cm), Weight (g), Volume (cm^3), Colour (white, light green, dark green, yellow), Shape (pyriform, subpyriform, ovoid, obovoid, flattened, spheroid, elliptic), Spines (absent,a few, intermediate, many), Distribution of spines (no spines, entire fruit, apex), Furrows (absent, shallow, intermediate, deep), Ridges (absent,a few, intermediate, many), Lenticels (absent, a few, intermediate, many), Texture of pulp (solid, soft), Taste of pulp (simple, sweet, insipid, tasty, salty), Fiber of pulp (absent, a few, many), Days to harvest and Fruits/plant.

Diversity and Genetic Resources

A few cultivated species of *Sechium* produce fruits with diverse morphology, mainly with regard to form, size, surface texture and, to a lesser degree, colour. In addition to this morphological diversity, yield also varies considerably. This reflects the fact that this genus appears to have enormous potential for genetic resources. Very little efforts are made to conserve and study these resources. The endocarpic and precocious germination of chayote seed seriously hinders its conservation. It has to be conserved in field genebanks which are expensive and complicated to administer and maintain, or *in vitro*.

The attempts to classify the diversity of cultivated chayote were so far neither systematic nor the variation observed was completely representative. However, genebanks of chayote housing diverse accessions are maintained in Nepal; at CATIE, Turrialba, Costa Rica and at Regional University Centre in Huatusco, Veracruz, Mexico.

Maximum genetic diversity of chayote is in Central America, where this diversity is declining due to forest destruction (Dutta, 1994).

In vitro Conservation

Seed propagation in chayote does not allow the production of morphologically homogeneous fruits (colour, shape and epidermal texture), which results in high percentage of rejected fruits during grading and packing. The establishment of a protocol for vegetative propagation of this crop will help to solve the heterogeneity problems. Wang *et al.* (1997) successfully regenerated chayote plants from hypocotyle on MS medium supplemented with 0.1 mg IAA, 1 mg BA [benzyl adenine] and 200 mg LH [luteinizing hormone]/litre. Among meristems, apices and shoot segments with an axillary bud, the shoot apices showed the best growth response (Somarribas *et al.*, 1991). Venutolo (1990) established micropropagation protocol based on the use of 5 mm long microcuttings cultured on supplemented MS medium. The *in vitro* conservation treatment under an osmotic pressure of 6 per cent sucrose and at a temperature of 18°C caused the least damage to explants and no morphological alterations.

Biochemical Markers

Abdelnour and Rocha (2008) established protocols for the analysis of genetic diversity in chayote by using isozyme markers and determined the level of genetic diversity present in 42 accessions of chayote from Costa Rica. Eight of the 14 loci examined were polymorphic.

Varieties and Hybrids

No named varieties of chayote are available in India. Local types such as Long white, Round white, Broad green, Pointed green, Creamy green, etc. are grown.

Future Breeding Needs

Much information are lacking on the genetic potential of wild, related species and the different landraces of chayote with respect to pests and diseases resistance. Some of the wild species can be a potential source to breed for disease and pest resistance. The taxonomic relationships between the species of *Sechium*, as well as many other biological aspects of the genus need to be explored.

The representative collections of the genetic diversity of all the species in the genus *Sechium* should be explored and promising materials identified. Efforts should be made for maintenance and handling of such collections. The *ex situ* and *in situ* conservation of wild species must receive attention. More studies need to be carried out on how the wild species are related to cultivated chayote and what is their potential as genetic resources.

Genetic erosion of chayote and related wild species is occurring. The genetic diversity of this crop is currently more or less being saved by traditional agriculture. If traditional farmers abandon their traditional production methods or decide to produce other crops, then it will impose a great threat to the genetic diversity in this crop. Therefore, preservation of genetic diversity in the centre of origin of the species should receive high priority.

Attempts should be made to explore not just food but also for the use of the fruit pulp in industrial processes. Research on the medicinal uses and other aspects of chayote should be looked into. The principal limitation for the industrial development and export of chayote is its short storage life. Chayote is viviparous and has a tendency to sprout under postharvest storage conditions. The precocious sprout development shortens storage life, lessens fruit quality (texture) and diminishes its value for sale and consumption. Tests should be conducted to determine the optimum temperature and humidity conditions in cold chambers for long term storage of the fruits. Currently, various parts of the plant are only processed on a small-scale. Research on various methods of processing chayote is required.

References

Abdelnour, A. and Rocha, O.J., 2008. Genetic characterization of a collection of chayote, *Sechium edule* (Jacq.) Swartz, in Costa Rica by using isozyme markers. *Genetic Resources Crop Evol.*, 55(1): 163–170.

Aung, L.H., Ball, A. and Kushad, M., 1990. Developmental and nutritional aspects of chayote (*Sechium edule*, Cucurbitaceae). *Econ. Bot.*, 44(2): 157–164.

Browne, P., 1756. *Civil and Natural History of Jamaica*. London, England.

Bukasov, S.M., 1981. Las Plantas Cultivadas de México, Guatemala y Colombia. CATIE–GTZ, Turrialba, Costa Rica.1a Ed. [In Spanish] (J. León, Transl.). 173 pp.

Chakravarty, H.L. 1990. Cucurbits of India and their role in the development of vegetable crops. In: *Biology and Utilization of the Cucurbitaceae* (D.M. Bates, R.W. Robinson, and C. Jeffrey, eds.). Cornell University Press, Ithaca, NY. pp. 325–334.

Cook, O.F., 1901. The chayote: A tropical vegetable. US Dept. *Agric. Div. Bot. Bull.*, 18: 1–31.

Donato, M.de., Cequea, H. and De-Donato, H., 1994. A cytogenetic study of six cultivars of the chayote, *Sechium edule* Sw. (Cucurbitaceae). *J. Hered.*, 85(3): 238–241.

Dutta, O.P., 1994. Genetic diversity in cucurbits: chow–chow. *Indian Horticulture*, 38(4): inside front cover, 36.

Engels, J.M.M., 1983. Variation in *Sechium edule* Sw. in Central America. *J. Am. Soc. Hort. Sci.*, 108: 706–710.

Eswarappa, G., Kuberappa, G.C., Sreenath, T.N., Gavi-Gowda and Sreeramulu, K.R., 2005. Relative abundance, foraging activity and pollinating efficiency of different pollinators visiting chow–chow (*Sechium edule*). *J. Ecobiol.*, 17(6): 509–516.

Flores, E., 1989. El chayote, *Sechium edule* Swartz (Cucurbitaceae). *Rev. Biol. Trop.*, 37(Suplemento 1): 1–54.

Jeffrey, C., 1990. Appendix: An outline classification of the Cucurbitaceae. In: Biology and Utilization of the Cucurbitaceae (D.M. Bates, R.W. Robinson and C. Jeffrey, eds.). Cornell University Press, Ithaca, NY. pp. 449–463.

Kabitarani, A. and Bhagirath, T., 1991. Powdery mildew of cucurbits in Manipur. *Indian Phytopathol.*, 44(1): 137–139.

Lira, R., 1988. Cucurbitaceae de la Península de Yucatán: Taxonomía y Etnobotánica. Tesis Maestría en Ciencias (Ecología y Recursos Bióticos), Instituto Nacional de Investigaciones sobre Recursos Bióticos. 329 pp.

Mandal, B., Mandal, S., Sohrab, S.S, Pun, K.B. and Varma, A., 2004. A new yellow mosaic disease of chayote in India. *Plant Pathol.*, 53(6): 797.

Newstrom, L.E., 1990. Origin and evolution of chayote, *Sechium edule*. In: *Biology and Utilization of the Cucurbitaceae* (Eds.) D.M. Bates, R.W. Robinson, and C. Jeffrey. Cornell University Press, Ithaca, NY. pp. 141–149.

Newstrom, L.E., 1991. Evidence for the origin of chayote *Sechium edule* (Cucurbitaceae). *Econ. Bot.*, 45(3): 410–428.

Ordonez, A.A.L, Gomez, J.D, Vattuone, M.A. and Isla, M.I., 2006. Antioxidant activities of *Sechium edule* (Jacq.) Swartz extracts. *Food Chemistry*, 97(3): 452–458.

Rai, N., Yadav, R.K. and Yadav, D.S., 2005. Genetic variability studies in chow–chow (*Sechium edule* Sw.). *Vegetable Science*, 32(2): 184–186.

Siciliano, T., Tommasi, N.de., Morelli, I. and Braca, A., 2004. Study of flavonoids of *Sechium edule* (Jacq) Swartz (Cucurbitaceae) different edible organs by liquid chromatography photodiode array mass spectrometry. *J. Agricultural Food Chem.*, 52(21): 6510–6515.

Singh, R.K., Verma, S.K., Arya, R.R. and Muneem, K.C., 2002. Genetic variability in chow–chow (*Sechium edule*). *Progressive Horticulture*, 34(1): 92–94.

Sobti, S.N. and Singh, S.D., 1961. A chromosome survey of Indian medicinal plants. Part I. *Proc. Indian Acad. Sci.*, 54: 138–144.

Somarribas, G., Sandoval, J. and Muller, L., 1991. *In vitro* vegetative propagation of chayote (*Sechium edule* (Jacq.) Sw.). Establishment phase. *Turrialba*, 41(4): 538–544.

Swartz, O., 1800. *Flora Indiae Occidentalis*. Leiden.

Vargas, E., 1991. Chayote *Sechium edule* (Jacq.) Swartz, Cucurbitaceae. In: Aspectos Técnicos sobre Cuarenta y Cinco Cultivos Agrícolas de Costa Rica. Ministerio de Agricultura y Ganaderia, Dirección General de Investigación y Extensión Agrícola. San José de Costa Rica. pp. 327–336.

Venutolo, S.A., 1990. Micropropagation and study of some options for germplasm conservation of chayote (*Sechium edule* Jacq. Sw.) *in vitro*. xvii +133 pp.; 13 pp.

Villegas-Rosas, M.L.O., Geissler, G., Handal–Silva, A. and Gonzalez–Vergara, E., 2003. Immobilization of a peroxidase from chayote (*Sechium edule* (Jacq.) and its potential use in the removal of phenolic compounds from contaminated water. *Revista Internacional de Contaminacion Ambiental*, 19(2): 73–81.

Wang, X.S., Li, B.X. and Wang, G.D., 1997. Regeneration of plants from hypocotyl of *Sechium edule* Swartz. *Rep. Cucurbit Genet. Coop.*, 20: 65.

Wu, T.H., Chow, L.P. and Lin, J.Y., 1998. Sechiumin, a ribosome–inactivating protein from the edible gourd, *Sechium edule* Swartz. Purification, characterization, molecular cloning and expression. *European J. Biochem.*, 255(2): 400–408.

Yadav, R.K., Rai, N., Yadav, D.S. and Asati, B.S., 2005. Genetic variability and correlation studies for fruit characters in chow-chow (*Sechium edule*). *Horticultural J.*, 18(2): 106–109.

Yang, S.L. and Walters, T., 1992. Ethnobotany and the economic role of the Cucurbitaceae of China. *Econ. Bot.*, 46: 349–367.

2011, Future Crops Vol. 1 **Pages 12–21**
***Editor:* Professor K.V. Peter**
***Published by:* DAYA PUBLISHING HOUSE, NEW DELHI**

Chapter 2

CLUSTER BEAN (*Cyamopsis tetragonoloba* L.)

H.R. Mahla[1], *B.R. Choudhary*[2] *and Dheeraj Singh*[3]
[1]*Central Arid Zone Research Institute, Regional Research Station, Jaisalmer – 345 001*
[2]*Agricultural Research Station, Rajasthan Agriculture University Mandor, Jodhpur*
[3]*Programme Co-ordinator, Krishi Vigyan Kendra, CAZRI, Pali-Marwar*
E-mail: dheerajthakurala@yahoo.com

Cluster bean (guar) (*Cyamopsis tetragonoloba* L.) is a major arid legume traditionally cultivated as a system sustaining crop under poorly endowed situations with the least management and aftercare in arid regions. Besides, it is a subsistence source of vegetable, food, fodder and feed security for both human and animal under harsh environments, where other crops fall short even to negotiate and produce biomass, needless to emphasize its role in improving soil fertility through biological nitrogen fixation and addition of foliage to the soil. Guar has diversified industrial utilities in an array of value added products. Guar gum and its derivatives are widely used in various industries like food, textiles, mining, cosmetics, pharmaceutical, nutrition, health care, oil well drilling and paper industry as thickeners, stabilizers, binders, gelling agents, explosives and corrosive materials. Recently, its use as an alternative of agar in tissue culture media was reported. The guar gum industry is a major industry in guar growing areas of Rajasthan (Jodhpur is known as hub of guar industry), Haryana and Gujarat providing employment as well as earning sizeable foreign exchange. It earns an annual foreign exchange worth Rs. 13390 million which has more potential to grow in volume due to untapped utilities (Mahla *et al.*, 2010).

Some Facts About Guar

- ✰ India is the largest producer of guar in the world, constituting more than 80 per cent world's guar acreage. World market for guar gum is estimated 3.0 lakh tonnes and 2/3 of which is being contributed by Indian guar gum industry.
- ✰ Rajasthan has more than 80 per cent area under gum of the country. More than 80 per cent of this are contributed by arid western parts of the Rajasthan (Barmer, Jaisalmer, Jodhpur, Jalore, Nagaur, Sikar, Churu, Bikaner and Hanumangarh).

- ☆ In rainfed *kharif* crops, guar has the maximum area in arid western Rajasthan. These districts are characterized by low and erratic rainfall distribution, sandy soils with poor water holding capacity and high evapo-transpiration losses resulting in very poor productivity (< 250 kg).
- ☆ Thus area and productivity of guar in Rajasthan have direct impact on national production. It is of great concern that area and productivity are fluctuating over the years (area: 15-30 lakh ha and productivity: 50-500 kg) due to erratic rainfall distribution and frequent droughts.
- ☆ The production is directly related to monsoon. In Rajasthan, the rainfall fluctuates between years and thus results in high volatility in production and consequently on prices (correlation between production and prices is -0.858 which reflects fundamentals of production and prices). While the demand is almost constant over the years, supply varies largely between years.
- ☆ The physical market of the commodity involves speculators and stockists. The commodity is subjected to a long storage period based on demand and market prices. There are no government rules and regulations governing production, distribution, marketing, exports or imports of the commodity and the market forces determine the prices. That is why, besides extreme variation in production over the years, its export in terms of quantity as well as value increased over the years.

Guar Scenario in Western Rajasthan

The guar scenario in Western Rajasthan is presented below from 1990 onwards for major guar growing districts. Arid districts of Western Rajasthan contribute lesser than 80 per cent in total guar production of Rajasthan besides having more than 90 per cent acreage. This ratio of 0.88 (production/ area) is reflected in the productivity and hence the contribution of Rajasthan remained static. Churu has the maximum area (385700 ha) with a coefficient of variation of 25 per cent while the C.V. for production is more than 62 per cent. The statistics clearly explains that average deviation in guar area

Table 2.1: Area, production and productivity of guar in Western Rajasthan (Average of 18 years from 1990-91 to 2007-08).

Districts	*Area (000 ha)*		*Production (000 t)*		*Productivity (kg/ha)*	
	Mean	*CV (per cent)*	*Mean*	*CV (per cent)*	*Mean*	*CV (per cent)*
Churu	385.7	24.91	66.23	62.35	167.6	63.76
Barmer	333.2	31.15	49.96	94.70	136.9	90.36
Ganganagar	277.0	39.03	145.0	72.40	499.8	50.14
Bikaner	216.0	88.69	40.97	148.87	151.2	62.11
Jaisalmer	193.5	55.50	23.22	117.51	81.3	100.20
Nagaur	157.8	25.05	52.64	59.34	345.4	63.90
Jodhpur	148.0	37.05	34.00	104.40	254.2	90.34
Sikar	101.0	21.36	39.50	79.87	386.5	70.75
Jalore	78.80	39.76	20.98	92.26	274.6	100.14
Total arid zone	1891.0	49.49	472.5	71.42	247.0	118.17
Rajasthan	2080.6	28.58	593.3	56.20	271.7	49.33
% of Rajasthan	90.88	–	79.64	–	–	–

is about 50 per cent while production deviates more than 70 per cent over the years compared to whole Rajasthan where the deviation is 28.58 and 56.20 per cent for area and production, respectively. The production of guar in major districts is not consistent and varies from 60–150 per cent (value of coefficient of variation). Bikaner, Jaisalmer, Jodhpur and Barmer districts are very much inconsistent with very low productivity. Moreover, these districts have more than 9.0 lakh hectare area under guar cultivation over the years. The productivity of guar is minimum at Jaisalmer followed by Barmer, Bikaner and Churu besides being cultivated over 11.3 lakh ha in these districts. This is due to the typical desertic situations in these districts characterized by very low and erratic rainfalls during growing period and sandy soils very low in organic matter and water holding capacity which lead to sparse plant stand and slow initial growth. Maximum productivity is exhibited in Ganganagar and Hanumangarh districts as guar is being grown as irrigated crop to some extent.

National Guar Scenario and Contribution of Rajasthan

The north-western parts of the country which are part of the Indian Thar Desert comprising of western Rajasthan (Barmer, Jalore, Jaisalmer, Jodhpur, Nagaur, Sikar, Churu, Bikaner, Ganganagar and Hanumangarh districts), North Gujarat (Banaskantha, Patan and Kutch districts) and adjacent districts of Haryana and Punjab receive scanty rainfall through south-west monsoon. India dominates the world Guar seeds production- contributing around 75–80 per cent (around 1.0 million tonnes). In

Table 2.2: Area, production and productivity of guar in Rajasthan and India and export of guar gum from 1990–91 to 2006–07.

Years	*Area (000 ha)*		*Production (000 tons)*		*Productivity (kg/ha)*		*Export of Guar Gum*	
	Rajasthan	*India*	*Rajasthan*	*India*	**Rajasthan*	*India*	*Quantity (tons)*	*Value (Rs. in Lakh)*
1990-91	2089.6 (86.9)	2402.9	946.2 (80.5)	1175.9	453	490	49852.12	5814.53
1991-92	1558.5 (88.3)	1764.8	204.0 (59.0)	345.8	130	200	61426.86	9255.06
1992-93	1822.2 (83.6)	2178.6	583.4 (73.2)	797.4	321	370	62373.97	10318.61
1993-94	1897.1 (90.3)	2101.1	287.2 (58.6)	489.9	151	230	74492.59	14082.09
1994-95	1959.4 (85.1)	2301.6	708.3 (75.4)	939.2	362	410	67748.12	14283.27
1995-96	1774.7 (84.4)	2103.0	274.2 (65.5)	418.8	155	199	83283.40	22720.24
1996-97	1818.8 (84.3)	2158.2	739.9 (81.2)	911.6	407	422	95169.93	35612.60
1997-98	1985.3 (86.3)	2301.2	733.5 (76.2)	962.7	370	418	102728.62	54498.81
1998-99	1611.9 (83.9)	1922.1	319.6 (65.4)	488.5	198	254	91989.99	72769.90
1999-00	2648.5 (90.3)	2933.9	231.6 (61.7)	375.1	87	128	113746.82	81476.62
2000-01	3056.3 (87.4)	3497.4	481.2 (73.0)	658.8	157	188	129530.85	60295.16
2001-02	2412.6 (83.1)	2903.1	763.3 (70.0)	1089.9	316	375	117883.01	40309.06
2002-03	556.5 (57.1)	974.1	27.90 (13.8)	202.6	50	208	111826.36	48615.39
2003-04	2278.3 (79.8)	2854.0	1163.2 (76.9)	1513.4	511	530	1205561.24	50789.55
2004-05	2363.8 (82.4)	2867.4	344.6 (38.1)	903.3	146	315	131299.96	68947.71
2005-06	2454.7 (83.1)	2955.5	600.6 (56.7)	1059.0	245	358	186718.40	104923.30
2006-07	2821.0 (84.4)	3343.7	663.1 (56.7)	1169.3	235	350	189304.38	112579.21
Average	2065.2 (84.5)	2444.9	533.63 (67.20)	794.2	258	325	–	–

*Figures in parentheses show the per cent contribution of Rajasthan.

India, Rajasthan dominates guar production contributing around 70 per cent, followed by Haryana (12 per cent) and Gujarat (11 per cent). The perusal of 17 years data shows that the guar is being cultivated over an area of 24.45 lakh ha in India and Rajasthan contributes 84.5 per cent while the production is around 8.0 lakh tons with only 67.2 per cent contribution from this major producing state (Henry *et al.*, 2008). The contribution of Rajasthan in production varies over the years and most of the times it hovers around 70 per cent which reached to 80 per cent in certain good monsoon years (1990-91 and 1996-97) only. In the worst drought years, the contribution of Rajasthan was barely 40 per cent and consequently the national production was influenced adversely (2002-03 and 2004-05).

In recent past, the contribution of Rajasthan in national basket decreased as the guar cultivation in Haryana expanded in irrigated area and realizing productivity levels of more than one tone/ha. The average area under guar seed and production in Haryana for the period of 6 years from 2003-04 to 2008-09 were 2.98 lakh ha and 3.11 lakh tonnes respectively. In 2008-09, the area was at its peak at 4.0 lakh ha with the highest production of 4.80 lakh tonnes. In the state, there is considerable increase in the production over the period of 6 years from 2003-04 to 2008-09. The level of yield since 2004-05 is increasing, while it recorded the lowest level of 435 kg/ha in 2003-04 and reached at peak in 2008-09 (1200 kg/ha). The average yield was 1044 kg/ha for the period of six years from 2003-04 to 2008-09.

The acreage under guar seed and production in Gujarat averaged at 2.15 lakh ha and 0.95 lakh tons respectively for the period of 9 years from 2000-01 to 2008-09. In 2000-01, the area covered under guar was at its peak at 2.75 lakh ha and arrived at ever lowest of 1.50 lakh ha in 2008-09. There was a declining trend in the production of guar seed from 2003-04 to 2006-07 which recovered in 2007-08 due to pleasing monsoon. The production was at its high in 2003-04 at 2.04 lakh tons. There was sharp decline in the yield of guar seed from 2003-04 to 2006-07 and further regained in 2007-08 (662 kg/ha) and drastically dropped again in 2008-09 (350 kg/ha). The yield was at its high in 2003-04 at 766 kg/ha. The average yield for the period of 9 years from 2000-01 to 2008-09 stood at 444 kg/ha.

Reasons of Low Productivity of Guar in Rajasthan

Besides development of high yielding and early maturing genotypes of guar in last 15 years,production and productivity of the crop are still very low compared to its yield potential. In spite of great significance of this crop in regional, national and climatological context, its poor productivity is still a continued concern. Several controllable and uncontrollable reasons can be cited for stagnation in poor productivity of this important legume. Some of the reasons for low productivity in Rajasthan are:

1. Erratic and low rainfall account for low and unstable production of the crop.
2. Cultivation of the crop on marginal and low fertile arid zone soils, high temperature and erratic, uneven and low rainfall during growing season lead to very poor productivity.
3. Frequent droughts, aberrant rainfall situations and poor economic conditions of farmers in Rajasthan discourage farmers to use adequate amount of recommended inputs for higher productivity.
4. Farmers are using local genotypes having very low productivity and long maturity duration due to inadequate supply/availability of quality and certified seeds of improved genotypes.
5. High risk due to biotic and abiotic factors and large price fluctuations and no minimum support price from government side also discourage farmers to grow this important legume.
6. Further, guar is a crop of secondary importance if there are timely rains. Farmers prefer to grow pearl-millet, sesame and mung on fertile and good soils instead of growing guar/moth/cowpea.

Production Technology for Higher Yield

Guar is mostly grown on marginal and sub-marginal lands without following the improved package of practices and there is wide gap in average grain yield at farmer's field and experimental yield potential on research farms. The yield can easily be increased substantially by adopting the latest improved production technology and cultivars suitable for the soil and climatic conditions of the region.

Soil and Climatic Requirement

Guar performs well under wide range of soil types. Deep well-drained sandy loam or sandy soils with moderate alkalinity (pH 7.5–8.0) are optimal for guar. Care should be taken when growing guar on heavy clay soils where the crop can be exposed to wet conditions. Soil crusting can reduce seedling emergence and water logging increases the likelihood of root diseases. Rainfall and its distribution have significant effect on guar productivity. A well distributed rainfall of 300–500 mm (4–6 rain events) during growth period is essential for higher productivity. Long dry spell during vegetative growth stage results in early senescence of leaves which ultimately leads to poor grain yield while excessive rains lead to more vegetative growth, lodging and poor yield. Adequate soil moisture is essential at flowering and pod formation stage while humid conditions during grain development as well as harvesting at high grain-moisture contents can both result in weathering of the grain. This grey or black discolouration reduces the commercial value and quality of produces.

Seed and Sowing

In most of the guar growing areas, first fortnight of July is the optimum sowing time for getting good yield. Early sowing results in more vegetative growth and higher incidence of *Bacterial Leaf Blight* disease while late sowing ends with low biomass and consequently poor yield. Guar for fodder and vegetable can be grown successfully from March to June depending on the irrigation facility and rainfall pattern of the region while green manuring crop should be sown as early as possible with the onset of monsoon.

Un-branched, early maturing and fodder type varieties should be sown at 30 cm row spacing whereas sowing of vegetable type, branched and medium duration varieties should be done at a row spacing of 45 cm and 15-20 cm intra-row spacing. For green manuring, sowing may be done by broadcasting the seed uniformly. Seed should be inoculated with specific *Rhizobium* strain to increase nodulation and efficient nitrogen fixation. Optimum seed rate for guar crop raised for different purposes is as follows:

Purposes	*Seed Rate (kg/ha)*
Seed and vegetable	12–15
Green manure crop	30–35
Fodder crop	35–40

Guar works well in rotation with wheat, barley, mustard, cotton, bajra and sorghum. Earlier, guar was grown as a mixed/inter crop with bajra, mung bean and moth bean in varying proportions but now a days, it is mainly grown in lines/rows as a pure/sole crop. However, where chances of guar failure are higher due to uncertain weather conditions, guar may be intercropped with bajra in 3:1 ratio, which reduces the risk against failure of crop. Intercropping with bajra or sorghum in 1:2 ratio for green fodder increases the nutritive value of the fodder (Singh *et al.*, 2002).

Cultivars

Guar is mainly grown in poorly endowed situations with the least care and management in north western parts of the country receiving scanty rainfall during the crop season. Under these poor growing conditions, secondary inputs generally show poor response and infestation of insect pest and diseases

Figure 2.1: Single stem.

Figure 2.2: Synchronous maturity.

Figure 2.3: Vegetable type.

is subsidiary, thus clearing the path for putting more concentrated effort for raising production through improved cultivars. It is obvious that the farmers find it difficult to utilize financial resource for supporting other technologies but improved seed (Kumar, 2008). The characteristics of the cultivars released for cultivation for guar growing areas of the country shown in Table 2.3.

Table 2.3: Characteristics of released varieties.

Variety and Year of Release	*States of Adaptation with Districts*	*Maturity (Days)*	*Grain Yield (kg/ha)*	*Specific Traits*
HG–884 (2007)	Hisar, Rewari, Bhiwani (Haryana); Jhunjhunu, Sikar, Churu, Ganganagar (Rajasthan)	90–95	1400–1500	Medium early, 30-32 per cent gum content, higher grain yield
RGC–1038 (2006)	Jaipur, Ajmer, Tonk, Swaimadhopur, Nagaur, Alwar (Rajasthan)	95–100	1200–1500	Branched, suited to both summer and kharif seasons, heavy podding and higher grain yield
RGM–112 (2005)	Jhunjhunu, Ganganagar, Hanumangarh, Sikar, Nagaur and Jodhpur (Rajasthan); Banaskantha (Gujarat) and Chittor, Ananthpur (AP)	85–90	1200–1300	Early maturity, moderately resistant to BLB, leaves light green in colour, higher yield potential
RGC–1031 (2005)	Jaipur, Ajmer, Tonk, Swaimadhopur, Alwar (Rajasthan)	110-114	700–1400	Suitable for irrigated condition, branched, white grain, field tolerance to BLB and alternaria blight
RGC–1017 (2002)	Jodhpur, Ganganagar (Rajasthan); Bathinda (Punjab); Gwalior, Bhind, Morena (MP), Banaskantha (Gujarat)	92–95	1200–1400	Wider adaptation, branched, leaf margin show slight cut, moderately resistant to BLB, PM and root rot
RGC–1002 (1999)	Jhunjhunu, Ganganagar, Hanumangarh, Sikar, Nagaur and Jodhpur (Rajasthan), Banaskantha and Kutch (Gujarat)	90–95	800–1300	Higher yield and more gum content, wider adaptation, heavy podding behaviour, less disease infestation
RGC–1003 (1999)	Jaipur, Alwar, Tonk, Ajmer (Rajasthan); Banaskantha (Gujarat)	90–95	1100–1200	Early maturing, branched, medium height, bold grains
RGC–986 (1999)	Jaipur, Ajmer, Tonk, Swaimadhopur, Alwar, Nagaur, Jodhpur (Rajasthan)	110- 115	1100 -1200	Suited to irrigated conditions and better soils, input responsive, quite late
HG–365 (1998)	Hisar, Rewari, (Haryana); Jhunjhunu, Sikar, Jodhpur, Hanumangarh, Ganganagar (Rajasthan) and Bathinda (Punjab)	85–90	1200–1400	Earliest maturing guar variety, high yielding, brisk podding behaviour, most popular in Haryana
HG–563 (1998)	Hisar, Rewari, (Haryana); Jhunjhunu, Sikar, Jodhpur, Hanumangarh, Ganganagar (Rajasthan)	85–90	1200–1300	Rich in gum content and viscosity, brisk podding behaviour, suited to Haryana area
RGC–936 (1991)	Jhunjhunu, Sikar, Jodhpur, Hanumangarh, Churu, Bikaner, Barmer, Jaisalmer (Rajasthan); Banaskantha and Kutch (Gujarat)	85–90	900–1150	Early, branched, dwarf, wider adaptation, drought hardy, high yielding, medium sized grain with light pink colour
RGC–197 (1990)	Jhunjhunu, Churu, Sikar, Hanumangarh, Alwar, Ganganagar (Rajasthan)	110- 115	900–1100	Suited for inter-cropping and irrigated conditions, un-branched, requires close spacing

There are also some old varieties and recent releases confined to particular states (Table 2.4).

Table 2.4: Old varieties and state released varieties.

Variety (Year of release)	*Characteristics*
Pusa Navbahar (1967)	Un-branched, glabrous, vegetable type
FS–277 (1974)	Un-branched, pubescent and deeply serrated leaves, late, fodder type, susceptible to *Bacterial Leaf Blight*
HG–75 (1981)	Branched, pubescent and smooth leaves, late maturing, creamy seed colour
HG–182 (1981)	Branched, pubescent with little serrated leaves, Late maturing and grey seed colour
Suvidha (1984)	Branched, early and suitable for all guar growing areas
PLG–85 (1984)	Un-branched, medium late maturity, suitable for all guar growing areas.
Maru Guar (1986)	Suitable for late sowing in western Rajasthan, branched
HFG–258 (1986)	Branched, pubescent and smooth leaves, late maturing
Naveen (1986)	Early, suitable for guar–wheat sowing
GAUG–34 (1987)	Branched and late (Gujarat)
GG–1 (1991)	Resistant to *Bacterial Leaf Blight*, late maturing, suitable for Gujarat
Bundel Guar–3 (2000)	Late, branched, dual purpose
GG–2 (2005)	Release for Gujarat state, early maturing, branched, white seeded
RGC–1055 (2007)	Released for Rajasthan state only, maturity duration 100–105 days
RGC–1066 (2007)	Un-branched, high yielding, maturity 100-105 days suitable for Rajasthan state

Nutrient Management and Weed Control

Being a leguminous crop, guar does not require much nitrogenous fertilizer except a starter dose of 15–20 kg N/ha. Application of 30–40 kg P_2O_5/ha was beneficial and it should be drilled before sowing. Application of 20 kg sulphur and 5 kg zinc (25 kg $ZnSO_4$)/ha at sowing is beneficial. Weeds compete for moisture, nutrients, light and space and thus affect the yield considerably if not controlled effectively. First hoeing should be done at 25–30 days after sowing and second at 40–45 days after sowing for effective control of weeds and better crop growth.

Irrigation

By and large, guar is being grown as a rainfed crop and due to extensive deep root system it utilizes moisture very effectively. A protective irrigation if available, at pod filling stage is beneficial in case there is severe moisture stress. Guar for fodder purpose requires 2–3 irrigations and for vegetable purpose 4–5 irrigations during summer may be applied.

Plant Protection

There is not much problem of insects in the guar crop. Sometimes Bihar hairy caterpillar and jassids attack the crop. Bihar hairy caterpillar can be controlled by spraying 1.25 litres of Endosulfan 35 EC in 1000 litres of water. Attack of jassids can be checked by 2-3 sprayings of 600-1000 ml of Malathion 50 EC in 600 in 1000 litres of water/ha depending upon the height of the crop with a manually operated sprayer at 15 days intervals.

Major diseases of economic importance are bacterial leaf blight, *Alternaria* leaf spot, root rot,wilt and powdery mildew. Losses in crop yield vary depending on nature of pathogen and severity of attack in the prevailing weather conditions. Bacterial leaf blight is the most important, causing losses to the extent of 50 -70 per cent, followed by *Alternaria* leaf spot causing losses to the extent of 25–30 per cent. Major symptoms and management of certain important diseases are given in Table 2.5.

Table 2.5: Symptoms and control measures of important diseases

Diseases	*Symptoms*	*Control Measures*
Powdery mildew [*Oidiopsis taurica* (Lev.) Sain, *Erysiphe taurica* (Lev.)]	All above ground parts of plant are attacked. White powdery growth on leaf, stem and pod. Severe defoliation takes place and pods remain smaller in size.	Spray Calixin (0.05 %), Bavistin (0.1 %) or Sulfex (0.2 %).
Bacterial leaf blight (*Xanthomonas axonopodis pv. cyamopsidis*)	All above ground parts are attacked but characteristic symptoms develop on leaves as dark brown, round to irregular spots. In the early stages, spots are water soaked which later turn grey to dark brown with concentric zonations. Whole leaf may give blighted appearance when several spots coalesce. Defoliation is very fast. Infected seeds are shrivelled and discoloured.	Spray Dithane M-45 (0.2 %) or Difolatan (0.3 %). Treat the seed with Agrosan GN (0.2 %) or Captan (0.2 %).
Alternaria leaf spot (*Alternaria cucumerina var. cyamopsidis*)	Water soaked spots on the leaves. Spots are olive coloured and oily on lower surface. Fully developed spots are jet black in colour with chlorotic halos. Several spots coalesce and form bigger blotches. Infection spreads from leaves to stem. Cracks are formed on stem.	Use disease free seeds. Treat seeds with Bavistin (0.2 %). Spray Difolatan (0.2 %) or Bavistin (0.1 %).
Anthracnose (*Collelotrichum capsici f. cyamopsicola*)	Small, dark, depressed spots appear on the leaves, stem and pods. In later stages, these spots enlarge and girdle the stem and pods. Sometimes die-back symptoms develop on well developed plant. Black to brown spots appears on infected seeds.	Treat the seed with Ceresan dry (0.3 %) or Captan (0.3 %).
Root rot complex (*Macrophomina phaseolina*)	Seedling mortality at early stage and in later stages root rot causes dry rot symptoms as it leads to decomposition of root tissues and prevents normal development of plants.	Crop rotation. Grow resistant variety (HG 75). Soil solarisation in May month is most effective.

Harvesting and Threshing

Seed purpose crop should be harvested when natural defoliation starts and the colour of pods turns golden yellow. Guar crop matures in 100–120 days depending on the variety and growing conditions. In light sandy soils, the crop matures early (75–80 days) with very low yield of poor quality. Delayed harvesting after maturity leads to shattering of pods and consequently reduced grain yield. After sun drying on threshing floor or in small heaps in the field, threshing may be done manually or with the power thresher.

A good crop in normal monsoon year can yield 0. 7–1.0 tone grain yield/ha while under protective irrigation and better management grain yield of more than 3.0 tons/ha can be attainable.

References

Henry, A., Choudhary, S.P.S. and Mathur, B,K., 2008. Clusterbean. Souvenir, 3rd National Symposium on "*Enhancing Productivity, Nutritional Security and Export Potential through Arid Legumes*" organized by Indian Arid Legumes Society, at CAZRI, Jodhpur during June 28–30, pp. 53–74.

Kumar, D., 2008. Promising varieties of arid legumes. *National Network Project on Arid Legumes,* Central Arid Zone Research Institute, Jodhpur, pp. 35.

Kumar, D., 2009. Genetic improvement of arid legumes: Perspective research activities of arid legumes in India (Eds). D. Kumar and A. Henry. Indian Arid Legumes Society, CAZRI, Jodhpur. Pp. 31–57.

Mahla, H.R., Kumar, D., Henry, A., Acharya, S. and Pahuja, S.K., 2010. Guar: Present status and future prospects in arid zone. *Journal of Arid Legumes,* 7(2): 1–5.

Singh, J.V., Arora, R.N., Saini, M.L., Grewal, R.P.S. and Yadav, B.D., 2002. *Guar Production and its Utilization in India* (Eds). Kumar, D. and Singh, N.B.. All India Co-ordinated Research Project on Arid Legumes, Central Arid Zone Research Institute, Jodhpur, pp. 50.

2011, Future Crops Vol. 1 ***Pages 22–39***
***Editor:* Professor K.V. Peter**
***Published by:* DAYA PUBLISHING HOUSE, NEW DELHI**

Chapter 3
DATE PALM
(*Phoenix dactylifera* L.)
Under Hot Arid Region

R.S. Singh, R. Bhargava and D.G. Dhandar
Central Institute for Arid Horticulture, Bikaner – 334 006 (Rajasthan)
E-mail: ciah@nic.in

Date palm (*Phoenix dactylifera* L.) is an important fruit tree of semi-arid and arid regions of the world. It is also known as *khajoor, kharek* etc. It grows well in poor desertic soils due to its hardy plant characteristics and deep root system. It is also one of the most suitable fruit trees, which can be easily grown under saline conditions. In arid region, crop production is a risky preposition, where date palm cultivation contributes to achieve food security, high nutritive value food, crop diversification, desertification control, income generation and foreign exchange earnings. It can be grown at such places where adequate irrigation facility is available besides other climatic requirements (Chandra *et al.*, 1992).

The fruits have better keeping quality under normal conditions and can be stored for a long period under refrigeration. Besides, doka dates, chhuhara (dry dates) and pind Khajoor (soft dates) are in high demand for raw eating (Vishalnath and Singh, 2003). Fruits are also used for making pickles, chutney, beverage, jam and syrup. Its delicious fruits at doka stage are eaten as fresh and also used for processing to different value added products (Singh *et al.*, 2004).

Origin and Distribution

Date palm is indigenous to countries around the Persian Gulf. The date palm is probably the most ancient cultivated fruit in the world. In the Indus Valley, date palm is believed to be introduced by the soldiers of Alexander the Great in the fourth century B.C. and during the Muslim invasions at the beginning of the eight century AD. After the partition of India,almost all the areas were left in Pakistan. In Rajasthan, date palm was brought by Maharaja Shri Ganga Singhji in his dynasty. The cultivation of date palm can be a boon to fruit growers particularly in arid areas where irrigation facilities are adequate. However, some wild groves of seedling date palm are found on the coastal belt from Anjar to Mandvi in Kachchh region of Gujarat. It is believed that Turk settlers introduced date palm to Kachchh. The commercial varieties of date palm were introduced from USA, Pakistan and Middle East countries during 1955-1962. The performance of some of the cultivars were found more beneficial under arid irrigated conditions at Abohar, Punjab. The success opened up the possibilities of commercial date palm cultivation in India. New thrust was given to date palm research since 1978 under the All India Coordinated Fruit Improvement Project of the ICAR at (AICRP on AZF) Bikaner and Jodhpur in Rajasthan, Abohar in Punjab, Hisar in Haryana and Mundra in Kachchh, Gujarat.

Area and Production

Date palm requires specific climate for fruiting and fruit quality. Hence, in India it is grown in the states of Rajasthan, Punjab, Haryana and Gujarat. However, commercial date palm growing countries of the world are Iraq, Saudi Arabia, Algeria, Iran, Egypt, Morocco, Tunisia, Pakistan and Sudan besides many other Near East and North African countries. Date palm has played a vital role in the history of mankind by providing food and shelter to millions of people. Pakistan is the fourth largest date producing country, contributing 11 per cent to world production (Zaid *et al.*, 2007). In India, date palm production is meager and maximum dates are imported from date-growing countries of the world to meet our requirements. India also exports small quantity of dates to neighboring countries like Nepal, Myanmar, Bhutan, Maldives and Sri Lanka.

The extremely dry hot areas comprising Jaisalmer, Barmer, Bikaner and Jodhpur in Rajasthan, Bhatinda, Fazilika, Abohar in Punjab, Sirsa, Hisar, Mahendergarh, Bhiwani in Haryana and Kachchh region in Gujarat are the potential areas for date palm cultivation in India. In Rajasthan, acreage under date palm cultivation is very less. An estimated area of 8, 973 ha. with a production of 53,836 tonnes was reported during 2000–01, increased to 12,493 ha with an annual fruit production of 85,352 tonnes (2004-05) of date palm in the Kachchh region (Bhuj, Anjar, Khedoi, Mundra, Mandavi etc.) as reported by Muralidharan *et al.* (2008). The maximum date palms are grown natural and as grove in depressions, on riverbank, *nala* and boundary of the farms (Singh *et al.*, 2004).

Looking to its potential, some new plantations are established in Kachchh region of Gujarat. There is a large chunk of potential area for the development of date palm groves in arid region of North- West India. The total area suitable for date palm cultivation is about 5.5 million ha. This includes low-lying stagnant and saline lands, natural wastelands and *Khadins* (natural water harvesting areas), canal command areas and well-irrigated lands. This potential area can be brought under date palm cultivation where other crops may not be remunerative.

At present, India imports date fruits from Gulf countries which involve about US $7.23 million. India also exports dates to neighboring countries valued US$ 2,18, 000.

Table 3.1: Import of date palm to India.

Units	2000	2001	2002	2003	2004	2005	2006	2007
Quantity (tonnes)	192619	244367	171523	193755	247875	240399	286317	253341
Value US$1000	41,554	52,786	27,798	33,011	46,407	55,753	74,760	72,359

Source: FAO Stat database.

Import of date palm can be reduced to some extent by expansion in area and quality of production, as 5.5 million. ha. area is suitable for planting date palm in Rajasthan, Gujarat, Punjab and Haryana.

The main constraints in expansion of date palm in India are non-availability of quality planting materials at reasonable cost. Standardization of agro-techniques for different ecological conditions and sufficient irrigation water are required for quality fruit production.

About 34 cultivars imported from different countries are successfully grown at Date palm Research Station, Bikaner but only the varieties Halawy, Barhee, Khuneizi, Medjool, Zahidi, Khalas and Sewi have shown better performance. Under the Rajasthan agro-climatic conditions, the cultivar should have (a) non-astringent and early maturing fruits for fresh eating (b) early maturing,high sugar and rain tolerant fruits for processing for preparation of *chhuhara* and *pind khajoor*, (c) tolerance to *Graphiola* disease besides high yield potential and dwarf growth habit.

Owing to heterozygosity and dioecious nature of the palm, offshoots are always used for its multiplication for commercial plantation. But 8-20 off-shoots can only be obtained during its whole life up to 10–12 years and none thereafter. This is obviously a slow rate of multiplication. The offshoots formation also varies from variety to variety. A few varieties produced more number of suckers in their life span. Absence of fast multiplication techniques for date palm is thus, a prominent bottleneck in the extension of area under its plantations. Some cultural practices were adopted to produce more suckers in date palm but are not efficient. Tissue culture technique is an appropriate option for fast multiplication and production of large scale planting materials in date palm.

Irrigation is absolutely necessary for date palm, because it is grown in hot and dry, low rainfall areas. The water requirement of date palm is quite high and under such circumstances, date palm cultivation could not be extended due to lack of irrigation facilities in most areas of arid zone.

Nutritive Value and Uses

Date fruit is a highly nutritious and favorite fruit throughout India. Fruits of date cultivars *viz.* Halawy, Khalas, Khuneizi, Chip chap, Braim and Barhee are eaten as fresh (doka stage). Besides fresh consumption, several value added products are prepared from fruits. This fruit can supplement the dietary needs of the desert people where very less nutritious food is available. Dates are cured as dry dates (chhuhara) and soft dates (pind khajoor) which have great demand and market in the country. In fact, it was the only staple food available to the desert people in the Near East and North African region till oil was discovered there. Even today, it forms a prominent component of daily diet of people in these countries. The per capita consumption of dates in Saudi Arabia, Iraq, Libya and Egypt is 28.9, 17.3, 28.0 and 10.8 kg/year, respectively. Date fruits provide abundant quantities of sugar, iron, potassium, calcium and nicotinic acid. Small amounts of protein, copper, magnesium, chlorine, sulphur, vitamins A, B_1 and B_2 are also present in date pulp. The flesh of dates with a moisture content of 20 per cent contains 60-65 per cent sugar, about 2.5 per cent fiber, 2 per cent protein and some what less than 2 per cent fat, mineral matter and pectic substances and provides about 3,550 calories/kg (Nixon and Carpenter, 1978). Fresh date fruit(100g) contains 59.2g moisture, 1.2g protein, 0.4g fat, 1.7g minerals, 3.7g fibre, 33.8g carbohydrates, 22.0 mg calcium, 38.0 mg phosphorus while 100g dry dates (chhuhara) provide 2.5 g protein, 2.1g minerals, 3.9g fibres, 75.8g carbohydrates, 120mg calcium, 50mg phosphorus, 7.30mg iron, 0.02 mg riboflavin, 0.9 mg niacin, 3.0 mg vitamin C (Gopalan *et al*,1985).

Date palm leaves are used for making handicrafts, mats, temporary huts, broom, etc. Its seed kernel is used in the preparation of cattle feed. In Gulf countries, several value added products are made from date palm plant parts and fruits. Many date products are exported to other countries where cultivation is not being done.

Botanical Description

The *Phoenix* genus consists of twelve species, out of which some are grown for edible fruits and others for ornamental purposes. *P. dactilyfera* has 36 chromosomes (n=18, 2n=36).The two wild species *Phoenix sylvestris* (L.) Roxb and *P. canariensis* are the most widely available, both are crossable to *Phoenix dactylifera.* Besides date palm, five species *viz. P.atlantica* Chev., *P.reclinata* Jaeq., *P.farinifera* Roxb. Synon. *P. pusilla* Gaertn., *P. humilis* Royle and *P. acaulis* Roxb. also bear edible fruits (Chandra *et al.*, 1992; Zaid,1999).

The date palm belongs to family Palmae. The latin name of date palm, *Phoenix dactylifera* is derived from a Phoenician name "*Phoenix*", which means date palm and *dactylifera* derived from a Greek word "daktulos" means a finger. In India, generally two species of *Phoenix* are found, out of which, *Phoenix sylvestris* Roxb. called "desi khajoor" is found throughout the country but it produces inferior quality of fruits. The fruit contains a large stone with less pulp. From its crown parts, a liquid is extracted known as "neera" or taddi, which is used as fresh drink and also for preparation of jaggary. The tree of desi khajoor (*Phoenix sylvestris*) is found in Bundel Khand region of the country and is a suitable tree component for agro.forestry system (Tewari *et al.*, 2001). *Phoenix dactylifera* is commercially cultivated through out the world for fresh fruits and processing.

It is a monocotyledonous plant with un-branched stem. Some times date palm shows a branching phenomenon due to several reasons. Branching in date palm is the result of either dichotomy or auxiliary bud development or polyembryony or attack by diseases (Zaid, 1999). Date palm roots go very deep. The roots can withstand wet soil for many months but not for longer period. Stem is tall straight (30-35m) bearing crown of leaves. Leaves are pinnate and spines are present. Depending on variety and age of palm, leaves of date palm vary from 3-6 meter long. Since, being a dioecious plant, male and female flowers are borne on separate palms.In general, male flowers are borne a few days earlier than female flowers. The fruit is a drupe and edible portion is called pericarp. Date fruit is a single seeded drupe in red, yellow and green colour. The fruit maturity is defined in three stages *i.e.*, doka, dang and pind stage.

Genetic Resources and Crop Improvement

All the commercial date cultivars have arisen by selection of chance seedlings based on local needs. A rich genetic diversity is available in costal belt of Kachchh region of Gujarat in India which should be exploited (Singh *et al.*, 2004).

Introduction

Most of the cultivars of date palm were introduced from different countries time to time *e.g.*, Halawy, Barhee, Medjool, Khalas, Sayar, Zahidi (USA), Khadrawy (Iraq), Barshi, Khuneizi, Nagal, Khashab (Oman), Hatemi, Tayar, Ruziz (Saudi Arabia), Amri, Sakloti, Agloni, Siwi and Amhat (Egypt), Chipchap and Braim (Iraq).In Sri Ganganagar area of Rajasthan, date palm was introduced from Pakistan.

Selection

From the rich genetic diversity of nearly 1.4 million palms developed from seeds in the coastal belt of Kachchh district of Gujarat, 20 promising palms were selected, most of which yield non-astringent fruits at doka stage. One of them bears coconut shape fruit. These selections flower twice an year. An early ripening date seedling was identified at Abohar.

No breeding work has so far been done on date palm in India except evaluation of cultivars against rain damage and selection of promising female seedlings from the Kachchh region. At Abohar, Zahidi cultivar was found resistant to rain damage, Barhee is more tolerant than Shamran. It was also found that Medjool or Mejhool was resistant to rain damage, as it withstood rains during fruit ripening (Pareek, and Vishal Nath, 1996). It is a late maturing cultivar.

A large number of varieties and promising selections (Sel.- 9, Sel.-13, Yaqubi, Kotho, Trofo, Gulchati, Madhepura, Khedoi-7, Sopari, Saidy,) are made in India from natural populations in Kachchh region. Some selections are also made by the farmers of the region. Further, these yellow and red fruit colour types are suitable for making different processed products (Singh *et al.*, 2004; Pareek and Sodagar, 1986). A genotype with green fruit at doka stage was identified in Kachchh region.

Cultivars

Date varieties were developed by thousands of years of selection of seedlings. More than 1000 varieties of date palm are found all around the world. There are about 400 varieties in Iran followed by Iraq (370) and Tunisia (250). In India, about 50 cultivars were introduced for evaluation at different research stations. The descriptions of important varieties are as follows:

Halawy

An early maturing, regular bearing, high yielding and suitable to grow in large areas for fresh fruit and processing. Its fruits are yellow and sweet in taste at doka stage with average weight 9.6g; size 3.56 x 2.10 cm and TSS 31.0 per cent. It yields 70- 80 kg doka fruits/tree/year.

Barhee

Its origin is Basrah, medium to late maturing cultivar and good for fresh consumption. The fruit at khalal stage (doka) has excellent flavour, and attractive yellow fruit colour and sweet taste. The stone is very small in size with high pulp: stone ratio It has high market potential in the world's market. The yield potential of this variety is 80-120 kg/plant.

Medjool

The origin of this variety is Morocco from where it was introduced to other countries of the world. It is No. 1 variety in the world. It is a very high yielding cultivar due to its bigger attractive coloured fruits (size 4.5 x 3.2 cm). It also matures late and is highly suitable for dry dates preparation. Its doka fruits are astringent in taste. The colour of fruit is yellow-orange at doka stage. It has high commercial value and is an important variety for export. Average yield is 80-100 kg/palm.

Khalas

It is a medium maturing type cultivar. Its berry is big in size at doka stage and sweet in taste. The yellow colour doka fruit is suitable for fresh eating and processing. This variety is a medium to heavy yielder.

Khuneizi

It is an early to mid season maturing variety. The fruit colour is red and suitable for fresh consumption at doka stage. The doka fruits are very sweet in taste. The yield potential of this variety is 60-80 kg/tree.

Zahidi

It is a late maturing variety and suitable for pind preparation. Its doka fruits are astringent in taste and solid with hard skin. The size of fruit is small with attractive yellow colour. It is a heavy yielder (125 kg *doka* fruits/tree). The bunch of this variety is compact with 18 to 24 fruits/strand.

Shamran

It is a mid season to late maturing high yielding variety. The average fruit weight is 6.5 g and fruit size 3.50 x 2.18 cm. The colour of fruit is light yellowish red and astringent at doka stage. This variety is grown commercially. It is suitable for processing to dry dates and other value added products. It is a prolific bearer as fully-grown tree of Shamran has average yield potential of 100 kg *doka* fruits/tree.

Sewi

The trees are tall and erect with moderate canopy. The fruits are yellowish green in colour at *doka* stage and oblong in shape. The average fruit weight is 6.1 g and fruit size 2.9 cm x 1.90 cm. It has a pulp thickness of 0.6cm having 40.0 per cent TSS at full *doka* stage. The fruits are sweet at full *doka* stage and therefore suitable for both raw eating as well as soft dates. It is late in ripening as fruits of this variety reach full *doka* stage by the first fortnight of August at Bikaner. The fully-grown tree of Sewi has an average yield potential of 50 kg *doka* fruits/year.

Chip Chap

This variety was introduced to India from Iraq in 1997 and is suitable for fresh and for processing purpose. Its fruits are yellow, bigger in size and sweet in taste at *doka* stage. It is average in yield and maturity period is also less in arid region. It is suitable for dry dates preparation.

Braim

This variety was introduced to India from Iraq in 1997. It is a medium maturing variety suitable for fresh use and processing. The berry is yellow in colour, juicy and sweet in taste at doka. (Singh *et al.*, 2005). It is a medium yielding variety.

Zagloul

The trees are vigorous in growth having erect habit. The fruits are red and large in size. The fruit bunches are bigger in size. The average fruit weight is 9.5g and fruit size is 3.90 cm x 2.20cm. It has a pulp thickness of 0.64 cm having 28.0 per cent of TSS at *doka* stage. The fruits are astringent at *doka*. The *dang* and *pind* fruits are relatively less sweet but have pleasant taste. It is suitable for soft dates and for preparation of fruit products like date juice and preserved products. It is medium in ripening as fruits of this variety reach full *doka* stage by the beginning of second fortnight of July at Bikaner. It is a prolific bearer and average yield potential is 150 kg *doka* fruits/year.

Khadrawy

The plants of this variety are dwarf in stature. The fruits at *doka* stage are greenish yellow in colour and oblong in shape with broad apex. The average fruit weight is 7.7 g and fruit size is 3.10 cm x 2.10 cm. It has pulp thickness of 0.66 cm having 30.0 per cent TSS at *doka* stage. The fruits are highly astringent at full *doka* stage and are badly spoiled due to rains at the time of ripening. This variety is suitable for preparation of both soft dates and dry dates. It is medium in ripening as the fruits of this variety reach full *doka* stage by end of July. Average yield potential is 40-50 kg *doka* fruits/palm.

Climate

The climatic factors which decide the commercial possibility of date palm production are: (a) long dry summer and sufficient heat unit accumulation for development and ripening of fruit, (b) sufficient water resources for irrigation, and (c) production technology suitable for different agro-climatic zones. The heat unit available at some of the locations in arid zone counted from the time of flowering (February) indicates that the most of these regions are suitable for date palm cultivation (Pareek, 1984). For successful date fruit maturation, nearly 3000 °C heat units are required, most of which accumulate by July in all these locations. The faster rate of heat unit accumulations in India compared to that in Near East and North African region, result in poorer quality of fruit. The ideal mean temperature during flowering and fruit ripening vary from 25-40 °C depending upon cultivars. The rate of maturity of fruit and quality depend upon maximum temperatures (45- 48 °C). The rainfall pattern also influences the cultivation of date palm. The annual average rainfall is less (240 mm) and erratic under Bikaner condition. However, it can be grown in areas receiving 150- 450mm annual rainfall.

In date palm growing countries with completely rainless period from February to September, date fruit maturation (up to *pind* or *tamar* stage) takes about 180 days. The number of days available for fruit development under Indian conditions, however, is cut short to 122-170 days (Pareek, 1984) as the growers are obliged to harvest early to protect from fruit spoilage owing to rainfall and high humidity, even though the fruits are not yet fully matured. Maximum date fruits are harvested at *Doka i.e.*, Khalal stage under Indian conditions.

Soil

Date palm thrives well in light to heavy soils, which have 2.0 to 3.0 m soil depth and good drainage system. The soil of Thar desert is suitable for planting of date palm. The pH of soil from 8.0 to 8.6 is suitable for date palm cultivation. Deep sandy loam and sandy soils are the best for date plantations, which are available in parts of Rajasthan, Haryana, Punjab and Gujarat. It tolerates salinity up to maximum level (30 $dSm^{-1)}$. The water logging conditions for longer period are not suitable for date palm trees. The soils of coastal belt of Gujarat are suitable for growing dates. These soils are light gray sandy and deep. The sandy and light grey sandy soils are suitable for growing date palm.

Propagation

Date palm is propagated both by seeds as well as vegetative methods but commercially multiplied through suckers and also by tissue culture techniques. The rooted suckers of appropriate size are suitable for better establishment and growth of plant.

Seeds

In date palm, variability and sex ratio are the main problems of seed propagation. The seed germination in date palm is 80-90 per cent when seed sowing takes place in July- August. Moreover, growth in seedlings is very slow at initial stages. It was observed that seed soaking enhanced germination and seedling growth by pre-treatments of seeds with cow dung for 24 hours in *Phoenix* species (Singh and Bhargava, 2009). Seedlings varied considerably with respect to vegetative and reproductive characteristics (Al-Ghamdi *et al.* 1988). Date palm propagation through seedlings can provide ideal material as seedlings palms may perform better sometimes (Chandra *et al.*, 1992). In Kachchh region, the seedlings are used for planting on farm boundary due to lack of other planting materials (Singh *et al.*, 2004).

Suckers/Offshoots

Presently, date palm is multiplied through suckers/offshoots. The offshoots bear fruits 2-3 years earlier than seedlings. Production of suckers is a limiting factor because a palm tree produces 10-20 suckers in whole life. Out of them, some are aerial offshoots and have no roots and fail to establish after detachment from mother palm. Survival of offshoots is 40-50 per cent under field conditions. Work on initiation of roots in suckers and better survival is also very important for increasing the area of cultivation in date palm. Sodagar *et al.* (1996) observed that the best time of planting is April–May for higher survival and rapid growth in Kachchh region.

In date palm, basal and aerial offshoot suckers are produced. Basal offshoot generally has well developed root system compared to poorly developed roots in aerial suckers. After 4-6 years of plantation, suckering starts and 2-3 suckers can be removed every year from each mother palm tree for a period of 8-10 years as observed at Abohar.

The offshoots should be well rooted and of proper weight for better survival in the field. Well rooted offshoot of medium weight (10-12 kg) is better in survival and performance in the field compared to poorly rooted small sized or extra bigger sized offshoots (Singh *et al.*, 2003). The offshoots weighing 8 to 12 kg give maximum rooting (Mohammed, 1978). High offshoots, which emerge from stem, perform poorly because of weak root development. Such offshoots of cultivar Medjool even after dipping in IBA (1000 or 3000 ppm) and planting in nursery failed to strike roots under arid irrigated condition at Abohar, Punjab. In an experiment when such offshoots were covered at the base with moist moss grass enclosed in polythene and attached to mother palm treated similarly with wooden shavings, there was 44.9, 7.5 and 22.0 per cent increase in average root length, root thickness and root weight, respectively (Vij *et al.*, 1977). The treatments of suckers with organics like cow dung solution, cattle urine etc enhanced the survival of plants.

Tissue Culture Techniques

Date palm is multiplied through micro-propagation also called *in vitro* propagation in Gulf countries.. By both organogenesis and embryogenesis, plantlets are raised. In India, date palm tissue culture work is limited. Tissue culture has many advantages in date palm cultivation like:

- ✰ Production of genetically uniform plants, a limiting factor at present situation.
- ✰ Propagation of healthy selected female cultivars.
- ✰ No seasonal effect on plants as they can be multiplied under controlled conditions in the laboratory throughout the year.
- ✰ Clones to be propagated are from elite cultivars or from F1 hybrids or previous selections or seed-only originated palms.
- ✰ Ensure easy and fast exchange of planting materials.
- ✰ Economically reliable when large production is required
- ✰ Large scale multiplication.

The physiological performance of these plants is at par with those of suckers (Bhargava *et al.*, 2003). Tissue culture methodology for mass multiplication of date palm mother plants is standardised In addition, supply of genuine plant material to farmers is another responsibility of scientists. There is need to standardize the biochemical markers for identification of superior plantlets produced from tissue culture.

Tissue culture is the only tool to produce a large number of plantlets in date palm to meet out the requirement. In India, groves of seedling date palm are found on the coastal belt in Kachchh region of Gujarat. In Kachchh area, some promising selections have also been made (Pareek and Sodagar, 1986). The promising selections should be multiplied on large scale, as some selections are better than exotic varieties.

At Central Institute for Arid Horticulture, Bikaner, 60 exotic as well as indigenous cultivars/ genotype of date palm are under evaluation. In order to develop phyto-chemical markers for delimitation of date palm cultivars, a study was undertaken using five cultivars. The foliar flavonoids were extracted and separated on TLC plates coated with cellulose. The spots thus developed were compared and subjected to cluster analysis using computer software. The results demonstrated that date palm cultivars can be identified using either specific spot or spot combinations. The results of the cluster analysis demonstrated that the selected five cultivars can be grouped into three distinct groups on the basis of their genetic distances. The group A is represented by cultivars having closer genetic distance with Sewi and the group C was produced with the group A and B. Within these three groups, the cultivars Khuneizi and Dayari have the closest genetic distance (42 per cent).

Tissue culture plants of eight varieties procured from M/s A.V. Thomas Co, Cochin, Kerala, are also under field evaluation and showed good vegetative growth under arid conditions. Only a few tissue culture plants of cultivars Dhamas, Nagal and Khuneizi have started flowering and fruiting after five years of planting. Moreover, it was observed that the number and size of bunches produced on tissue culture plants are comparatively lesser and smaller in size than that of offshoots plantations. Further, evaluation of tissue culture plants is in progress. A private company has initiated planting of tissue culture plants of date palm in Western part of Rajasthan imported from UAE and it is under initial stage of establishment.

Planting

The preparation of pits is very important aspect for planting of suckers. Sucker requires lot of efforts and care for proper establishment. The pits should be at least 1x1x1 m in size and filled with equal proportion of well rotten manure, pond silt and top 30 cm fertile soils. For effective control of termites, Methyl parathion-5 per cent dust (100g/pits) should be added with filling mixture. Planting at distance of 8 x 8 meter is suitable for date palm (Singh and Dhandar, 2007). However, in Kachhch region of Gujarat, date palm suckers are planted at 9 x 9m or 10 x 10 m distance. This space seems to be excessive in the initial years but it is appropriate in view of the long bearing life of date palm orchard even up to 50 years. Secondly, it will also facilitate intercultural operations like ploughing, hoeing, removal of weeds and also will provide adequate space for intercropping. Date palm orchard should be planted in square system. Date palm being dioecious requires plantation of male suckers for proper pollination along with the female offshoots. These should be 5-10 per cent of the total plantation. However, Muthana *et al.* (1976) recommended plantation of male suckers along with female suckers at the rate of one male plant for about 30-40 female plants.

At the planting time, suckers should be treated with Carbendazim (Bavistin) @ 0.1 per cent solution to control fungal infections. Planting of date palm suckers should be done either in spring season *i.e.*, during February-March or during August–September in arid regions. However, May is the best time of planting of suckers for better survival and growth of plant in Kachchh region of Gujarat (Sodagar *et al.*, 1996). The offshoots should be planted in such a way that their tops get straight or they lean very slightly if at all towards the South. Too deep planting of offshoots may be damaged by irrigation water. The crown of the planted offshoots should be at least 10-15 cm above ground level so that the irrigation water does not damage it.

Irrigation

Irrigation plays an important role in establishment of suckers during initial stage. The date palm trees require frequent irrigation for better survival and growth. Water stress conditions during this period may be detrimental for survival of suckers because at the time of plantation, suckers have generally poor root growth and it can not take moisture from deeper soil layer. After planting, the suckers should be irrigated daily during the first month of planting and thereafter these may even be irrigated at weekly intervals. Frequency of irrigation largely depends upon soil texture, season and age of the tree. However, it is known that water requirement of date palm is very high particularly after fruit set till the growth and development of the fruits. During this period moisture stress condition should be avoided. Weekly irrigation should be given because high temperature prevails during April to July when fruiting takes place in date palm. During winter season, irrigation to date palm plantations may be given at an interval of 20 -30 days. On heavy soil, the requirement of water is reduced by 50 per cent. The soil should be kept moist up to a depth of 2.0-2.5 m.The saline water can be used for irrigation purpose as it is a salt tolerant plant.

Mulching

Mulching in date palm with locally available grasses and weeds *viz.* bui (*Aerva persica*) and shaniya (*Crotolaria burhia*) helps in retention of soil moisture even when irrigation interval was prolonged up to one month during April-May. The available soil moisture was 1.52 per cent with bui mulch followed by 0.99 per cent in shaniya mulch and 0.002 per cent in control. In addition, bui mulch produced fruits of better quality in cv. Halawy having 5.2 mm pulp thickness, 47.5 per cent TSS and 0.13 per cent total acidity at doka stage (Chandra *et al.*, 1990). Mulching with black polythene also helps moisture retention, checks weed population and improves fruit yield and quality under hot arid conditions. Pruned leaves can also be used as mulch in date palm to retain soil moisture.

Manures and Fertilizers

Date palm plantations should be provided with adequate farmyard manure, which not only results in better growth of trees but also improves physical properties of soil and productivity. Considering poor soil fertility status of potential date palm growing area in Western Rajasthan where the soils are coarse fine textured, the application of farm yard manure (FYM) may largely help in improving moisture retention characteristics. Application of nitrogen, phosphorus and potassium may be helpful in better vegetative growth and fruit yield of better quality. The application of farm yard manure (50-60 kg/tree) and 400 g N/palm/year results to maximum yield in cv. Khadrawy under irrigated conditions of Abohar, Punjab. The phosphorus and potash should be given @ 250–500g/ tree according to age of plant. In Kachhch, Gujarat, the manure and fertilizers are applied @10 kg FYM and 100 g N,P and K/plant/year with increasing age of plant up to 10 years (Patel *et al.*, 2001). Under Bikaner conditions, application of 1500 g N/tree/year was optimum with respect to better yield and fruit quality in full-grown date palm cv. Halawy.

Training and Pruning

Being monocotyledonous, date palm trees have single stem and do not branch. Diseased, dried and damaged leaves must be removed during winter. Very old leaves must be removed, as these are less efficient in photosynthesis. The leaves should be cut as close to the base as possible. Unwanted leaves should be cut down before manuring period *i.e.* September–October. Cultivars, which have short stalks, should be pruned more severely to provide ventilation to bunch. If possible, spines on the leaves and old stalks should be removed to avoid physical injury to the developing fruits.

Besides removal of leaves, dethroning is also essential to fruit bunches as it helps in pollination, wire net fixing, thinning of bunches, spraying and harvesting of fruits. After fruit set, twisting of bunch is done during April and May, which helps to withstand bunch breakage due to high wind velocity. Removal of one third of their central strands after fruit set shows better development of fruits and hastens ripening and improves fruit quality. Pruning date leaves leaving 6 leaves/bunch and also removal of one-third strands from centre have given good quality date production (Anon., 2002). It is suggested that nearly 70-100 leaves/palm are required for optimum productivity.

Weed Management

Weeds pose serious problem under irrigated conditions in arid region of North–West India. Weed competes for nutrients and moisture, besides causing diseases and insects. Mulching can also reduce weed population and loss of soil moisture. Hoeing of plant basins is required to control weed population and soil aeration. Post emergence application of Glyphosate (Paraquat) @ 3.75 litre/ha controlled both the monocot and dicot weeds under Abohar, Punjab conditions.

Inter-cropping

Information on the effect of intercropping in date palm plantation is lacking. Intercrops may be taken in strips between rows without disturbing the physiology of date palm trees when trees are planted at wider spacing of 8x8m.Citrus tree can be grown in the partial shade of date trees. At Abohar, citrus plantations *viz.* Mosambi, Kinnow gave good results under shade of date palm (Anon., 2001). Sapota tree is planted in interspaces of date palm orchards in Kachchh region (Singh *et al.*, 2004). In initial years of establishment, different annual crops *viz.* cluster bean, moth bean, cowpea can be taken in between rows of plantations without any loss. The crop cultivation under tree's shade is economical and suitable for per unit production. During *rabi* season, interspaces can be effectively utilized for growing mustard and taramira in arid regions. Some shade loving crops like turmeric, seasonal flowers etc. are also grown as inter crops under date palm plantations at Abohar.

Spathe Emergence and Flowering

Flowering takes place from January to March depending upon agro-climatic conditions of the region and varieties. Proper pollination is required for better fruit set. It is dioecious in nature and identification of sex (male or female) of seedling palm trees is possible only after the trees start flowering and fruiting and thus has equal possibility of male and female sex of seedling palm trees. However, 5-10 per cent male plants are required of the total population for pollination of female palm trees.

In North India, flowering starts after winter season *i.e.*, with the start of spring season. Flowering is preceeded by spathe initiation. Date of spathe initiation is governed by various factors like cultivar, place of cultivation, temperature and other climatic conditions prior to spathe initiation. Under Bikaner conditions, spathe emergence starts from last week of January and completed in February at Abohar, flowering takes place by middle of March. Flowering in male palm trees was observed 15 days earlier than female trees, yet fresh pollens were available in abundance for pollination. Most of the spathes, however, emerged during March. Spathe opening starts by the fourth week of February and continues upto third week of April. Similar observations are also reported by Jawanda and Kalra (1972) at Abohar. The anthesis of flower is completed between fourth week of March and second week of April. Mertia and Vashishtha (1985) observed variation in spathe emergence in different years in cultivar Halawy. It was noticed from 3rd to 28th February during 1981, 7th to 23rd February during 1982 and 7 to 9 March during 1983 at Chandan in district Jaisalmer. Vashishtha (1987) reported variation in initiation of flowering during 1985 and 1986 in different date palm cultivars at Jodhpur. Under

Bikaner conditions, Chandra *et al.* (1989) reported that out of 24 cultivars maintained at Date Palm Research Centre during 1987, the spathe initiation commenced early in cultivars Medjool, Muscat, Khadrawy and Bint-a-isha while it was delayed in cultivars Gizaz and Barhee. The maximum duration of flowering was in cv. Khadrawy (35 days) followed by cv. Medjool and Khalas (32 days). A palm tree usually carries 6-12 clusters and one cluster may carry 50-100 strands with 16-31 fruits (Anon, 1989).

Pollination

Pollination in female palm trees is done manually and for that strands of freshly opened male inflorescence are cut from the tree and 3-4 strands are placed between the strands of female flower cluster in an inverted position as soon as flower anthesis in female takes place. It is necessary to tie the pollinated inflorescence with a knot to hold male strand in pollination. The second method of pollinating female clusters consists of dusting of pollens on it within 2 days of anthesis. The pollens may be collected in abundance as a creamy powder from the freshly opened male inflorescence. It should be done in morning hours for obtaining better berry set. If pollens are in short supply, it may be mixed with talcum powder. Pollen storage of date palm can help in pollination of female palm trees where spathe opening takes place much after flowering. in male trees. The pollens can be stored at room temperature or in a refrigerator in a dry air tight container which can be used in later part of pollinating period (Kalra, 1976). The pollen stored in refrigerator (9°C) gave higher fruit set (42.9 per cent) as compared to that stored at room temperature (36.8 per cent) (Gupta and Thatai, 1980). The pollens can be stored either in refrigerator or at room temperature for a period of four weeks after collection without any loss in viability but thereafter viability decreases under room temperature. Under refrigeration, the pollens can be stored for one year but viability of pollen is decreased.

Fruit Growth and Development

The fruit development in date palm passes through various developmental stages. Fruit initiation was the earliest in Sewi cultivar (Ist week of April) followed by Medjool, Bint-Aisha and Sedami cultivars (II nd week of April). It was the latest in cultivars *viz.* Zahidi, Barhee and Nagal Hilali (IInd to IIIrd week of April) under Bikaner conditions during 1987 (Chandra *et al.*, 1992). After pollination, the fruit set stage is known as Hawabuk, when un-pollinated or unfertilized flowers are dropped.

The first stage of fruit is known as pre-doka or gandora or kimiri stage in which the fruits are hard and remain green in colour. The fruits at this stage attain growth and size. Second stage is known as *doka* or *khalal* stage in which the fruits are fully grown, hard and attain yellow, red or its combination depending on varieties characteristics. This stage may further be subdivided into two stages *viz.* early *doka* stage (colour changes to yellow reddish) and late *doka* stage (colour fully changes to yellow/red). Early *doka* stage may also be called as colour break stage. The third development stage during ripening of *doka* berries is known as *dang* or rutab stage, characterized by softening of date berries. When the berry becomes completely translucent or soft, it is known as pind khajoor. At Abohar, Ganganagar, Churu, Sirsa and Hisar, the fruits are usually harvested at doka stage due to early commencement of rains and also due to other climatic reasons whereas the late doka and dang stages can be obtained at Bikaner, Jaiselmer and Jodhpur areas of Western Rajasthan. However, dang to early pind/full pind stages can be obtained in extremely dry areas like Jaisalmer and Western parts of Barmer, Bikaner and Jodhpur (Pareek, 1984).

Therefore, it indicates suitability of date palm cultivation in Thar desert of Western Rajasthan where good quality date palm fruits are harvested. Kalra (1976) reported that under Abohar conditions, *doka* stage is normally attained in July but afterwards most of the fruiting cultivars are seriously damaged due to rains and humidity' during July and August.

Fruit Thinning

Fruit thinning increases the fruit size, fruit quality and enhances ripening of dates. This can be achieved either by even up of the bunches (by cutting back the strands in their length) or by removal of some strands from the bunch or by fruit thinning. The bunch thinning through removal of 1/4 or 1/3 strands from the centre improves size of the fruit. Fruit thinning is essential in date palm owing to several factors *viz.*, (*i*) Excessive crop load may cause shrivelling of fruits, (*ii*) spathe stalks may break down due to excessive fruiting, (*iii*) excessive crop load may cause delay in ripening, alternate bearing and fruits of reduced size and poor quality and (*iv*) excessive load of fruit on palm tree may be more prone to damage by rains and humidity.

Fruit Ripening

Fruit matures when colour turns either yellow or red according to cultivars from green. In general, yellow and red coloured cultivars are available for cultivation. One genotype has observed green fruit colour even at full maturity stage in Kachacch region of Gujarat. Application of ethrel improves quality of dates at colour break stage. Different concentrations of ethephon.(0, 250, 750 and 1000 ppm) as spray were tried at colour turning stage in cultivars, Hayani and Khadrawy. Ripening was advanced by 20 or 28 days by spray of ethephon(1000ppm) in Hayani and Khadrawy, respectively. It also increased fruit weight, reducing and total sugars. Besides, improving colour development in fruits, total acidity was also reduced Spray of ethrel (1000 ppm) at colour turning stage also increased berry weight (18.83 g compared to 8.87 g in control) at Bikaner (Chandra *et al.*, 1992). Average fruit size also increased. Pulp thickness increased from 0.64 cm in control to 0.85 cm, in ethrel sprayed fruits. Maximum TSS (50 per cent) was recorded in ethrel(1500 ppm) application at late doka stage. Meena *et al.*, (2009) reported spray of ethephon(1000ppm) suitable for early ripening in date palm cultivars under arid conditions.

Diseases

Like other fruit crops, date palm plantation is also not free from attack of diseases. In fact, survival of offshoots is a limiting factor for large-scale establishment after planting.

Graphiola Leaf Spot

The major disease of date palm plantation is *Graphiola* leaf spot, caused by *Graphiola phoenicis* and is characterized by numerous dark coloured black pustules with full of yellow spores, formed on both sides of pinnae and on the rachis. There occurs premature leaf fall in infected trees. Foliar application of bavistin or benlate @ 0.2 per cent is suggested as control measure to the disease. Spray of copper oxy-chloride (COC) @ 0.4 per cent solution was effective to control *Graphiola* leaf spot.

Besides leaf spot disease, fruits of date palm are also spoiled by *Aspergillus* sp. This type of spoilage is more in cultivar Medjool where the fungus penetrates deep up to the stone of the fruit resulting in blackening of pulp. Application of Blitox @ 0.2 per cent after fruit set may help in minimizing the incidence of diseases.

The main problem in date palm plantation is survival of offshoots due to rotten main leaf sheath and thus the tree collapses even after a year. For the control of such disease, suckers should be taken from healthy palm. The offshoot should be of proper weight *i.e.*, between 10 and 12 kg with well-developed root system. At the time of planting, suckers should be treated with Carbendazim (Bavistin) solution (0.1 per cent) for 2-5 minutes and in Indole Butyric Acid (1000 ppm) solution to promote rooting and better establishment. After planting, suckers should be drenched with fungicide

(Carbendazim 0.1 per cent or Ridomyl 0.1 per cent) at monthly interval to check crown rot diseases in suckers under field conditions

Insects

Scale insects, red palm weevil and termites are major insects, which have posed problems for date palm cultivation in arid region.

Termites

Termites (*Odontotermes obesus*) feed on the root zone of newly transplanted suckers and gradually tunnel their way upwards and galleries are formed inside the main trunk. The tree may ultimately wither and die away. The termites infest suckers as well as palms. It is the main problem in establishment of plants in arid region. Endosulfan dust (50g/pit) or Methyl parathion dust (2 per cent) should be mixed in the filling mixture in each pit before planting to check the termites attack. After planting, apply Chloropyriphos @ 0.1 per cent at the time of irrigation to control termites.

Scale Insects

It is one of the most dreaded insects for date palm and its incidence may even wipe out the whole crop. This pest is reported causing damage to date palm orchard from each and every part of potential date palm growing areas. *Parlatoria blanchardi* Targ. attacks date palm plantations. The nymphs and adults of scale insect cause damage by sucking the sap from leaves. This insect covers the entire surface of foliage and also feeds on the white succulent tissues at the base of leaf stalks. With increase in its population, they move to the pinnae, which ultimately withers and dries up. To control this insect, removal and burning of all the infested leaflets in severe incidence are required. Spraying with malathion (0.08 per cent) controls the pest population. Sprays of monocrotophos (0.03 per cent), dimethioate (0.05 per cent), parathion (0.05 per cent) or diazinon (0.02 per cent -0.04 per cent) are very effective when done at an interval of 20 days.

Date Stone Beetle (*Coccotrypes dactyliperda* F.)

Both the adults and grubs damage the fruit. The adult beetle penetrates the unpripe fruits and reaches the stone and feeds inside the pulp. They make a circular hole through the pulp. The fruits thus infested ultimately drop. In early ripening variety like Halawy and Khadrawy, the infestation was 22 and 17 per cent, respectively while late varieties like Medjool and Thoory escaped from infestation. The pest was active from June to August. The adult beetle is small and brown coloured and the full-grown larvae is creamy white. Spraying the plants with malathion(0.05 per cent) is effective.

Dried Fruit Beetle (*Haptoncus luteolus* Ericl.)

The adults of beetle are oval, slightly oblong, yellowish brown and 2 mm long. Eggs are laid on fruit. It was reported from Abohar centre. The larvae on hatching penetrate the fruit and feed on the fruit pulp till fully grown and are white or yellow in colour. The pupation takes place in the soil. The pupaa are robust, oval, white or light yellow. Both the adults and larvae feed on the ripe date usually soft and sour dates are preferred by them.

Red Palm Weevil (*Rhynchophorus ferrugineus*)

It attacks only on stem and the grubs are harmful. Due to hollowing of trunk from inside, the top of young tree dries up and rots giving out bad smell and finally trees break and drop off. For control of insect population, dead and infected palms should be destroyed.

Bird Management

The period of fruit growth and development from May to July in date palm coincides with the period when there is almost no field crop available for birds for feeding in the areas where date can be grown as in Rajasthan. Birds cause severe damage to date palm fruits. The bunches at the time of ripening are very much attacked by birds like bulbuls, sparrows, mynas, parrots, purple sunbird and crows. The number of damaged fruits is more during later half of June to July when the fruits are in full *doka, dang* and *pind* stages. Doka fruits are lesser preferred than dang and pind fruits. For control of the birds, individual bunches should be covered with wire netting (3x3mm mesh) after thinning of bunches in a conical shape. It provides maximum ventilation for proper fruit development though it is costly but durable for 8-10 years (Singh and Dhandar, 2005). In Kachchh region, bunches are covered with transparent alkathene sheet (600 gauge) to protect from birds and rain damage. In case of alkathene covering, lower ends should be opened for proper ventilation.

The maximum damage to the fruit is caused at full *doka* and *dang* stages. Protection from bird attack is very much essential during this period. The protection from bird damage can also be done by covering fruit bunches with wire gauge (3x3 mm mesh) at the start of *doka* stage till harvesting.

Harvesting

The date palm fruits are harvested before monsoon. Monsoon occurs from June to September, which coincides with fruit-ripening season and thus harvesting needs to be done before attaining complete ripening of fruits. Pareek (1984) has reported that dates can be harvested after 110-170 days compared to 180-220 days or more required for complete ripening. At Hisar, Abohar, Sri Ganganagar, Churu, fruits are harvested at *doka* stage and in Jodhpur-Bikaner area at late *doka* or *dang* stage. In Jaisalmer-Barmer, fruits are harvested at dang or early pind stage, which is the most desirable stage of harvesting. Unusually, due to successive drought conditions at Bikaner during 1987, pind dates were obtained from most of the date palm cultivars except cv. Barhee, during the third and fourth week of July and the earliest pind was obtained from cultivar Sedami (Chandra *et al.*, 1992). Cultivars Halawy and Barhee have sweet fruits at doka stage and can be used as raw eating dates. Harvesting of fruits at doka stage is done by cutting the bunches and plucking of fruits from the bunch. It should be done at a stage when more than 80 per cent fruits in a bunch attain desirable quality. When fruits are harvested at dang stage for soft dates, only those fruits, which have attained 1/3rd to ½ dang, are picked and thus the harvesting is done of selected fruits. For fresh consumption and chuhhara preparation, fruits are harvested at full *doka* stage.

Yield

Date palm trees start commercial bearing after five to six years of planting. The trees remain productive for a very long period. The trees of even 40-50 years age bear heavily. The fruit yield varies according to age, cultivar and growing sites. About 50 to 60 kg *doka* fruits of good quality can be harvested from a full-grown tree provided the crop is protected against birds damage and adverse climatic conditions. Fruit yield may go up to 100- 200 kg/palm with proper cultural practices.

Postharvest Management

Date fruits are utilized as fresh for raw eating due to delicious, sweet taste and good flavour at *doka* (Khalal). However, *doka* fruit of a few cultivars are astringent in taste. Processing involves curing of dates for preparation of dry dates *i.e.* chhuhara and soft dates (pind khajoor). *Doka* dates are used for preparation of a number of processed products like jam, chutney, RTS, beverage, wine and bakery products.

Processing

For preparation of pind, fully ripened *doka* fruits are taken and dipped in boiling water for 60 seconds which vary with cultivar and then dried in air circulating oven. Katyal and Dutta (1976) reported dipping of fruits in boiling water for about 5 minutes and then drying the fruits either in air circulating oven at 48°C or in sun. Halawy, Medjool, Khadrawy and Thoory cultivars are suitable for making dry dates. According to Kalra (1976), good quality chuharra is prepared at Abohar by dipping doka fruits of Medjool for 10 minutes in boiling water and drying the berries at 50°C for about 120 hr in temperature controlled air-circulating oven. Excessive boiling causes splitting of fruit and loss of sugars. Best quality dry dates (chhuhara) can be obtained from cv. Medjool due to bigger size of berry and recovery (per cent).

Godara and Pareek (1985) reported that the best quality of chuhhara could be prepared from doka fruits of cv. Khadrawy by dehydration in oven at 50°C after 5 minutes boiling water treatment which gave a recovery percentage of 33.3 and organoleptic rating of 33/36 followed by cultivar Shamran which gave a recovery of 35.3 per cent and organoleptic rating of 28.8. At CAZRI Jodhpur, preparation of dry dates (chuhhara) from *doka* fruits of date palm involves boiling of *doka* berries for 20 minutes followed by treatment with potassium meta-bisulphate(1500ppm) for 2 minutes. These fruits are then dehydrated at 50°C for 70 hr. Fruits can be dehydrated in oven fitted with hot air circulating fan (Anon., 1989).

Soft dates are also prepared in North India after curing. At Abohar, soft dates were prepared in cultivars *viz.* Halawy, Khadrawy and Medjool from 1/4, 1/2 and full dang fruits. When the fruits were dried in the sun, 39.9 to 63.7 per cent of the original fresh weight was obtained and it was maximum in cv. Halawy fruits cured at 1/2 dang stage (63.7 per cent). The time taken was 153 hr for full dang fruits of cultivar Khadrawy, whereas 338 hr were required by advanced doka fruits of cultivar Medjool dried in the sun.Kalra (1976) reported that doka fruits of Khadrawy and Shamran were successfully converted into dang by smearing with 1.5 to 2.0 per cent common salt on the berries and after that keeping them for 24 hr in wooden box lined with newspaper.

Soft dates (pind khajoor) were prepared from cultivars Halawy and Medjool and the fruits were picked at full doka to 1/4 dang and kept for 48 hr at room temperature. During this period, 30-80 per cent berries turned 1/3 to 1/2 dang which were removed and used for making pind khajoor by drying at 42 to 48°C in the dehydrator. This process could be completed in 60-65 hours.

Doka fruits are also used for making delicious drinks, chutney, jam, biscuits and pickles. A large number of processed products of date palm are available in the world market. Date biscuits are prepared by adding 10 and 20 per cent dried date pulp in wheat flour, ghee, sugar and it was found suitable on score basis by judges (Singh *et al.*, 2009). Date products can be preserved for a long period by using preservatives and citric acid. At Mundra, recipe for date chutney was standardized. Its green mature fruits can be used for making vegetable and pickles also. Delicious drinks can also be prepared from *doka* fruits of local red and yellow type's berries of date palm from Kachchh region of Gujarat.

References

Anonymous, 1989. Biennial Report. All India Coordinated Research Project on Arid Zone Fruits, NRCAH, Bikaner.

Anonymous, 2001. *Annual Report*, Central Institute for Arid Horticulture, Bikaner.

Anonymous, 2002. *Annual Report*, Central Institute for Arid Horticulture Bikaner.

Al-Ghamdi, A.S., Hassan,A.L. and Jahjah, M., 1988. Evaluation of eight date plam seedling male and their effect on fruit characteristics of three female cultivars. *Arab Gulf J.Sci. Res.*, 6: 175–189.

Bhargava, R., Singh, R.S., Vashishtha, B.B. and Dhandar, D.G., 2003. Photosynthetic parameters in tissue culture and sucker plants of date palm cultivars. Paper presented in 2nd *International Congress of Plant physiology*, Jan.8–12, New Delhi.

Chandra, A., Swaminathan, R., Choudhary, M.L., Manohar, M.S., Pareek, O.P. and Chandra, A., 1990. Performance of date palm cultivars in Thar desert. *Indian J. Hort.*, 47: 1, 28–33.

Chandra, A., Anju Chandra and Gupta,I.C., 1992. *Date Palm Research in Thar Desert*. Scientific Publishers, Jodhpur p.116.

Godara, N.R and Pareek, O. P., 1985. Effect of temperature on storage life of ready to save date juice beverage. *Indian J. Agric. Sci.*, 55: 347–349.

Gopalan, C.; Ramasastri, B.V. and Balasubramanium, S.C., 1985. *Nutritive Value of Indian Foods*. National Institute of Nutrition, ICMR, Hyderabad, p. 87.

Gupta, M.R.and Thatai, S.K., 1980. Receptivity of pistillate flower in date palm cultivare Shamran. *J. Research, (P.A.U.)*, 17(4): 366–368.

Jawanda, J.S. and Kalra, S.K., 1972. Date improvement work at Abohar. *Punjab Hort. J.*, 12: 33–38.

Kalra, S.K., 1976. Date palm research in India. *Indian Farming*, 26: 11–12.

Katyal, S.L. and Dutta, C.P., 1976. Scope and problems in date industry in India. *Ind. Farming*, 26: 3–5.

Meena, R; Yadav, P.K. and Singh, R.S., 2009. Response of different date palm varieties to ethephon spray on yield and yield attributes. *Prog.Agric. (communicated)*.

.Mertia, R.S. and Vashishtha, B.B., 1985. A note on performance of date palm cultivar Halawy at Chadan, Jaiselmer. *Annals of Arid Zone*, 24: 263–64.

Mohammed, S., 1978. Problems in date palm propagation. *Indian Hort.*, 23(3): 15–18.

Muthana, K.D., Puri, D.N. and Singh, U., 1976. Date palm cultivation in Arid region. *Indian Hort.*, 21(1): 10.

Muralidharan, C.M., Tikka, S.B.S. and Piyus Verma, 2008. Date palm cultivation in Kachchh. *Tech. Bull.* S.D. A.U. Mundra, Kachchh p. 36.

Nixon, R.W. and Carpenter, J.B., 1978. Growing dates in the United States. *Agric. Inf. Bull.*, No. 207, USDA, Washington, D.C., pp 63.

Pareek, O.P., 1984. Date palm growing potential of Indian arid zone. *Indian Hort.*, 29: 2–5, 8.

Pareek, O.P. and Nath, Vishal, 1996. Coordinated Fruit Research in Indian Arid Zone–A Two Decade Profile 1976–95. NRCAH, Bikaner.

Pareek, O.P. and Sodagar, N.N., 1986. Date palm grooves in Kachchh. *Indian Hort.*, 31: 21–27.

Patel, R.C., Sodagar, N.N. and Ramdevputra, M.V., 2001. *Kharek* (in Gujarati). Directorate of Extension, Date palm Research Centre, GAU,Sadar Krushinagar, p. 50.

Singh, R.S., Samadia, D.K and Vashishatha, B.B., 2004. Genetic diversity in date palm (*Phoenix dactyliferra* L.). *Indian Horticulture*, 49 (3): cover page II and 24.

Singh, R.S. and Dhandar, D.G 2007. *Khajoor ki Kheti*. Bulletin published by Directorate of Extension, Min. of Agric. (Govt. of India), New Delhi. p 21.

Singh, R.S., Bhargava, R. and Dhandar, D.G. 2003. Status and prospects of Date palm propagation, *Current Agric.*, 27(1–2): 9–16.

Singh, R.S., Bhargava, R. and Dhandar, D.G., 2005. Performance of date palm introductions under hot arid conditions. *Indian J. Plant Genetic Res.*, 18 (1): 65–66.

Singh, R.S. and Bhargava, R., 2009. Effect of Seed treatments on Germination and Growth behavior of *Phoenix* species under hot arid conditions. *J. Tropical Forestry*, 25: (1) 42–48.

Singh, R.S., Bhargava, R., Singh, S.K. and More, T. A., 2009. Postharvest utilization of date palm fruits. In Book of Abstracts "*International Conference on Nurturing Arid Zones for People and the Environment: Issues and Agenda for 21st Century*", November 24–28, 2009, CAZRI, Jodhpur, India, p. 310–311.

Sodagar, N.N.; Murlidharan, C.M. and Ramdeoputra, M.V., 1996. Date palm propagation problems and prospects with special reference to Kachchh. *J. Applied Hort.*, 2: 102–05.

Tewari, R.K. Solanki, K.R., and Bajpaie, C.K., 2001 *Krishi vaniki ke liye upyukt desi khajoor, Phal–Phool*, Jan.–March issue, p. 37–38.

Vashishtha, B.B., 1987. Evaluation of date palm germplasm. Proc. of 4th National Workshop of AICRP on Arid zone fruits. TNAU, Madurai, 16–18 September, 1987.

Vishal Nath, and Singh, R.S., 2003 Boost up date cultivation. *Intensive Agric.*, 41: 3–7.

Vij, V.K.; Kalra, S.K. and Bajwa, M.S., 1977. Studies on rooting of high offshoots in date palm. *Punjab Hort. J.*, 17 (3&4): 135–138.

Zaid, A., 1999. *Date Palm Cultivation*. FAO, Rome. pp 238.

Zaid, A., Hegarty, V. and Kaabi, H.H.S., 2007. *Proceedings of the III International Date Palm Conference. Acta Horticulturae*, 736, pp. 586.

2011, Future Crops Vol. 1 ***Pages 40–54***
***Editor:* Professor K.V. Peter**
***Published by:* DAYA PUBLISHING HOUSE, NEW DELHI**

Chapter 4
GRAIN AMARANTH
(*Amaranthus* spp.)

K. Krishnakumary
Associate Professor,
Kerala Agricultural University, Vellanikkara – 680 656
E-mail: krishnakothottil@rediffmail.com

In the developing world, the painful fact is that around 192 million children and 2000 million others experience micronutrient deficiencies, particularly in pregnant women, leading to infant mortality (Gaddagamath, 2002). Food based strategy to combat malnutrition prevalent among vulnerable segments of the population involves identification of optimal and good sources of nutrients and ensuring their availability. Grain amaranth was selected as one of the four crops from among 400 selected by National Academy of Sciences, USA Select Panel as an underexploited tropical plant with promising economic values (NAS, 1980). A survey conducted by Theisen *et al.* (1980) identified amaranth as one of the five crops which should receive immediate attention as a food crop better adapted to environmental stresses. Both of these studies show concern for identifying broader varieties of plant species, which have potential as major food sources and for developing these for more efficient food production for our world economy.

Grain amaranth is a unique, nutritionally rich non-cereal crop capable of combating malnutrition and is identified as an alternate crop to the present grain crops. It is a multipurpose crop where both leaves and grains are utilized. Commercialization of under exploited vegetables like grain amaranth will help in solving the problem of malnutrition to a larger extent.

Botany, Taxonomy and Physiology

The name amaranth hails from the Greek word for "never-fading flower." The plant is an annual herb, not a "true" grain and is a relative of pigweed, a common wild plant also known as lamb's-quarters, as well as the garden plant we know as Cockscomb. Aside from amaranth being such an attractive plant, it is extremely adaptable to adverse growing conditions. It resists heat and drought, has no major disease problems, and is among the easiest of plants to grow.

Amaranth (*Amaranthus spp*.L.) is a false cereal belonging to the dicotyledonous class in the genus *Amaranthus* and family Amaranthaceae. The genus *Amaranthus* consists of approximately 60 species. However, only a limited number are of the cultivated types, while most are weedy species. The genus Amaranthus has two sections *viz.*, Amaranthus and Blitopsis. Most of the grain amaranth types are under the section 'Amaranthus' with dibasic chromosome number (x=16 and17), terminal panicles and pentamerous flowers. The three principal species considered for grain production include: *A. hypochondriacus*, *A. cruentus* and *A. caudatus*. Mallika (1987) developed a simplified key for the identification of the different species based on inflorescence characters and reported that inflorescence of *A. hypochondriacus* is stiff, *A. cruentus* is moderately stiff and *A.caudatus* is smooth and drooping type. Each plant is capable of producing 40,000 to 60,000 seeds.

Grain amaranth types show considerable variation in morphological characters. The plants can vary from 0.6 to 2.4 m tall and have a main stem axis that terminates in an apical large branched inflorescence. The plant is branched to unbranched; leaf, stem colour range from purple, maroon to light green and leaf shape from lanceolate, rhombic to elliptical. The flowers are unisexual, purple, orange, red or gold in color, and are developed on branched flower clusters (glomerules). A glomerule is described as a dichasial cyme that forms large flowering panicles. Amaranth seeds are born in an utricle, which are classified as dehiscent,semi dehiscent or indehiscent types (Brenner and Hauptli 1990).The amaranth seed is quite small (0.9 to 1.7 mm diameter) and seed numbers vary from 1000 to 3000 seeds/g. Seed colour varies from cream to gold and pink to black. Actual stature of the amaranth plant will vary significantly and depends upon species and environment.

Amaranth, whether grain or leaf, constitutes a veritable solar factory. Being a C_4 plant, it is one of a few dicots in which the first product of photosynthesis is a four carbon compound. The combination of anatomical features in amaranth and C_4 metabolism, results in increased efficiency to use CO_2 under a wide range of both temperature and moisture stress environments, and contributes to the plant's wide geographic adaptability to diverse environmental conditions. Many amaranth types are sensitive to day length, useful in developing varieties.

Environmental Adaptation of the Crop

Grain amaranth grows well in hot dry regions since it uses lesser amount of water than conventional cereals and millets. For seeds to germinate and establish root, amaranths require well moistened soil, but once seedlings are established, grain amaranths do well with limited water. Infact they grow the best under dry, warm conditions. The drought tolerant capacity is due to its robust appearance, hardy and fibrous nature and taproot system which help in soil and water conservation. Grain amaranths are grown in dry land in areas receiving as little as 200 mm of annual precipitation whereas vegetable amaranths are routinely grown in areas receiving 3,000 mm of annual rainfall.

Amaranths can be cultivated in environments ranging from semi arid lands to high Himalayas and generally cultivated within 30° latitude of the equator.

Some of the grain species of amaranth are sensitive to day length. Some strains of *A. hypochondriacus* do not set flowers in summer, however, they do mature during short day conditions of winter. *A. cruentus* sets seeds with long day conditions. *A.caudatus* thrives well in high hills and is known to be a short day species. The leaf area, number of branches and chlorophyll content increase with increase in light (Eze, 1987).

Amaranth grows satisfactorily from sea level to above 3000m amsl, but only *A.caudatus* thrives at altitude 3000m in Andes region and Himalayas. In the tropics, altitudes above 1000m are considered the best.

The grain species *A. hypochondriacus* and *A.cruentus* tolerate high temperature and they are not frost hardy. *A.caudatus* is more resistant to chilling than the other two grain species. The plant grows the best when the daily high temperature is atleast 21° C. Optimum germination was found at temperature varying 16° to 35° C. Growth of the plant ceases at about 8° C and plants are injured below 4° C.

Origin and History

Amaranth is native to South and Central America, but over 400 varieties are found throughout the world in both temperate and tropical climates.

Grain amaranth is a pseudo cereal that played an important role as human food in the ancient civilizations of America. Amaranth was cultivated for more than 8,000 years, dating back at least to the Mayan civilization of South and Central America. It was a staple in the diets of pre-columbian Aztecs who believed in its supernatural powers and incorporated into their religious ceremonies. Many of their ceremonies included the formation of an image of one of the Gods, made with a paste of amaranth grains mixed with honey. In 1516, the conquistadors prohibited the growing of amaranth. But, even now much of the genetic base are maintained because amaranth has continued growing in these areas as a weed.

For the Aztec people, the grain amaranth represented a value on therapeutic, ritual and nutritional level. On the nutritional level, the grain and leaf entered into the preparation of numerous dishes. On the therapeutic level, the Aztec healers used both the seeds and the leaves of the plant to remedy several health problems. On the religious level, this plant is considered as a very sacred plant. During certain religious festivals, figurines fashioned from amaranth dough were offered up to the Pantheon of the Aztec Gods.

Although production of amaranth in Latin America diminished dramatically subsequent to Spanish rule, the positive attributes of the crop led its adoption in other areas of the world. By the 1700s, amaranth had spread throughout Europe as herb and ornamental. In the late 1800s, amaranth is reportedly being grown in mountain valleys of Nile and parts of East Africa. During the 20th century it was grown in China, India, Africa and Europe as well as North and South America. Although US is the leading producer of grain amaranth used in retail food products, the largest production area in the last decade is in China. In India, grain amaranth is cultivated in north Indian hills and also in the plains of Gujarat. In south India, cultivation is steered in hilly tracts of the Western Ghats in tribal areas and considered as the food for poor and marginal peasants. Now, the scientists and institutionalists are rediscovering the importance and potential of these promising grains as source of nutrition for the future.

Nutritional Value of Grain Amaranth

Amaranth is quite nutritious and the interest in amaranth is recently aroused due to it's high nutritive value. Nutritional value of grain amaranth and it's supplementary value with other cereals were described by Reddy *et al.* (1992) and Bhuvaneswari *et al.* (2001). Amounts of vitamin C, iron, carotene, calcium, folic acid and protein are especially high .

Smitha (2006) evaluated the biochemical content of twenty three grain amaranth accessions and the ranges of different nutrients in these accessions are given in Table 4.2.

Table 4.1: The proximate composition of grain and raw leaves of amaranth (100g portions).

Components	*Leaves*	*Grain*	*Components*	*Leaves*	*Grain*
Moisture	86.9 g	9.0 g	Protein	3.5 g	15.0 g
Fat	0.5 g	7.0 g	Total carbohydrates	6.5 g	63.0 g
Fiber	1.3 g	2.9 g	Calories Kcal	36	391
Phosphorus	67 mg	477 mg	Iron	3.9 mg	–
Potassium	411 mg	–	Vitamin A	6100 I.U.	0
Riboflavin	0.16 mg	0.32 mg	Niacin	1.4 mg	1.0 mg
Ascorbic acid (Vit. C)	80 mg	3.0 mg	Thiamin (B1)	0.08 mg	0.14 mg
Ash	2.6 g	2.6 g	Calcium	267 mg	490 mg

Table 4.2: Nutritive value of twenty three grain amaranth accessions.

Components	*Leaves (100g)*	*Grain (100g)*
Protein	2.85–3.81 g	15.15–18.65 g
Starch	4.95–7.1 g	49.75–67.17 g
Fiber	7.57–14.0 g	1.15–4.70 g
Iron	10.42–22.13 mg	33.78–44.58 mg
Calcium	199.50–460.67 mg	559.50–600.75mg
ß-carotene	59995.7–10007.80 µg	134.17–158.37µg
Vitamin C	96.70–500.60mg	3.53–8.32 mg

The amaranth seeds have a unique quality in that the nutrients are concentrated in a natural "nutrient ring" that surrounds the center, which is the starch section. For this reason, the nutrients are protected during processing.

Amaranth seed is high in protein (15-18 per cent) and contains respectable amounts of lysine(5 per cent) and methionine, two essential amino acids which are not frequently found in cereals(Raju, 1990). Joshi and Rana (1991) identified that amaranth protein has twice the lysine content of wheat protein, three times that of maize and the same as found in milk. Using amaranth in combination with wheat, corn or brown rice, results in a complete protein as high in food value as fish, red meat or poultry. The carbohydrates in amaranth grain consist primarily of starch made up of both glutinous and non-glutinous fractions. Gaddagamath (2002) reported that supplementation of the diet with amaranth seeds is the economical way of combating protein energy malnutrition.

The protein is also relatively rich in sulphur-containing amino acids, which are normally limiting in pulse crops. The "protein complement" of amaranth grain is very near to the levels recommended by FAO/WHO and it has a protein score of 67 to 87. By comparison, wheat (14 per cent protein) scores 47, soybeans (37 per cent) scores 68-89 and maize (9 per cent) scores 35. Although amaranth is theoretically close to the ideal, combining it with another grain increases the quality to very close to the FAO/WHO standards. Nutritionist's protein value score chart (100 is considered ideal) for amaranth is 75, cow's milk 72, soybean 68, barley 62, wheat 60, peanuts 52 and corn 44. The protein value of amaranth grains is highlighted when amaranth flour is mixed with other cereal grain flours. When amaranth flour is mixed 30:70 with either rice, maize, or wheat flour, the protein quality (based on casein) rises from 72 to 90, 58 to 81, and 32 to 52, respectively (Bressani 1989). Amaranth seed protein also differs from other cereal grains by the fact that 65 per cent is found in the germ and 35 per cent in the endosperm, as compared to an average of 15 per cent in the germ and 85 per cent in the endosperm for other cereals.

The complete amino acid pattern of *A.hypochondriacus, A.cruentus and A.caudatus* are shown below and comparison of various grains in amino acid content isalso given. The results indicate that methionine, lysine and tryptophan contents of amaranths were significantly higher than cereals.

Table 4.3: Protein amino acid composition (8/16 g of N) of grain amaranth species.

Amino Acids	*A. caudatus*	*A. hypochondriacus*	*A. cruentus*
Lysine	5.3	5.5	5.1
Histidine	2.5	2.5	2.4
Threonine	3.5	3.6	3.4
Cystenine	2.3	2.1	2.1
Methionine	2.4	2.6	1.9
Valine	4.1	4.5	4.2
Isoleucine	3.6	3.9	3.6
Leucine	5.3	5.7	5.1
Tyrosine	2.8	3.3	2.6
Phenylalanine	3.4	4.0	3.4
Serine	5.9	6.3	5.4
Glycine	6.9	7.4	7.0

Table 4.4: Essential aminoacid composition of various grains (per cent protein).

Protein Sources	*Tryptophan*	*Methionine*	*Threonine*	*Isoleucine*	*Valine*	*Lysine*	*Leucine*
Amaranth	1.5	4.6	3.0	3.2	3.8	5.5	5.0
Barley	1.2	3.2	3.2	4.0	4.6	3.2	6.5
Buckwheat	1.4	3.7	3.9	3.8	5.2	5.9	5.8
Maize	0.6	3.2	4.0	4.6	5.1	1.9	13.0
Oats	1.2	3.4	3.1	4.8	5.6	3.4	7.0
Rice	1.0	3.0	3.7	4.5	6.7	3.8	8.2
Soybean	1.4	3.1	3.9	5.4	5.3	6.3	7.7
Wheat	1.2	3.5	2.7	4.1	4.3	2.6	6.3
FAO/WHO Recommendation	1.0	3.5	4.0	4.0	5.0	6.5	7.0

The fibre content of amaranth is three times that of wheat and its iron content, five times more than wheat. It contains two times more calcium than milk. As per the report in American Health and Nutrition (1973), one fourth cup of amaranth flour supplied 60 per cent of the RDA of iron. The availability studies of iron and β-carotene from amaranth indicate that these are excellent sources of iron and β-carotene. Daily inclusion of amaranth in the diet of children alleviates their iron and vitamin- A deficiencies which cause blindness in thousands of children each year. Besides, it is an excellent source of folic acid which increases the blood haemoglobin level.

The amaranth foliage is edible with biological value comparable or even superior to common vegetables of popular consumption consisting of less oxalates and nitrates. Mallika (1987) reported that grain amaranth types have only lower levels of oxalate and nitrates as compared to vegetable amaranth.

Amaranth grain consists of approximately 5 to 9 per cent oil generally higher than other cereals. The lipid fraction of amaranth grain is similar to other cereals, being approximately 77 per cent unsaturated, with linoleic acid being the predominant fatty acid. The lipid fraction is unique due to the unusually high squalene content (5 to 8 per cent) of the total oil fraction. Squalene and its hydrogenated squalene are relatively high value lipids. Squalene is an expensive terpenoid compound, derived primarily from shark and whale liver oils. It has many applications in the pharmaceutical and cosmetic industries. Also present in the amaranth oil fractions are tocotrienols. These are rare forms of vitamin E shown to inhibit the key regulatory enzyme in cholesterol biosynthesis known to effect lower cholesterol levels in mammalian systems. The major fatty acids in *Amaranthus* oil consisted of palmitic acid (19.1-23.4 per cent), oleic acid (18.7-38.9 per cent), and linoleic acid (36.7-55.9 per cent).

Pant (1985) studied the effect of heat processing (popping) on the nutrient composition, amino acids content and protein nutritional quality of grain amaranth. He reported that although popping did not affect the proximate composition of the grain, there was significant loss in total lysine content of the protein of the popped grain. This loss was more in home popped grains (35 per cent) which included some amount of heat damaged grains than in commercially popped grains (25 per cent). The net protein ratio (NPR) of home popped grain was consistently lower compared to that of commercially popped grains.

Table 4.5: Approximate fatty acid composition of oil (per cent by wt).

Fatty Acid	*Per cent by Weight*
C16: 0	13.4
C18: 0	2.7
C18: 1	20.4
C18: 2	62.1
C18: 3	1.1
C18: 4	0.7
Squalene	6.96

Extrusion is one of the best methods to maximize the nutritive value of amaranth. The extruded amaranth diet at 24 per cent moisture reached higher corrected protein efficiency ratio (corrected-PER) than all the other treatments. Each of the extruded amaranth diet reached greater net protein ratio (NPR) values than casein control and raw and toasted amaranth diets. The differences between extruded and raw amaranth samples were probably due to destruction of toxic factors in the raw grain, in combination with increasing palatability, improvement of the carbohydrate use and it was not due to increased protein digestibility. The results indicate that extruded amaranth contains very high quality protein.

Amaranth is highly suitable for utilization in infant formulas and cooked amaranth has a total digestibility of about 90 per cent and because of this ease of digestion, it has traditionally been given to those recovering from an illness or ending a fasting period.

The presence of rather high amounts of oxalic acid and nitrates places some limitation on the quantity of amaranth leaves which can be consumed daily. Excessive amounts may result in a level of oxalic acid that begins to reduce the availability of calcium in humans. This is especially a concern if calcium intake levels are low to begin with. Nitrate in vegetable portions of amaranth is a concern because it is hypothesized that nitrates may be chemically changed in our digestive tracts into poisonous nitrosamines. According to scientists, over 100 g/day may be an unsafe amount to eat.

Applications and Uses

Amaranth has a wide variety of applications in the food industry. It can be puffed or ground into whole grain flour. Amaranth can be used in a number of food products including breakfast cereals, breadings for meat, fish, poultry, or vegetables, confectionery products, salad condiments (in croutons, or puffed grain sprinkled directly on salads), baked products (breads, muffins, cookies, biscuits), extruded snack foods and chips, pastas, health foods, soups and dietetic products. It is used in the preparation of weaning foods. Rathod and Udupi (1991) developed weaning mixtures using melted, roasted and puffed amaranth grains, which could be used at the community level to produce low cost nutrient rich weaning mixtures. Amaranth Food Company released "Amarlac" and "Amarameal" for children and pregnant women in Guatemala. Amaranth seeds can also be popped like popcorn, expanding to about 10 times the original volume. The popped amaranth has a toasted, nutty flavour, and can be used in confections bound with sorghum, molasses or honey and in high-energy granola bars. Because amaranth has very little gluten, it is ideal for use in gluten free diets. It is recommended to blend amaranth flour to whole wheat flour in sweet dough. When amaranth seed is boiled in water and then chilled, it develops gelatinous texture possibly due to a waxy starch similar to cornstarch. Therefore, it can be used to prepare jam-like fruit spread using no pectin and very little sweetener. Because of the unique characteristic and nutritional profile of the seed, amaranth is identified as a new potential crop for certain specialized applications, health foods and baked goods.

The unique aspect of amaranth grain starch is that the size of starch granules is much smaller than found in other cereal grains. Due to the unique size and composition of amaranth starch, it is suggested that the starch may possess unique gelatinization and freeze/thaw characteristics which could be of benefit to the food industry (Becker *et al.*, 1981and Lehman, 1988). Several considerations for the use of amaranth starch in food preparation of custards, pastes, and salad dressing are published (Singhal and Kulkarni 1990a, b,c). Starch also finds application in high quality plastics, cosmetics and other industries. The other new uses of amaranth are natural dyes, pharmaceuticals (laxative) and squalene, a high priced material found in amaranth grain but normally obtained from shark livers and used in cosmetics (National Research Council, 1984).

The crop is a tribute to beauty and genetic diversity. Variegated leaves and showy inflorescence make the crop attractive and it can be grown in groups, rows or mixed with other plants. Amaranth is also used in specialty artesian dyes. Red dye from amaranth leaves is used to colour food and beverages(Hauptli and Jain,1980).

Amaranth is used in various cultures in some very interesting ways. In Mexico, it is popped and mixed with a sugar solution to make a confection called "alegria" (happiness), and milled and roasted amaranth seed is used to create a traditional Mexican drink called "atole.". Peruvians use fermented amaranth seed to make "chicha" or beer. In Ecuador, the flowers are boiled, then the colored boiling water is added to "aquardeinte" rum to create a drink that "purifies the blood," and regulates the menstrual cycle. In certain areas, the flowers are used to treat toothache and fevers and as a food colourant.

In India, grains are used in different culinary preparations, white grains are popped and mixed with milk and sugar (sweet porridge, Kheer), popped grains in the form of sweet puddings (Halwa), white grains mixed with maize and finger millet flour for making chapattis. Popped grains are mixed with sugar syrup to make sweet balls (laddoo). In North India, popped grains mixed with honey to make flat round breadings, soaked and grinded seeds mixed with wheat and finger millet flour and salt and spices to make delicious paratha and very crisp pakoras are also made of amaranth flour (Thomas, 2005). In Nepal, amaranth seeds are eaten as gruel called "sattoo" or milled into flour to make chappatis. In Himachal Pradesh, the decoction of grains is very useful and effective in treatment of foot and mouth disease of animals. As per Parker (1985), the stem and leaf juices of grain amaranth are used in the treatment of kidney stones by the tribes in Tamil Nadu and Kerala hills. The amaranth leaves are used as a vegetable either boiled or fried.

Crop Improvement

Germplasm Collection, Breeding and Research Centres

Current interest in amaranth resides in the fact that it exhibits a high nutritional value, a C4 photosynthetic pathway, a great amount of genetic diversity and phenotypic plasticity (Dowton 1973; Hauptli and Jain1977; Jain *et al.* 1979; Kauffman 1982; National Research Council, 1984). Amaranth germplasm is available in 11 countries (Sauer, 1967; Toll and Von Sloten, 1982).

There are presently three main centres for amaranth germplasm collection and breeding.

1. USA: Rodale Organic Gardening and Farming Research Centre, Kutztown and the National Seed Laboratory, Fort Collins, Collorado,USA
2. Nigeria: National Horticultural Research Institute and Other Institutions at Labadon.
3. India: National Bureau of Plant Genetic Resources, New Delhi-12

In India, initial efforts to study grain amaranths were made in sixties by the Division of Plant Introduction, Indian Agricultural Research Insititute, New Delhi (now NPBGR),which then held small germplasm collection(less than 50) at Shimla station, assembled from indigenous sources and from a few foreign countries *viz.*, Nepal and USA. From 1978 onwards, more efforts were made for germplasm collection and evaluation of the existing genetic resources to enrich these collections and to analyse and study the potential of this dual purpose crop (Joshi, 1981 a, b).

Himalayas offer a wealth of diversity in terms of landrace populations of grain amaranths. These crops are cultivated sporadically in a large stretch of land of higher hills of Himachal Pradesh, Uttarakhand, Jammu and Kashmir and North Eastern hill regions for ceuntries. They offer a great deal of variability in terms of various traits of economic importance for collection and utilization.

NBPGR, Regional Stations, Delhi and Shimla have comprehensive collections of genetic resources of grain amaranth from different areas. Collections are also made by the Vivekananda Parvatiya Krishi Anusandhanshala (VPKAS) Almora; GB Pant University of Agriculture and Technology, Ranichauri; and ICAR North Eastern Hill Complex, Shillong.

Cultivars and Varieties

The improved varieties of grain amaranth used in the U.S. were developed at the Rodale Research Center in Kutztown, Pennsylvania. Initially, many farmers grew one of the shorter, earliest maturing Rodale line, called K432.

The first line of amaranth registered by The Crop Science Society of America was Montana-3 (MT-3) (Schulz-Schaeffer *et al.*, 1989a). MT-3 is *A. cruentus* from a selection of RRC-1041 obtained from the Rodale Research Center, Emmaus, PA. MT-3 is light green flowered, produces white seeds, and was selected for uniform height and high yields. In 1989, Montana-5 (MT-5), an *A. cruentus* variety was registered from a selection of RRC-425 (Schulz-Schaeffer *et al.* 1989b). MT-5 is green flowered and has white seeds. MT-5 was selected for advanced dry-down characteristics, with the seed head dry down simultaneous to stalk dry down. The third amaranth selection released by Montana State University was named 'Amont' (Schulz-Schaeffer *et al.* 1991). 'Amont', a selection from MT-3, is green flowered with white seeds and has resistance to lodging. The plant height of `Amont' varies 91 to 244 cm depending upon the moisture availability throughout the season and plant population.

In 1992, a cooperative amaranth breeding programme between Rodale Research Center and the University of Nebraska resulted in the release of the cultivar 'Plainsman', an interspecific hybrid between *Amaranthus hypochondriacus* L., a gold seeded selection from Mexico, and *Amaranthus hybridus* L., a black seeded selection from Pakistan. Single plant selection in earlier generations (F_2 to F_5) and mass selection in advanced generations (F_6 to F_7) were used (Baltensperger *et al.* 1992). Plainsman, is usually 1.6 to 2 m tall in Missouri. Plainsman has a single unbranched stem, with a large mass of tiny maroon flowers clustered in an inflorescence at the top of the plant and golden seeded. Grain heads of Plainsman can range from 12 to 36cm long and from 6 to 24 cm wide. Seeds are small, about 3/25 cm. Plainsman's maturity is very early in Western Nebraska, reaching maturity in 110 days after planting. Evaluation tests in Nebraska, Colorado, Missouri, Minnesota, and South Dakota indicated that 'Plainsman' was one of the most promising amaranth cultivars for the United States and most farmers have switched to Plainsman, a release of the University of Nebraska Experiment Station, which is a selection of Rodale's K343.

Genetic Variability and Promising Types in India

In TNAU, Coimbatore, Mohideen *et al.* (1983) collected and evaluated grain amaranth types in field and noted significant variations in plant morphological characters. Variations for many characters among different species of amaranths were reported by Das *et al.* (1991), Devadas (1982), Verma (2002), Priya (1998) and Hossain and Rehman (1999). Grain amaranths were free from leaf spot disease, an important disease in rainy season in Kerala,India(Krishnakumary,2000)

In the process of germplasm evaluation, several promising types were identified. One of the top yielders, selection IC-42258-1(*Amaranthus hypochondriacus*), appropriately named 'Annapurna', was released by NBPGR, Regional Station, Shimla (Joshi, 1985). It's average grain yield is 22q/ha. It has great promise for the rainfed land of the hills as well as to the drought prone areas and can be recommended for cultivation under Himalayan watershed. Annapurna gave 68.9 per cent increase in grain yield over VL-21, a promising selection from Almora centre. This variety is tall, medium late maturing with dark green broad leaves. The inflorescence colour is yellowish green with long compact terminal and lateral spikelets. The inflorescence length is 74 cm. The seed shape is medium bold, round bulging at centre and tapering at periphery. Seed colour is creamish white. The popping quality of the seed is excellent and the protein content of dry seed is 14.5 per cent. The other varieties of grain amaranths suited for cultivation in different parts of the world are RI04, 20 USA, Jumla, VL-21 and S.K.Nagar.

In Kerala, field evaluation of 23 grain amaranth accessions revealed wide variations in important morphological and biochemical characters (Smitha, 2006). The plant height ranged from 53 to 138 cm and the grain yield/plant from 49 to 128 g. The accessions GA-25, GA-31 and GA-20 emerged as high yielders in Kerala. Biochemical analysis revealed that GA-15 and GA-22 had high protein content in grains where as GA-31 had high iron and beta carotene content in leaves. Bhatiwada (2007) prepared value added products by substituting grain amaranth flour with other flours and found that chapathi prepared by substituting wheat flour up to 30 per cent grain amaranth flour was acceptable. In biscuits and puttu, even upto 50 per cent substitution of grain amaranth flour was highly acceptable.

Cultivation

The plant is easy to grow sprouts quickly and needs very little special care. Although the best nurtured by good soil and moisture, it is capable of surviving both in drought conditions and poor soil, a fact probably responsible for it's lasting through intervals of near neglect, and has earned it the appellation of "never fading."

Planting

Grain amaranth is a shade sensitive crop and yields are comparatively high in sunny days. The crop is generally sown after the monsoon showers in July- August. Traditionally seeds are broadcasted, but for better crop stand, transplanting is practiced. Transplanting results in efficient use of seeds and allows the growing areas to be weeded just before the seedlings are transplanted. In direct sowing, due to the very small size of seeds, some may be washed way. Though extra work is involved in transplanting, seed quantity can be saved and a two-week jump on the weeds can be gained. Planting in a nursery also reduces risk of loss due to diseases like damping off. Direct seeding involves much less labour, but incurs a greater risk of poor stand due to diseases and predators of young seedlings and competition with weeds in the crucial initial couple of weeks. If direct seeding is used, sowing should probably be in rows to facilitate cultivation. The optimum planting depth depends upon soil type.

Nursery

To initiate amaranth production, the farmer should prepare a seedbed, preferably on soils having a pH above 6.0. A firm moist seed bed with soil temperatures above 15°C is required to establish a good plant stand. Seeding rate of 1.5-2.0 kg/ha is recommended. Many producers have successfully planted amaranth with a standard grain drill. Field establishment is satisfactory with soil temperature between 18.5 and 24°C (Weil and Belmont,1987)

Whether sown in the nursery or field, seeds need to be planted 3–4½ cm deep for good germination. Since seeds are very small, they should be mixed with sand and sown in rows at a depth of 1 cm. Seed germinates within 4-5 days after sowing and needs maximum care. Because of the shallow depth, special care must be taken to prevent drying out of the soil until plants are established. Transplanting or thinning may be done in about two to three weeks when plants are 25-30 cm tall. Delay in transplanting for even one week can reduce total yield. Recommendations for plant spacing vary widely for grain amaranth. One recommendation is to space 20 cm between plants and 75 cm between rows.

Field Preparation and Fertilizer Application

The field must be well levelled and two to three ploughings are sufficient for sowing of amaranth seed. Before sowing, application of 10-25 tons of farm yard manure/ha has given very good yield. Fertiliser recommendation varies with locality. Application of 150 kg SSP and 80-100 g K_2SO_4/ha are also recommended. A dose of 70-90 kg/ha of urea should be top dressed in three split doses. Under hilly condition in India, NPK at the rate of 20:30:20 kg/ha meet the full requirements of the crop (Rana and Rameshwar, 2003).

Interculture

During the growth period, crop is made free from weeds and must receive adequate moisture. One or two weedings or hoeings are sufficient for controlling weeds. Hoeing between the rows not only checks weeds, but also reduces number of irrigation.

Harvest and Yield

Amaranth grains mature earlier and the plant dries up quite late. If the heads are allowed to remain till the plants dry up, heavy shattering of grains is noticed which leads to heavy grain loss. It is important to harvest at the right stage. Rub the flowers with your hands and if lots of seeds are released easily, it is ready to harvest. Seeds can be chewed to test whether they have passed beyond the "dough stage".The flowers should still be quite damp at this point, and not at all dried out. The heads are cut from the stalk and side branches, when the plant is still somewhat green and start weathering and kept for sun drying for 6-7 days in the threshing yard. Closer plant spacing may be advisable for mechanical harvesting. Once dry, the seeds are knocked from the heads, sifted through an ordinary window screen, and winnowed to remove chaff. Threshing is done by beating the produce and winnowed like other cereals. If the seed is very damp, spread it thinly on a tray in the sun and dry for a day or so. Then winnow the seed by pouring it from one bowl to another in a steady breeze. Grain should be dried to about 9 per cent moisture for safe storage. It is reported that grain remains viable for up to seven years.

Amaranth can be effectively harvested by grain combines which are seed tight and by reducing the cylinder speed just high enough to effectively thresh the seed heads. Amaranth grain yields are extremely variable depending upon the cultivar, growing season and soil moisture availability. Yields of 450 to 700 kg/ha in dryland and 900 to 2,000 kg/ha under irrigated or high rainfall would be considered reasonable.The varieties introduced into USA by the Rodale Institute and other Centres, like Plainsman and K432 produce on an average 2 tons/ha. Yields up to 6 tons/ha have been achieved on certain experimental plots.The seeds of the amaranth plant are incredibly tiny and a gram will contain 1000–3000 seeds.

Crop Protection

Hymenis recurvalis, *Cletus* sp., lygus bug (*Lygus lineolarius*), spinach flea beetle, the amaranth weevil (*Conotrachelus seniculus*) and tarnished plant bug are presently the most important insects pests of grain amaranth. Lygus bugs are cosmopolitan pests. They feed on hundreds of plant species. Lygus bugs damage grain amaranth by feeding on the meristematic tissue, developing floral buds, immature blossoms and developing embryos. It causes substantial yield losses both by preventing flowers from developing into seeds and by reducing seed weight. The amaranth weevil causes severe damage to the roots resulting in lodging and predisposition to root diseases. Stem borer is another problem in Africa and Asia which causes high degree of lodging in plants. Seed beds should be guarded against ants and termites. Root knot nematodes also slightly affect *Amaranthus sp.*

Damping off, leaf blight, white rust, mycoplasmal and viral are identified as the serious diseases of grain amaranth in India. Reddy *et al.* (1980) reported a bacterial leaf spot disease of amaranths in India caused by *Xanthomonas amaranthicola*. A severe mosaic disease caused by a strain of cucumber mosaic virus in *Amaranthus caudatus* was reported in Himachal Pradesh by Sharma and Chowla (1987) damaging the crop. The other virus affecting the crop is ivy vein clearing virus, alfalfa mosaic virus and beet western yellow virus.

Productivity Constraints in Grain Amaranth

Compared to the cereals, yield of grain amaranths is very low and the major productivity constraints of the crop are given below:

1. Non availability of improved variety(s) for cultivation
2. Lack of information on agro techniques for high seed production
3. Heavy lodging and seed shattering
4. Little or non utilization of seed for various domestic and agro industrial products by mass consumers
5. Lack of public awareness about the organoleptic taste and quality of both grains and greens
6. Grain amaranth is generally grown in lands of marginal productivity without fertilizers and pesticides. Lack of information on scientific cultural practices results in poor yield potential

Conclusion

Eighty two(per cent)of Indian population live in rural areas and are primarily engaged in agricultural pursuits.Low cost cereals and vegetables form the major constituents of their diet. An increase in the production and consumption of amaranth, particularly in tropical regions of the world, should be considered due to it's excellent nutritional value particularly high content of quality protein, rich in amino acids. It is a good source of Fe and β-carotene and daily inclusion of amaranth in the diet of children and pregnant women alleviates/eradicates malnutrition which ultimately results in nutritional upliftment of the society. Just like any alternative crops, cultivar improvements, production and utilization research etc. are needed. The colourful appearance of the crop and it's chequired history generated interest in this crop. It's good nutritional characteristics combined with it's variety of potential uses illustrated the importance of continued work with this rediscovered crop. The crop can be grown in dry and drought prone areas and hence it becomes an ideal choice for the future, when the much feared global warming and other weather abnormalities become a reality in the near future. This crop ensures nutritional and lively food security for people inhabiting in marginal, hot and dry regions of the world.

References

American Health and Nutrition. 1973. http: //www.organic harvest. com. limg/pdt. icon. grain amaranth flour.

Baltensperger, D.D., Weber, L.E. and Nelson, L.A., 1992. Registration of 'Plainsman' grain amaranth. *Crop Sci.*, 32: 1510–1511.

Becker, R., Wheeler, E.L., Lorenz, K., Stafford, A.E., Grosjean, O.K., Betschart, A.A. and Saunders, R.M., 1981. A composition study of amaranth grain. *J. Food Sci.*, 46: 1175–1180.

Bhatiwada, N., 2007. Standardization and quality evaluation of grain amaranth flour supplemented food products. *M.Sc (Home Science) Thesis*, Kerala Agricultural University, Thrissur.

Bhuvaneswari, G., Sharada, G.S. and Patil, U.C., 2001. Nutrient composition of grain amaranth varieties. *Karnataka. J. Agric. Sci.*, 14(3): 869–870.

Brenner, D. and Hauptli, H., 1990. Seed shattering control with indehiscent utricles in grain amaranth. Legacy 3: 2–3 Amer. Amaranth Inst., Bricelyn, MN.

Bressani, R., 1989. The proteins of grain amaranth. *Food Rev. Int.*, 5: 13–38.

Cole, J.N., 1979. *Amaranth: From the Past, for the Future.* Rodale Press, Emmaus, USA.

Das, P.K., Dey, G. and Ghosh, S.C., 1991. Genetic variation for quantitative traits and yield components in grain amaranth (*A. hypochondriacus*). *Indian J. Agric. Sci.*, 35(3): 197–201.

Devadas, V.S., 1982. Screening for non bolting type(s) of amaranths suited for year round planting. *M.Sc. (Hort.) Thesis*, Kerala Agricultural University, Thrissur, p.85.

Downton, W.J.S., 1973. *Amaranthus edulis*: A high lysine grain amaranth. *World Crops*, 25 (1): 20.

Eze, J.M.O., 1987.Growth of *Amaranthus hypochondriacus* under different day light intensities in the dry season in southern Nigeria. *Experimental Agriculture*, 23(2): 193–200.

Gaddaganath, P.B., 2002. International vegetable conference: The philosophy and thereafter, Souvenir, *Int. Conference on Vegetable*, Bangalore No.11–14, p.111.

Hauptli, H. and Jain, S.K., 1977. Evaluation of agronomic potential and genetic variation in amaranth collection. *Agronomy Abstracts*, American Society of Agronomy, 58: 485–488.

Hauptli, H. and.Jain, S.K., 1980. Genetic polymorphism and yield components in a population of amaranth. *J. Heredity*, 71(4): 290–292.

Jain, S.K., Vaidya, K.R. and Joshi, B.D., 1979. Collection and evaluation of Indian grain amaranth. *Proceedings of the Second Amaranth Conference.* Rodale Press, Inc., Emmaus, PA, USA, pp. 123–128.

Joshi, B.D., 1981. *Catalogue on Amaranth Germplasm.* Regional Station, National Bureau of Plant Genetic Resources (NBPGR), Phagli, Simla, India.

Joshi, B.D., 1985.Annapurna–a new variety of grain amaranth. *Indian Farming*, 25(8): 29–31.

Joshi, B.D. and Rana, R.S., 1991.Monograph No. 3. *Grain amaranth–The Future Food Crop.* NBPGR, Shimla, p. 86, 115.

Krishnakumary, K., 2000. Genotypic and seasonal influence on leaf spot disease in amaranth. *Ph.D. (Hort.) Thesis*, Kerala Agricultural University, Thrissur.

Kauffman, C.S., 1982. Improved grain amaranth varieties and their yields. Rodale Research Report, Emmaus, USA pp. 45 *Agriculture*". (Ed). W. Lokeretz, Proyer Press, New York, p.289.

Lehman, J., 1988. *Carbohydrates of Amaranth.* Legacy 1: 4–8 Amer. Amaranth Inst. Bricelyn, MN.

Mallika, V.K., 1987. Genome analysis in the genus *Amaranthus. Ph.D. (Hort.) Thesis*, Kerala Agricultural University, Thrissur, p.103.

Mohideen, K.M., Muthukrishnan, C.R., Shanmugavelu, K.G., Rangaswami, P. and Vadivel, E., 1983. Evaluation of grain amaranth types at Coimbatore. *South Indian Hort.* 31: 11–14.

NAS (National Academy of Sciences), 1980. Under exploited tropical plants with promising economic value. Washington, D.C. pp.14–19.

National Research Council, 1984. Amaranth-modern prospects for an ancient crop. *National Academy Press*, Washington D.C. USA, pp. 80.

Pant, K.C., 1985. Effect of processing (popping) on protein and nutritional quality of grain amaranth. *Nutrition Reports International*, 32(5): 1089.

Parker, A., 1986. Amaranth: Origin and arrival in the old world reconsidered. *A working paper on amaranth*, Dept. Geography of Oregon, USA.

Priya, U.P., 1998. Screening amaranth genotypes (*Amaranthus spp.*) for yield, quality and resistance to biotic stress. *M.Sc. (Hort.) Thesis*, Kerala Agricultural University, Thrissur, p. 103.

Raju, S.N.P., 1990. Effect of water stress in grain amaranth species. *Amaranth Newsletter* (2): 12.

Rana, S.S. and Rameshwar, E.U., 2003. Response of grain amaranth to phosphorous and potassium under irrigated conditions. *Himachal J. Agri. Research*, 29(1/2): 11–15.

Reddy, B.N., Chakravorthy, B.P. and Shekawat, P.C., 1980.Nutritional requirements of *Xanthomonas amaranthicola*. *Indian Phytopathology*, 32(4): 581–586.

Reddy, M.V., Padmavathi, P., Vijayalakshmi, U. and Aruna, K., 1992. Nutritive value of grain amaranth. *J. Res.*, 20(1–2): 18–21.

Rathod, P. and Udipi, S.A., 1991. The nutritional quality and acceptability of weaning food incorporating amaranth. *Fd. and Nurt. Bull.*, 13: 58–64.

Sauer, J.D., 1967. The grain amaranths and their relatives: A revised taxonomic and geographic survey. *Ann. Mo. Bot. Gard.*, 54: 103–137.

Saunders, R.M. and Becker, R., 1983. Amaranthus: A potential food and faced resource. In: *Adv. Cereal Sci. Technol"*, Vol. VI, Cereal Chemists, St. Paul, USA.

Singhal, R.S. and Kulkarni, P.R., 1990a. Some properties of *Amaranthus paniculatas* (Rajgeera) starch pastes. *Starch/Starke*, 42: 5–7.

Singhal, R.S. and Kulkarni, P.R., 1990b. Utilization of *Amaranthus paniculatas* (Rajgeera) starch in salad dressing. *Starch/Starke*, 42: 52–53.

Singhal, R.S and Kulkarni, P.R., 1990c. Studies on applicability of *Amaranthus paniculatas* (Rajgeeraa) starch for custard preparation. *Starch/Starke*, 42: 102–103.

Schulz-Schaeffer, J., Stallknecht, G.F., Baldridge, D.E. and Larson, R.A., 1989a. Registration of Montana-3 grain amaranth germplasm. *Crop Sci.*, 29: 244–245.

Schulz-Schaeffer, J., Webb, D.M., Baldridge, D.E., Stallknecht, G.F. and Larson, R.A., 1989b. Registration of Montana–5 grain amaranth germplasm. *Crop Sci.* 29: 1581.

Schulz-Schaeffer, J., Baldridge, D.E., Bowman, H.F., Stallknecht, G.F. and Larson, R.A., 1991. Registration of 'Amont' grain amaranth. *Crop Sci.*, 31: 482–483.

Smitha, K.S., 2006. Morphological and biochemical studies in grain amaranth (*Amaranthus* spp.). *M.Sc. (Hort.) Thesis*, Kerala Agricultural University Thrissur.

Sharma, P.N. and Chowla, S.C., 1987. Studies on a mosaic disease of amaranth (*A.caudatus*) in Himachal Pradesh. *Indian Journal of Mycology and Plant Plathology*, 16(3): 349–350.

Theisen, A.A., Knox., E.G., Sprague, H.B. and Mann, F.L., 1980. Possibility of introducing food crops better adapted to environmental stress. Washington D.C., National Science Foundation, Report No. NSF/RA 780038.

Thomas, R., 2005. Amaranth–the solution to world hunger, malnutrition, diseases and poverty. *http: //intinite* play/themovie. Com/amaranth. Aspx.

Toll, J. and Von Sloten, D.H., 1982. Directory of germplasm collections. *Int. Board Plant Genet. Resour.* Rome, Italy.

Verma, P.K., Gupta, S.N., Deen, M.K. and Malik, B.P., 2002. Genetic divergence in grain amaranth. *Annals of Biology*, 18(1): 35–38.

Weil, R.R. and Belmont, G.S., 1987. Interactions between winged bean and grain amaranth. *Amaranth Newsletter*, 1: 4.

2011, Future Crops Vol. 1 ***Pages 55–78***
***Editor:* Professor K.V. Peter**
***Published by:* DAYA PUBLISHING HOUSE, NEW DELHI**

Chapter 5

KAIR

[*Capparis decidua* (Forsk.) Edgew.]

Dheeraj Singh[1], H.R. Mahla[2], M.K. Choudhary[1], H. Dayal[1], M.L. Meena[1] and H.S. Jat[3]

[1]KVK, CAZRI, Pali-Marwar, Rajasthan
E-mail: dheerajthakurala@yahoo.com
[2]Senior Scientist, Central Arid Zone Research Institute, Regional Research Station, Jaisalmer - 345 001, Rajasthan
[3]Senior Scientist, CSSRI, Karnal, Haryana

Kair belongs to Capparaceae (Kingdom: Plantae; Sub-kingdom: Tracheobionta; Superdivision: Spermytophyta; Division: Anthophyta; Class: Magnoliopsida; Subclass: Dilleniidae; Order: Capparidales). The family comprises of 39 genera and 650 species distributed through warm regions. The genus *Capparis* comprises 250 species including shrubs, trees and woody climbers. This genus is represented in Egypt by six species. There are reportedly 26 species of this genus occurring in India (Anon, 1992). *Capparis decidua* (Forsk.) Edgew. (Syn. *C. aphylla* Roth; *Sodada deciduas* Forsk) commonly known as Kair, is an important indigenous shrub found growing on farm boundaries, orans, gochars and wastelands, widely distributed in arid and semi-arid tracts of India. It is a densely branched shrub that reaches a height of 4-5 m, with a clear bole of 2.5 m. Its branches are tender waxy with rough corky, and gray bark. Kair has the ability to survive in various habitats having good soil binding capacity. It has fair tolerance to salinity and alkalinity, can help improve fertility of sand dunes and reduce alkalinity. Kair grows unattended and unprotected on barren lands. On account of its deep tap root system, scanty foliage, mucilaginous sap and tough conical spines, the plant had acquired special xerophytic mechanism and these features make the species suitable for cultivation on a large scale to combat soil and wind erosion on sandy wastelands (Gupta *et al.*, 1989).

This indigenous multipurpose shrub has played an important role in the rural economy of Western Rajasthan as it provides food (pickle and vegetable), fodder and fuel etc. The plants' mature fruits serve as valuable and integral source of nutrition for villagers of arid and semiarid tracts (Pareek *et al.*, 1998). Besides, it has medicinal value in cardiac and gastric troubles, is commonly used as a biofence and its termite resistant wood is used by rural people for making handles, cartwheels, etc. The immature fruits are collected from natural stands by the rural people is an additional source of income and nutrition for them (Saroj and Awasthi, 2006).

Ecology and Distribution

History of Cultivation

Hooker (1960) listed 234 species of Capparis in Index Kewensis. Of these, 26 species occur in India. In these species, *Capparis decidua* is an important constituent of desert ecosystem and plays a significant role in the rural economy of northwest arid regions of the Indian sub-continent. The natives of Thar Desert recognized the importance of this shrub long ago. *C. decidua* is one of the important multipurpose tree species of desert and arid regions of the Indian subcontinent, Africa and Saudi Arabia. It provides vegetative cover in dry, hot, sandy desert areas where little else grows and is an extremely hardy species (Von, 1986).

Natural Habitat

This species is common in dry tropical Africa, especially in the Sahel, where it sometimes constitutes lines of small trees in Wadi beds, as in Mauritania for instance. In West Africa, the area of distribution is identical to that of Cadaba farinosa; its southern limit corresponds to the northern loop of the Senegal river. In the Republic of Niger, it reaches the Konadougou. Its area includes Tibesti (West Chad), much of the Sudan (except the extreme South),the Arabian Peninsula, Jordan, India, Pakistan, Iran, the Mascarene Islands and Natal. It is tolerant to prolonged drought and is an interesting plant by reason of its excellent adaptation to arid conditions.

Geographic Distribution

Kair is indigenous to the tropical and subtropical regions of Chad, Egypt, Ethiopia, India, Arabia, Pakistan, Iran, Jordan, Mauritania, Niger, Nigeria, Senegal, Somalia, South Africa, Sudan and Egypt. It is a member of the mixed xeromorhic woodland and psammophytic scrub vegetation (Shankarnarayan *et al.*, 1987). Scrub vegetation consists of low trees and bushes and occurs in areas with very little rain. *Capparis decidua* is found in several states of India *viz.*, Gujarat, Western Rajasthan, Uttar Pradesh, Madhya Pradesh, Haryana and Punjab. It is locally referred as Kair in Rajasthan, Karil in Uttar Pradesh, Ker in Gujarat, Teent in Haryana, Della in Delhi and Punjab and Nepti in western Maharashtra. Its natural habitat is pediment plains all over the dry regions. It also grows in Central India and Deccan.It grows on all types of wastelands in arid zone and other soils. No authentic data on this aspect are available so far, as no systematic plantation of kair is done. However, as an estimate, kair is spread over 3, 540 km^2 pediment plains in Bikaner and Jodhpur districts of Rajasthan and giving an estimated annual production of 7,000 tonnes (Chandra *et al.*, 1994).

Biophysical Limits

- Altitude range: 300–1200 m amol
- Mean annual rainfall: 100–750 mm
- Mean annual temperature: 25–41°C

Soils

It prefers alkaline, sandy and gravelly soils, thriving on shallow, hard soils and rocky outcrops. In India, *kair* can be found in the dry regions of Rajasthan, Gujarat, Punjab, Haryana, Madhya Pradesh, and many parts of Tamil Nadu, Karnataka and Andhra Pradesh. Although it is well suited to areas with low rainfall (150 mm), *kair* survives in areas with rainfall as high as 600 mm (Gupta *et al.*, 1989). It can also survive in tropical and sub-tropical regions or in areas with an annual rainfall of 100- 400 mm and temperatures of 16-50°C. Singh and Singh (1994) reported that *C. decidua* grows well on soils with lower sodicity (pH <9, ESP <35, EC 4 dS/m).*Kair* usually grows in dry, exposed habitats like foothills and wastelands. It grows in very shallow soils, soils affected by saline irrigation water, or in stabilized sand dunes. *Kair* grows in association with *Anogeissus pendula* Edgew., *Calotropis procera*, *Maytemus emarginatus*, *Zizyphus nummularia*, *Salvadora oleoides*, *Prosopis cineraria*, etc. It is a common species in stabilized sand dunes (Shankarnarayan *et al.*, 1987). The population density of *Capparis decidua* on different land forms were reported by Gupta *et al.* (1989), as the population varied from 9-170 bushes/ha. They recorded the highest population density on older alluvial plains (81 plants/ha) and minimum on the periphery of runs (20 plants/ha). Kumar and Bhandari (1993) reported 0.50 per cent and 3.50 per cent cover of *kair* in fenced and open sand dune habitats, respectively. Research revealed that *kair* can also improve the fertility of sand dunes. It also has the tendency to reduce alkalinity very sharply (Gupta *et al.*, 1989). Moreover, a *kair* plantation can reportedly increase organic carbon up to 10 times within 20 years.

Botany: Tree Characteristics

Capparis decidua commonly known as kerda, karir, kair, kirir, karril etc is a small much branched tree or shrub, 4-5 m high, or occasionally a small tree found in Thar desert and other arid regions in Southern Asia with a mass of slender, leafless branches, the small caducous (tending to drop off, as leaves deciduous) leaves being found only on young shoots. Leaves small, linear-oblong, acute, spinous-pointed, very minute (2 mm long), with a very short life span on young shoots, so that the plant looks leafless most of the time.

Flowers

Flowers pink, red-veined, in small groups along the leafless shoots, in the axils of spines in many-flowered corymbs, from the old branches, or from short lateral shoots; pedicels slender, about 12 mm. long.

Sepals

The outer pubescent, ciliate, the lower one very saccate, acuminate, the upper much smaller, ovate-oblong, concave; sepals elliptic, acute, with floccose margins. Petals red, narrow-oblong, Gynophore up to 12 mm long.

Bark

Bark turns whitish-grey colour with age, but most branches and twigs are a glossy dark green. Small, light brown spines occur in pairs on the twigs at each node.

Fruit

Fruit globular,glabrous, beaked. many-seeded ovoid or sub-globulous, slightly mucronate pink berry of the size and shape of a cherry, becoming blackish when dry. The generic name is derived from the Arabic 'kapar', the name for *Capparis spinosa*. The pink fleshy berries are readily eaten by birds.

Seeds

Seeds are usually classified into two types. Reports revealed dimorphism of two distinct seed types. Small seeds had 20-seed weight of 375 mg, 4.67 x 3.67 mm size and were pale violet with white markings, while large seeds had 10-seed weight of 471 mg and 6.17 x 4.67 mm in size and were cream-coloured (Figure 5.1).

Root: Studies on development of roots in nursery showed that during initial months, a single tap root is developed and at 6-months stage it starts developing a few secondary roots. At 12-months stage, a number of secondary branches develop even though the tap roots continue to dominate. High shoot-root ratio was observed in nursery. Less growth in shoot and more growth in root can be an important adaptation for initial establishment in arid areas. Studies on root system show that roots penetrate upto nearly 4 metres. The major roots tend to go deep by growing directly downwards and the lateral branches spread to great distance. These roots are also not superficial and their terminal ends tend to go deep. It is for this very reason that *C. decidua* does not compete for moisture with crops whose root zone is upper one meter or so. The root biomass comprised of major portion of the plant biomass and has the shoot-root (w/w) in mature plant recorded as 0.68 (Singh *et al.*, 2005).

Flowering and Fruiting

Flowering occurs at the beginning of the dry season. The flowers of *kair* are red or pink (Figure 5.2), and sometimes yellow in lateral corymbs. Flowers occur in many flowered corymbs, on old branches or on short lateral shoots; pedicel slender. *Kair* flowers in three seasons: February to March (Ambe Bahar), July to August (Mrig Bahar) and October to November (Hast Bahar). Flowers appear on new and old shoot, but mostly on one year old shoots. Profuse flowering was observed on older shoots, while the new shoots bore afewer flowers. The inflorescence is terminal fascicle or corymbose raceme, Flowers are pedicelate, complete, usually brick red, but a few flowers are yellow in colour. In Ambe Bahar phase (February to March season), the flowering was observed upto the first week of April, the peak flowering period being from March 20 to March 25 (Shekhawat, 1994). In some of the trees, the flowering continues from March to November in varying intensities. Flowering takes place during summer and winter months. Winter flowering generally starts in September- October and goes upto November-December. There is wide variation in flowering pattern between plants and within the same plant over different years. The February-March flowering is profuse, producing fruits of up to 20 kg/plant (Anon., 2001). Fruit yield varies between 2-20 kg/plant depending on age, genetic potential, and flowering season. Each fruit has at least five seeds. It is observed that freshly mature seeds have more than 80 per cent germination.

Initiation of floral buds in kair was observed in the second fortnight of February.Bud development took 14 to 15 days; the appearance of the flower bud at inflation stage was greenish white, hairy at the base. The buds became dark brownish red with tinge of greenish white at fully mature stage. The size of flower bud increased gradually, throughout the period of development (Shekhawat, 1999).

Table 5.1: Flower bud development in kair.

Day of Observation (Julion)	*Average Length (cm)*	*Average Diameter (cm)*	*Size Index (LXD)*	*Colour Appearance*
70	0.62	0.27	0.167	Greenish white hairy at base
81	0.90	0.36	0.319	Greenish red at top, greenish white at lower portion
83	1.05	0.43	0.473	Dark brownish red with tinge of greenish white
85	1.30	0.47	0.580	Dark brownish red with tinge of greenish white
87	1.44	0.56	0.806	Dark brownish red with tinge of greenish white

Figure 5.1: Cut half of fruits depicting seeds.

Figure 5.2: Kair flower.

Anthesis

Duration of anthesis in kair is from 14:00 to 19:001 h (Figure 5.3). Dehiscence occurs 7 to 10 hours after anthesis. Process took about over four hours to complete. Maximum receptivity of stigma is exhibited just after anthesis, and it is considerably lost in 24 hours. The flowering is asynchronous and the transformation of buds into flowers on a raceme takes about six days (Shakhawat,1994). There were about three to four flowers on a raceme. The opening of buds is normally sequential; the buds close to the base open first. The pollen viability is high (88 per cent), it being slightly more in March flowering.

Figure 5.3: Flowers of Kair.

Pollination

The flowers are large, highly conspicuous and coloured, attractive and have large stamens of different sizes which bend slightly downwards. All the floral characters along with asynchronous flowering and foliation at the time of flowering apparently suggest that pollination is by insects. As the blooming takes place in March-June period when hardly any food is available, the visit by large number of insects occurs. Many of these are predators, and some of the hymenopteran and dipteran insects like bees and flies are likely to be pollinators. Black ants can be observed sucking nectar from flowers while squirrels feed on corolla parts, but their role in pollination is uncertain (Figure 5.4). Pollination by insects has also been observed in other Capparis species (Eisikowitch *et al.*, 1986; Dafni *et al.*, 1987). *Xylocopa finestra*, a carpenter bee found in *Capparis* is a useful pollinator. In Jodhpur

Figure 5.4: Ants involved in pollination in *Kair*.

district, this species hibernated from late November to early February and occasionally during warm spells (Sharma, 1981).

In an experiment, it was observed that when flowers are emasculated and left as such for open pollination, there is very poor fruit set. The little fruit set that has occurred may be due to random visits of insects or by wind. There was also low fruit set in both selfing and sibbing where individual flowers and inflorescences were bagged (Singh *et al.*, 2005). The setting of seed indicates that there is no self-incompatibility. At the same time low fruit set shows that degree of pollination is very limited in the absence of insects. Pollen sterility of about 35 per cent was observed in some plants (Rao and Rao, 1977).There was hardly any difference in cross-pollination by hand and in natural open pollination. The fruit set in the two cases was 27 and 29.52 per cent respectively (Shekhawat 1999). This value is much higher than that observed in other arid zone trees. Studies have shown very low fruit set in arid zone tree species like *P. cineraria* and *T. undulata* where on an average 1-2 per cent flowers develop into fruits. The high fruit set in *C. decidua* can be attributed to lack of competition of space due to long pedicel and gynophore of the fruit. Further, the green branches of stem may be able to provide more continuous supply of nutrients leading to better fruit set.

The variation observed in flower colour, morphological traits and fruit and seed related characters in nature also confirmed that the species is highly cross-pollinated. Some degree of self-pollination by insects while moving within the same shrub cannot be ruled out. Low fruit set in selfing is not an indication of less self-pollination but only indicates that selfing is possible and its extent is very low in the absence of pollinators. In practice,more self-pollination may be taking place as fruit set was observed even in totally isolated single shrub, with no other shrub in visible range.

Fruiting

The dull red berries or fruits are globose or ovoid and 1-2 cm in diameter (Figure 5.5). The seeds are 2-5 mm diameter, and are embedded in the pulp (Anon., 1992; Gupta *et al.*, 1989). Green fruiting in summer was observed in March-April in most of the plants. There was a second flush of green fruits in May-July. In some of the trees, however, green fruiting may appear to be continuous from March to July.

During winter, fruit set occurs in October and November. The trees showing continuous flowering from March to November-December did not produce fruits continuously. Generally in August-September, no green fruits were available. Some of the trees did not produce fruits in winter. Thus different trees produced green fruits once, twice or thrice within a year and pattern within same tree were not uniform over years.

Phenological observations indicate that there are high intra-population variations suggesting high degree of homeostasis in the species. Even though, *C. decidua* produces leaves, flowers and fruits two to three times in a year but the peak flowering occurs during summer and maximum mature fruits are available just before onset of monsoon. Apparently, it looks paradoxical that plant should produce flowers and fruits during the driest period when there is maximum water stress. But a close inspection reveals that this is an important adaptation for continuity of the race. By producing flowers about 1-2 months before the rainy season the species ensures that the seeds are available in ground when

Figure 5.5: Fruiting in *Kair*.

rainfall season starts. The plant can afford this behavior because of its ability to draw moisture from deeper layers (Singh *et al.*, 2005).

Chemical Composition

The ripe fruits of *C. decidua* contains 55 per cent of total pulp of the total fruit weight including moisture (56 per cent), crude protein (8.6 per cent), true protein (5.02 per cent), total sugars (1.8 per cent), reducing sugars (1.1 per cent), P (0.057 per cent), K (1.026 per cent), Ca (0.055 per cent), Mg (0.055 per cent) and ascorbic acid (7.81 mg/100 g) Thus, this fruit is rich in protein and mineral matter (Khera and Singh, 1975).

Table 5.2: Chemical composition of Kair fruits.

Sl.No.	*Chemical Constituents*	*Quantity*	*Sl.No.*	*Chemical Constituents*	*Quantity*
1.	Moisture (per cent)	28.00	2.	Crude protein (per cent)	14.88
3.	Ether extravatives (per cent)	7.43	4.	Crude fibre (per cent)	12.32
5.	Ash (per cent)	5.96	6.	Digestible carbohydrates (per cent)	59.41
7.	Ascorbic acid(Mg/100gm)	120.70	8.	Carotene(Mg/100gm)	5.40
9.	Calcium(Mg/100gm)	90	10.	Phosphorus(Mg/100gm)	179
11.	Iron(Mg/100gm)	3.50	12.	Zinc(Mg/100gm)	1.60
13.	Copper(Mg/100gm)	1.10	14.	Manganese(Mg/100gm)	1.90

Among locally available fruits and Vegetables of the arid zone of Rajasthan *viz.*, Dansra (*Rhus myserensis*), kachri (*Cucumis callosus*), kair (*Capparis decidua*) and sangri (*Prosopis cinararia*); the kair was the richest source of β-carotene in fresh and various processed states like blanched, dried and roasted (Chaturvedi and Nagar, 2001). The unripe fruits of *C. decidua* consisted of protein (14.88 per cent), fat (7.43 per cent), ash (5.96 per cent), crude fibre (12.32 per cent) and digestible carbohydrates (59.41 per cent DM). It also included (mg/100 g) b-carotene (5.4), ascorbic acid (120.7), calcium (90.0), phosphorus (179.0), iron (3.5), copper (1.1), zinc (1.6), manganese (1.9); phytic acid (304). Total protein fraction includes albumin 53 per cent, globulin 16 per cent, prolamine 11 per cent and glutelin 12 per cent (Chouhan *et al.*, 1986). Duhan *et al.* (1992) reported that among thirteen non-conventional foods; Khejri (*Prosopis cineraria*), pinju (*Capparis decidua*) and kachri (*Cucumis* sp.) contained considerable amounts of protein (15-18 per cent).

Analysed fruits, flowers and buds of *C. decidua* yielded 14 per cent surface wax (w/w) and consisted of a mixture of compounds, mainly straight chain saturated hydrocarbons and ketones with C 28 to C 32 chain lengths associated with N- and S-containing oils. Seeds contained 1.7 per cent sugar and 8.6 per cent protein with 20 per cent oil and 14 per cent oil in flowers (Rai and Rai, 1987). Seeds of *C. decidua* contain 20.3 per cent oil with 68.6 per cent of unsaturated fatty acids and 31.4 per cent of saturated acids (Rai and Shekhawat, 2000).

Table 5.3: Chemical analysis of *Kair* seeds and flowers.

Sl.No.	*Plant Part*	*Quantity (per cent Wt)*
1	Seeds	
	(*a*) Oil(Dry)	20
	(*b*) Ash	5.7
	(*c*) Moisture	55
	(*d*) Sugar	1.7
	(*e*) Protein	8.6
	(*f*) Vitamin	7.8 mg/100 gm of pulp
2.	Flowers	
	(*a*) Oil (Dry)	14
	(*b*) Moisture	72.3
	(*c*) Sugar	Glucose and galactose

The seeds contain glucocapparin from which the mustard oil methyl isothiocyanate is released when the plant is crushed (Gupta *et al.*, 1989). Extraction and isolation of 15 natural products from *C. decidua*, which are also used as emergency famine foods were reported by Rai (1987). On the basis of spectral and chemical data studies, two new spermidine alkaloids 14-N-acetyl-isocodonocarpine and 1 5-N-acetyl-capparisine were isolated from root bark of *C. decidua* (Ahmad *et al.*, 1992) and two compounds N-nonacosane and N-henteiacontane were isolated from the flowers (Garg *et al.*, 1993).

Foliation

Foliation in *C. decidua* populations occur up to three times in a year during June-July, October and January-March and can be referred as summer foliation, pre-winter foliation and winter foliation respectively. The winter seems to be the dominant period of foliation and all plants bear leaves during this period. Summer foliation occurs only in a few plants and that also not regularly in every year. Water stress and environmental factors are likely to affect the foliation patterns. There is vast variation for this trait between plants within a season and in the same plant during different years in this species, which is commonly described as a leafless shrub. It is not, however, possible to infer the contribution of leaves as the stem and branches are also green and some photosynthetic activity must be going on even when leaves are not there. Leaves are more of a reflection of moisture availability in the deeper layers and normally appear a month or so before flowering. As reproduction is a high energy consuming activity, it is possible that leaves supplement the food requirement and help to generate better food reserve (Singh *et al.*, 2005).

Seed Dispersal

In *C. decidua*, the seed dispersal appears to be through birds. The mature red fruits of *C. decidua* are palatable and hence dispersal can also be by fruit burst, squirrels, other animals or even human beings. The mature fruits are not marketed but are eaten by birds, insects, animals, children and grazers. The size, shape and other characteristics of the seed indicate that seed is not dispersed by wind.

Natural Variability

Under natural population, rich genetic diversity with wide range of variability is available for bearing habit, fruit size, colour of fruits, petals, pulp content, spiny habit, plant spread and compactness of canopy, time of flowering and fruiting, etc. in *C. decidua* but no systematic efforts are made so far to collect and conserve them and to popularize the promising types. There is a need to identify suitable types to select heavy yielder with large fruit size, high pulp content, rich in protein; proper TSS:acid blend, less acrid, small and soft seeds. Attempts were made by NBPGR, Regional Station, CAZRI, Jodhpur and CIAH, Bikaner to identify promising germplasm of kair from natural population. There is no named variety of kair.

While selecting the shrubs for commercial exploitation, shrubs with good growth and non-spiny or less spiny branches and open canopy were preferred. The primary criteria, in addition to the above were good fruiting. The late fruiting types were not preferred as the fruits set in hot months are likely to have poor quality. At present, *C. decidua* plants occur in orans, gochars and on wastelands in scattered pockets. Chundawat (1990) opined that if proper selection is made from available variability, *C. decidua* can make an excellent crop for extreme arid zones of Rajasthan and Gujarat where a few species can survive and can be domesticated.

In general,two distinct plant types of kair occur, one type having tree form whereas majority occur as bushes (Figure 5.6). It appears that plant attains tree form when it grows from seed and remains undisturbed. On the other hand, plants hich get exposed to biotic interference may tend to produce more shoots and also propagate through root suckers. This view is supported by the fact that *C. decidua* occur singly in tree form and mostly in clusters in bush form in orans and gochars. The plant can also be trained by allowing single stem and can be integrated as one of the components in different

Figure 5.6: Variability in tree type in *Kair*.

cropping models. It needs scientific evaluation to assess the compatibility and economic viability under different resource situations. Gautam and Bishnoi (1991) reported that aqueous extract of root and shoot of *Capparis decidua* had strong inhibitory effect on seed germination of *Calligonum polygonoides* and *Lasiurus scindicus*, thus while integrating in a system its allelopathic influences must be ascertained.

Figure 5.7: Variability in fruit shape and size.

Fruit is the economically useful trait in this species and the variability in fruit (Figure 5.7) related traits are of interest to the breeder. The coefficient of variation for pulp wt./fruit was 70.4 per cent, fresh ripe fruit weight 64.4 per cent, pulp-seed ratio 61.9 per cent, seed weight/fruit 49.5 per cent, and number of seeds/fruit 43.5 per cent. The high variability indicates that there is very good scope of selection for these fruit characteristics. The variation for height and spread was also ample; coefficient of variation being 43.8 per cent and 48.1 per cent, respectively for these two characters. If the species is to be developed on wastelands then fast growing types with better height and spread will be desirable. If the species is to be developed as a horticultural crop then selection parameters may be slightly different.

An interesting observation related to plant spread was that in natural population the spread was more in north-south direction than in east-west direction indicating that more growth takes place in north-south direction to avoid high heat load. In established populations, no difference in the spread was observed and it can be attributed to regular and close spacing. The coefficient of variation for 100-seed weight, internodal length and internodal angle were 20.0, 14.9 and 2.74 per cent, respectively. It appears that 100-seed weight is relatively less affected by the environment and has limitations of its range. Even though, variation for internodal length and internodal angle are less but these traits along with other traits contributing to compactness of shrub are very important (Singh *et al.*, 2005).

Table 5.4: Coefficient of variation in different traits of *C. decidua*.

Trait	*CV (per cent)*
Pulp wt./mature fruit	70.42
Fresh ripe fruit wt.	64.41
Pulp-seed ratio	61.90
Seed wt./fruit	49.47
Plant spread	48.13
Plant Height	43.77
No. of seeds/fruit	43.63
Fruit width	26.89
Fruit length	22.65
100-seed wt.	20.00
Internodal length	14.95
Internodal angle	02.74

Variability in flower colour in natural stands of *C. decidua* (Figure 5.8) was reported by Shekhawat (1999) wherein brick red and yellow coloured flowers were observed. In the present scenario, where there is no commercial cultivation of this species and green fruits are harvested by the rural people, it is obvious that more fruits will be picked up from the non-spiny open type shrubs having early fruiting. At present, the demand for *kair* is very high and mass harvesting of green fruits is likely to lead to skewed perpetuation of different plant types. The plant type from which it is most difficult to pick up the fruits is likely to pass on maximum seeds to the next generation.

Propagation and Management

This tree is over-exploited. There is wide genetic variability in the field population of *Capparis decidua* as it is open pollinated trees. Propagation is not a problem as *kair* can be grown from seeds and root suckers. If grown from seeds, mature fruits must be collected during May-June. The seeds have short period of viability. The seeds are then separated by washing out the pulp. They are then soaked in fresh water for 24 hours. Afterwards, the seeds are germinated in nurseries by sowing them in polythene bags filled with clay, sand and farmyard manure at 1:1:1 ratio. Sowing is the best done during August. Light irrigation must be applied everyday during the evenings. After six to seven months, the seedlings are ready for transplanting. This is usually done during early March or July-August at the onset of the monsoon season. One thousand plants of *Capparis decidua* can be planted in an acre of land with a spacing of 2 x 2 m. This is expected to yield an annual fruit production of 20 tons/acre, at the production rate of 2 kg/plant.

Figure 5.8: Variation in flower colour of *Kair.*

Vegetative Propagation

Vegetative propagation is by cuttings or tissue culture. Direct sowing, planting stock and wildings are used. *C. decidua* is reported to naturally propagate through root suckers which are thrown up to 5 m from the shrub. The ability to produce root suckers is a good indication of regenerative ability. The ability to produce shoots is also a fair indication of its power for vegetative propagation.

Efforts to propagate the species by collecting root suckers, separating the sprouts and raising them in the nursery met with a limited success. A few sprouts which produced shoots and survived up to 1 year in nursery failed to survive in the field beyond two months. It appears that the suckers survive in nature by remaining dependent on mother plant for prolonged period and there early separation during collections has been lethal.

The trial on identifying the best month for collection of cuttings showed that there is variation in sprouting of cuttings taken from different plants. In one case no sprouting was observed in any of the months whereas in another case 20-40 per cent sprouting was observed in seven out of 12 months. This may be a reflection of genetic variation or the food reserve available in the cuttings. February-March and July-September appear to be the suitable months. However, all collections may not behave in the same manner. This clearly indicates that *C. decidua* is a hard to root species and application of hormones is necessary.

Even use of hormones BA, PA, BNOA and commercial preparations like Seradex B_2 and B_3 did not yield very promising results. Hormones led to enhanced sprouting but the rooting response was, in general, poor. The earlier studies made by Meghwal and Vashishtha (1998) also showed that the time of planting had marked influence on rooting and sprouting and this species is very hard to root. The rooting response of the species was poor and the best result was obtained during rainy season.

The natural regeneration of *kair* takes place by seeds and suckers.Though, the seed multiplication is generally not recommended but suckers with some roots can be used as planting materials. Attempts to propagate kair through vegetative means yielded low rate of success so far. Vashishtha (1987) suggested that the best time for taking cutting is August-September and these cuttings should be treated with IBA(100ppm) before planting. In later study, Meghwal and Vashishtha (1998) suggested that July-August is the best time for taking the cutting under Jodhpur conditions. Shekhawat (1994) claimed the best response at IBA(5000 ppm)while propagating the kair through cutting under mist system of nursery shade net house. Bhargava *et al.* (2002) reported that semi hardwood cuttings taken in March, treated with IBA(7500ppm) + Thymine(1000ppm) in DMSO had given maximum sprouting (50 per cent). In another experiment, aqueous extracts of *Capparis decidua* cuttings inhibited *Nerium* sp. root growth and cluster bean sprouting. ABA-like substances (at 3.2μg/g) were observed in *Capparis decidua* aqueous extracts, which may be responsible for inhibition of sprouting and root growth in other plants (Bhargava *et al.*, 2000). Experiment by Bhargava *et al.* (2000) using cuttings of *C. decidua* of 15-20cm length and 1-1.5cm diameter and treated with ISA at 3000, 5000 and 7000 ppm alone or in combination with Thiamine(1000ppm) showed that maximum sprouting (25 per cent) occurs in BA(5000ppm).

Micropropagation

Capparis decidua is an open/cross pollinated out breeding tree that exhibits enormous genetic variability. The viability of seed is very short in *C. decidua* and the seeds are attacked/damaged by insects. For cloning of selected plants, tissue culture methods could be highly desirable as the conventional method of vegetative propagation is not effective in the multiplication of desired germplasm of capper.

In an investigation on micropropagation, various concentrations of BA (1, 3 and 5 mg/l) showed development of multiple shoots and shoot buds in the primary cultures of *Capparis decidua* within 30 days. With nodal explants,shoot multiplication was the best on medium supplemented with BA (5mg/l) and after a second and subsequent subculturing, further multiplication of shoot buds was the highest on the medium containing 3mg/I BA (Bhansali, 1994). Micropropagated plants were hardened in the green house. The hardened and acclimatized plants were transferred to pots and subsequently to field. The cloned plants are growing normal. The protocols defined are reproducible. These can be used for mass multiplication of selected clones and genetic improvement of these species. A method for micropropagation of *C. decidua* from mature trees using nodal explants was developed by Deora and Shekhawat (1995). *In-vitro* micropropagation was achieved from mature trees using nodal explants and seedling explants from cotyledonary nodes, cotyledon and hypocotyls on MS medium supplemented with bezyladenine (BA 1, 3 and 5mg/l). Rooting medium with IBA (1 mg/l) was proved the best for rooting. *In-vitro* hardened plants were transferred to pots, Bhargava and Saroj (2003) reported that both callus induction and shoot proliferation were the best in the medium containing MS = BAP(5mg/l) + IBA(1.5mg/l). However, rooting was not found in this medium, thus the plantlets were subcultured in modified MS medium and roots were induced in MS + 0.1 mg l^{-1} IAA(0.1mg/l) + BAP(1.0mg/l) and IAA(0.1mg/l) + ½ strength MS media.

Cultivation and Production

Harvesting

Hand picking is a common practice for harvesting of fruits (Figure 5.9). By hand picking only tender fruits are harvested and mature ones are avoided, hence there is no need for further grading of fruits. Moreover, plants are spiny in nature, thus careful harvesting should be done. Sometimes, harvesting with twigs is also seen but it affects the plant vigour and subsequent fruiting.

Figure 5.9: Harvesting of *kair*.

Maturity Indices

Kair comes into fruiting after 4-5 years of age. The fruits are harvested in April- May. Shekhwat *et al.* (1999) reported the flowering habit, duration of flowering, flower bud development, anthesis time, dehiscence, receptivity of stigma, fruit setting and fruit development in *C. decidua* grown for its edible fruits and medicinal properties. The fruits should be harvested at green tender at small pea stage for pickling and vegetable purposes The stage can be judged by size of fruits and also by pressing the berries.Meghwal (2002) suggested that the fruits should be harvested after 7- 10 days of fruit setting with 5-8 mm diameter during March-April to fetch better price in the market. After maturity, seeds are hardened and fruits are not preferred for consumption. The fruits of second flowering available in September- October are not harvested as the crop load is poor.

Postharvest Technology

The unripe/ripe fruits of kair are generally not used due to acrid taste, but this fruit can be converted into a variety of by-products after processing.The pickles are the most commonly and widely utilized postharvest product of ker. The processed fruits can be utilized directly for preparation of pickles or as vegetable or can be dehydrated for off-season utilization (Goyal and Sharma, 2009). The recovery is about one fourth *i.e.* one kg of fresh immature fruit yields about 200-240 g of processed dried fruit. Based on size, three relative grades of processed kair are available in the market *i.e.*, big, medium and small size. In fact, the basis of size grading is maturity of fruits. The smaller fruits are more tender and of better quality than bigger fruits. The processed fruits are stored either in pots or in plastic containers while processed dried fruits are stored in flexible polybags. The dried fruits can be stored in polybags for a year without any deterioration in quality. The dried fruits are sold in small packets of 250 g to 1 kg in polybags as retail packaging. Intensive cultivation can yield an annual income of Rs. 1, 20,000 at the selling price of Rs. 60/kg from an acre of *kair* plantation. The dried fruits fetch premium price of Rs. 500/kg in the local market. Such cultivation entails the planting of 1000 plants in an acre to produce 2000 kg of fruits annually. A five to seven-year-old bush of kair gives 1-2 kg of green tender fruits. Though there is high variation in fruit yield and sometimes the old kair plants of tree shaped give as high as 6- 8 kg fruits/tree. However, Pareek (1998) estimated fruit yield of 10 -20 kg/bush, as well as 20- 80 q/ha by giving proper management practices. One man can harvest about 8 -10 kg fruits/day. Fruit production increases as the plant gets older. It can even increase when plants with high genetic potential are cultivated. Scientific efforts towards developing genetically superior strains and standardization of optimum cultivation practices may increase the yield, and thereby increase income from the plantation. Additional income may be derived from the wood as building materials and fuelwood. These benefits make *kair*'s cultivation both economically rewarding and environmentally sustainable.

Uses

Food

The flower buds and immature green fruits of *kair* are pickled, cooked and consumed as vegetable (Anon, 1992; Harsh and Tiwari, 1998; Pareek *et al.*, 1998). The immature fruits are in great demand in different regions of India. Each plant yields approximately 2 kg. One kilogram of fresh fruit yields about 200 g of processed and dried fruit. The mature fruits, meanwhile, are consumed fresh. The fruits are rich in protein (8.6 per cent) and vitamin C (7.8 mg/100 g of pulp by weight). As the fruits mature, the amount of sugar increases from 1.7 to 3.0 per cent (Gupta *et al.*, 1989). Unripe fruit is pickled and has good commercial value. They are also cooked as vegetables along with the fruits of Prosopis cineraria (sangri) and seeds of Acacia Senegal (Kumat). Ripened fruits are also eaten by locals. Young green fruits are an important ingredient of the famous "Panch koot" mixtures of dry vegetables of Rajasthan which include *kair* fruits, *kumat* seeds, dry *kachri* (*Cucumis*), *sangri* (Khejari pods) and *kamal* dandi (lotus roots) (Hocking, 1993). The fresh or dehydrated fruits are used extensively as vegetables and pickles. The air-dried processed and air-dried immature fruits are delicious and very expensive.

Fodder

The fruit is relished by camels and also, wherever within their reach, by goats. The fruits are also consumed by man in Sudan.Vyas *et al.*, (2009) revealed that fruits are rich in carbohydrates (63.06 µ 4.20 per cent) and protein (23.34 µ 4.38 per cent) with high amount of potassium (3.23 µ 0.13 per cent). Its browse value is probably the most important asset, despite being low in nutritional value. In Sudan for instance, it is a major source of camel food as it can be eaten when little else is available. Green or dried leaves of *C. decidua* are used as diet supplement for goats, sheeps, camels and other animals during scarcity period of grass (IBPGR, 1989).

Fuel

It is used for charcoal and firewood in its native range. Quantitative analysis of fuelwood properties like calorific value, density silica, carbon, ash-biomass ratio, moisture, volatile matter and fuelwood index of *C. decidua* revealed its the best fuel wood characters (Jam, 1994).

Timber

The wood is very hard and used to make water pipes and water troughs. This provides hard, heavy, termite resistant/timber. Aside from the fruits and seeds, *kair*'s wood also provides varied uses. The wood is light yellow to pale brown, smooth, moderately hard and heavy. Each plant yields 625-775 kg/m^3 of wood(VOGT,1995). The wood's strength and durability are suitable for making small beams, rafts, knees of boats, tool handles, cartwheels, axles, and even combs. It also serves as material for making huts and fences. Estimation of herbage production under canopy cover of *C. decidua* was greater as compared to that in open field and is thus a suitable species for use as woody component in silvipastoral systems.

Oil

The seeds of *C. decidua* contain high quality edible oil. The seeds, meanwhile, contain 20.3 per cent oil. The oil consists of 68.6 per cent unsaturated fatty acids and 31.4 per cent saturated fatty acids. When processed, the oil is edible.

Medicinal Uses

Aside from economic uses, *kair* is also tapped for its medicinal qualities. The plant has a bad smell and taste; carminative, tonic, emmenagogue, aphrodisiac, alexipharmac, improves the appetite; good for rheumatism, lumbago, hiccough, cough and asthma (Satyanarayana *et al.*, 2008).

When pickled or cooked as vegetables, the immature fruits are used to cure stomach problems, especially constipation. Fruits possess antidiabetic activity. The alcoholic extracts of fruit pulp and root bark possess anthelmintic activity (Mali *et al.*, 1994). The extract of unripe fruits and shoots of *C. decidua* produced reduction in plasma triglycerides, total lipids and phospholipids (Goyal and Grewal, 2003); hence used as hypocholesterolemic. It appeared to operate through increased faecal excretion of cholesterol as well as bile acids (Agarwal and Chauhan, 1988). Capparidisine a new alkaloid from *C. decidua* has dose dependant depressant effect on heart rate and coronary flow. (Rashid *et al.*, 1989) *C. decidua* powder has hypoglycaemic activity, decreases lipid peroxidation and alters free radical scavenging enzymes like superoxidedismutase (SOD) and catalase (CAT) in erythrocytes, liver, kidney and heart in aged alloxan induced diabetic rats. *C. decidua* powder is used against alloxan induced oxidative stress and diabetes (Yadav *et al.*, 1997). The green immature fruits are good for antihelminthic, laxative, asthma, constipation, coughs, hysteria, worms and other psychological problems (Ghazanfar, 1994). The fruits and seeds of kair are used to cure cholera, dysentery and urinary purulent discharges. The fruit has a sharp hot taste astringent to the bowels, destroys foul breath, biliousness and urinary purulent discharges and good in cardiac troubles (Ayurveda). Dietary fibre content of *C. decidua* varied from 38.5 to 55.7 per cent. Hemicellulose in *C. decidua* given at 10 per cent of the diet to rats induced a greater resistance to hyperlipaemia than cellulose and has the most pronounced hypocholesterolaemic effect, which operated through increased faecal excretion of cholesterol as well as bile acids (Agrawal and Chauhan, 1988). *C. decidua* has diuretic properties and antidiabetic properties.

The flower buds and roots are used as renal disinfectants, diuretic, tonic and for arteriosclerosis and as compresses for eyes. Flower buds are also used as potherbs. The alcoholic extract of flowers, fruit husks and seeds show anti-bacterial potentials. Flowers contain the hydrocarbons Nonacosane and Triacontane. Capers *i.e.*, the flower buds of *Capparis decidua* Edgew, are pickled and used as condiment. Members of this family contain thioglucosides (known as glucosinolates) which release isothiocyanates (mustard oils) when the plants are damaged. Typically the plants yield methyl isothiocyanate from methyl glucosinolate (otherwise known as glucocapparin). Flowers and fruit husk contain phthalic acid (Dahot, 1993). The young shoots of *Peganum hurmala* and *kair* can be combined as an antifertility drug.

Oil from the seeds contains nitrogen and sulphur. This is used to cure skin diseases. The bark is also used in liver affections. Infusion of stem and root bark is used against diarrhea and febrifuge. Bark has an acrid, hot taste, analgesic, diaphoretic, akheteric, laxative, antihelminthic and good for cough and asthma, ulcers and boils, vomiting, piles and all inflammations (Anon, 1992). The alcoholic extract of root bark possesses significant antibacterial and antifungal activities (Mali *et al.*, 2004; Gaind *et al.*, 1969). Stem bark, meanwhile, is used for rheumatism and toothaches, including pyorrhea. The very bitter roots are used in the Indian and Farsi pharmacopoeia and the root bark is used to cure swollen joints. From the root bark, spermidine alkaloids like isocodonocarpine (Rai, 1987), capparisine (Dahot, 1993; Gupta and Ali 1997), capparadisine (Goyal and Grewal, 2003) and capparisinine (isomer of capparidisine) (Ahmad *et al.*, 1987) were isolated. The stem and root bark extracts contain isocodonocarpine and other alkaloids, effective in curing the asthama, inflammation, diabetes and cough. The aqueous extracts of roots of *C. decidua* have purgative activity (Gaind *et al.*, 1969). Two sterols, one diterpene alcohol, two aliphatic constituents and one diterpenic ester were reported from *C. decidua* root barks. (Ahmad *et al.*, 1987). Root bark is used as anthelmintic and purgative. In the traditional system of medicine, the bark is useful in the treatment of coughs, asthma and inflammation; roots in fever and buds in the treatment of boils.

In Unani, leaves act as appetizer, helps in cardiac troubles and fruits used in biliousness; Shoots along with shoots of *Peganum harmala* are used as anti fertility drug; ground stem and leaves are used in alveolaris and pyorrhea; Wood coal is used in muscular injuries (Kirtikar and Basu 1993; Sharma, 2003; Chopra *et al.*, 1999). In Sudan, *C. decidua* is used in swellings, jaundice and infection of joints (Ahmad *et al.*, 1985).The plant has a number of alkaloids, terpenoids, glycosides and some fatty acids. Ethanolic extract of aerial parts exhibited anti-inflammatory and analgesic activity (Ageel *et al.*, 1985). Isocodonocarpine was responsible for anti-inflammatory activity and anti asthmatic activity (Ahmad *et al.*, 1987). Antibacterial activity of seeds indicated that isothiocyanate aglycon had good antibacterial activity (Juneja *et al.*, 1970). It inhibited cell cultures of *Vibrio cholerae* Agave, *V. cholera* Inaba and *V. cholera* Eltor (Gaind *et al.*, 1972).

Uses from Rajasthan and Other States

Stomachache

Soak fruits in saline water for 10 days, dry and make a powder. Take a spoonful orally twice a day.

Diabetes

Take two spoonful of the powder of the deseeded fruit orally.

Constipation

Soak fruits overnight. Dry them and grind into a fine powder. Take a spoonful of powder orally every morning.

Uses from other States

Toothache

Put two drops of young shoot juice in the ear.

Rheumatism

Cook the chopped fruit (3kg) along with jaggery (500g) in ghee (500g).Take the preparation (30g) twice a day for a month.

Eczema

Make a paste from the bark of *kair* and leaves of *mamejava* (*Enicostema littorale* Bl.) taken in equal quantity. Apply the paste on the infected area.

Uses in Classical Codified Literature

The fruit is taken orally to reduce blood sugar. Plant extract is applied topically on skin diseases; decoction of the plant is taken orally to alleviate rheumatic pain. Tablet 'Jyoti kalash' a multi-herb drug acts as an antioxidant; 'Maharshi Amrit Kalash' is used against vomiting, anorexia and improving general well being. One patent is found out its cosmetic use.

In Punjab, the top shoots and young leaves are made into powder and used as blister. They are also used as antidote to poison and also in afflictions of joints. The fresh young leaves are crushed and soaked in water. The water is strained off. The residium is dried and allowed to solidify. A tiny piece of it is eaten with butter and gives relief from pains of bruise after a fall.

When crushed, the leaves are applied as a poultice on burns and inflammations. They are chewed to relieve toothaches, or consumed as healthy appetizers for those with cardiac problems. The young leaves and shoots, meanwhile, are dried and powdered and used as antidote against poison and as a cure for joint problems.

Ethnobotanical expeditions from Western Maharashtra revealed the use of root extracts of *C. decidua* in fisheries for catching predatory and weed fishes which compete with cultivated carp.

Services

Erosion Control

Though natural catastrophes, overgrazing and over utilization have threatened genetic resources since ages, but in it,is the fast pace of human intervention, a threat to such fragile ecosystems. The need is thus to understand and practice the conservation of desert wealth with the involvement of conservative desert dwellers and their indigenous technical knowledge (ITK) thereby promoting the conservation ethics and retaining the traditional cultural values. *Capparis decidua* can be used in landscape gardening, afforestation and reforestation in semidesert and desert areas. It provides assistance against soil erosion. *C. decidua* helps in controlling wind erosion in sandy areas, reduction of alkalinity and increase of organic carbon and available N, P and K and is the best species for shelterbelts to check the movement of sand and its morphometry plays an important role in sand dune stabilisation in the thar desert of India (Pandey and Rokad, 1995).

Shade or Shelter

One of its preferred uses in Sudan is as a shade and shelterbelt (Vogt, 1995).As it is drought resistant and withstands neglect, this species could be particularly useful in arid areas as a live hedge providing edible fruits (Booth and Wickens, 1998).

Biodiversity Conservation/Ecosystem

Over a large part of the State of Rajasthan, agriculture is a struggle against inhospitable nature. About a third of the state is stark desert–the Thar, which represents the harshest agricultural environment of India. Conservatism, as regard to agricultural practices and utilization of agricultural produce is deeply ingrained in the cultivators of Thar Desert and is the outcome of loosing everything year in and year out. In a bid to grow, multiply and utilize every blade of grass available, the desert dwellers have found out innovative means of sowing, managing, harvesting and utilizing agro-biodiversity around them (Baumer, 1983,). In fact, their livelihood and lives revolve around the genetic wealth,used extensively and resolutely (CABI, 2000). Traditionally, they are very conscious of

Figure 5.10: Biodiversity conservation by *Kair.*

conservation and have allowed multiplication and regeneration of plants even when utilizing it. Some of the practices of biodiversity conservation are unique in this land and have strengthened the life support system of the area for facing challenges of inhospitable hot deserts (Figure 5.10).

Associated Insects and Microflora

Stenozygum fieber, a beetle is long associated with *C. decidua* (Ahmad and Khan, 1983). *Bruchidus obscurus* (Coleoptera) was seen on the leaves of *C. decidua* in Dehradun forest in Uttarakhand but rarely seen in Rajasthan (Arora and Tewari, 1984).The highest density of termites was observed in *C. decidua* with 114.6 nests/acre in plant communities of eastern Baluchistan (Kayani *et al.*, 1985). Reports on investigation of locust by spider web (*Stegpduphus sarasinorum* Karsch) were studied at Bikaner. Web spun on leaves and branches of *C. decidua* trapped insects from Coleoptara, Lepidoptera, Diptera, Hymenoptera and others. Biology of *Pierid Belenois aurota* collected from *C. decidua* in Rajasthan was reported by Chandra (1988).

Diseases and Pests

In fact, the systematic information on pests and diseases have not been collected so far. Moreover, no attempt is made for management of pests and diseases of kair as bushes of kair are growing naturally. Similar to other arid vegetations, kair is also affected by several pests and diseases Abundance of termites can be seen with several natural growing kair bushes. Association of web of spider (*Stegpduphus sarasinorum*) with kair plants can also be seen The small hole in both raw and ripe fruits was observed but whether it is due to puncture of kair fruits by their own spines or due to some insect borer is yet to be investigated. Field establishment of *Capparis decidua* is poor. The attack of *Colletotrichum* and *Alternaria stem rot* are observed in vegetatively propagated saplings transplanted in the field. *Fusarium solani* and *Rhizoctonia root rot* are also recorded in young seedlings raised in polybags (Bhansali, 2002). The drying of twigs from top to bottom is a common phenomenon in kair and reasons are not known so far. Some of major diseases and pests of Kair are summarized as index.

Soft Rot of *Capparis decidua* L.

Fruit rot of *Capparis decidua*, spoils fruits to the extent of 7-10 per cent in the field, and 18-20 per cent in the market at Bharatpur (Rajasthan), Agra, Hathras (U.P). A range of 25-30º C temperature and 60-80 per cent R.H. prevailed in fields in summer season. The fruits exhibited soft rot symptoms and carried mixed infections. Most of the infected lesions were circular and olive-gray, when young and irregular, brownish-black or dark gray when old. Fruit rot disease on *C. decidua* caused by *Dichomera capparidis* in arid and semi-arid regions of western Rajasthan was reported by Gehlot *et al.* (2001). Different symptoms and intensity of rot induced by pathogen are described below.

Penicillium nigricans

Rot pathogen caused soft rot of *Capparis* through injury. The rotting was initiated in the forms of small, irregular, light green patches on the third and forth day of incubation which increased in area. With increasing incubation period (10 days), the spot became irregular, dark green lesions that developed below the infected ring. Internally the tissues changed in colour and produced bad smell. About 15-20 per cent fruits tissue got decayed within 8 days of incubation.

Aspergillus flavus

Rot pathogen caused soft rot of *Capparis* fruits but not on uninjured ones. The infected ring showed several irregular cracks and tissues beneath were macerated. Secretion of watery substance, with foul odour was also associated with this disease. The pathogen spoiled 25 per cent fruit tissues in 10 days. Several fungi caused postharvest disease on *Capparis decidua* but the present pathogens had so far remained undetected.

Fungi

Other fungi attacking *kair* are listed below:

- ☆ *Albugo capparidis* De By
- ☆ *Aschochyta capparidis* (Cast.) Sacc.
- ☆ *Botrytis*–gray mold
- ☆ *Camarosporium suseganense* Sacc. et Speg.
- ☆ *Cercospora capparidis* Sacc.
- ☆ *Cloeosporium hians* Peck et Sacc.
- ☆ *Hendersonia rupestris* Sacc. et Speg.
- ☆ *Leptosphaeria capparidis* Pass.
- ☆ *Phoma capparidina* Pass.
- ☆ *Phoma capparidis* Pass.
- ☆ *Phyllosticta capparidis* Sacc. et Speg.
- ☆ *Septoria capparidis* Sacc.

Viruses

Viruses are transmitted by mechanical inoculation; by grafting and vegetative propagation of cultivated varieties. Certain insect pests may also be vectors.

Two viruses: Caper Latent Carlavirus, and Caper Vein Yellowing Virus are reported in Puglia, Italy.

Insects Pests

- ☆ *Acalles barbarus* Lucas–weevil that attacks roots
- ☆ *Asphondylia capparis* Rubs.–a dipterian (Cecidomyiidae) that disfigures flower buds.
- ☆ *Calocoris memoralis* Sacc.
- ☆ *Cydia capparidana* Zeller–a lepidopteran that disfigures flower buds.
- ☆ *Eurydema ventralis* Kolen

Constraints

This tree is over-exploited. There is wide genetic variability in the field population of *Capparis decidua* as it is open pollinated. The seeds have short period of viability and propagation of this plant through seed is poor. Vegetative propagation through stem cutting is reportedly difficult. At present, the immature fruits continue to have high economic value. They are often harvested and sold at high prices. This practice puts seed production and propagation of *kair* at risk. Poor or lack of seed production continues to be the major cause for *kair*'s declining population. In fact, Pandey and Shetty (1985) listed the plant as one of the endangered species in India. It is also suppressed by *Prosopis juliflora*, which regenerates faster and grows aggressively. This plant, however, cannot be used as fodder. *Prosopis juliflora* is seriously threatening the survival of *Capparis decidua* and other indigenous species.

Future Research Areas

Kair has proven to be an economically viable plant. It provides varied medicinal uses, building materials, fuelwood and other income-generating opportunities. It contributes to environmental sustainability due to its soil-binding capacity and its ability to improve the soil fertility of sand dunes and reduce soil alkalinity. Extensive research and support activities are thus needed to maximize the production, propagation and utilization of this species to help contribute to rural livelihood and wasteland development. Some of these research and support activities include the following:

- ☆ Establishment of a germplasm bank for seeds or suckers of *kair* from the different regions of India to conserve the plant's genetic variability;
- ☆ Development and promotion of intensive cultivation practices for sustainable production and utilization of *kair*;
- ☆ There is need for developing methods for cloning of selected germplasms through vegetative means as the seed raised plants are not true-to-type.

References

Agarwal, V. and Chauhan, S.M., 1988. A study on composition and hypolipidemic effect of dietary fibre from some plant foods. *Plants Foods for Human Nutrition*, 38(2): 189–197.

Ageel, A.M., Parmar, N.S., Mossa, J.S., Al-Yahya, M.A., Al-Said, M.S. and Tariq, M., 1985. Anti-inflammatory activity of some Saudi Arabian medicinal plants. *Agents Actions*, 17(3/4):383–384.

Ahmad, I. and Khan, N.A., 1983. A revision of the genus *Stenozygum fieber* (Pentatomidae: Strachini) from the oriental and Australian regions with reference to zoogeography and phylogeny. *Australian Journal of Zoology*, 31(4): 581–605.

Ahmad, V.U., Arif, S., Amber, A.R. and Fizza, K., 1987. Capparisinine, A new alkaloid from *Capparis decidua*. *Liebigs Ann. Chem.*, (2): 161–162.

Ahmad, V.U., Arif, S., Amber, A.R., Usmanghani, K. and Miana, G.A., 1985. A new spermidine alkaloid from *Capparis decidua*. *Heterocycles*, 23(12): 3015–3020.

Ahmad, V.U., Nargis Ismali, Shoth A.and Aziz-ur-Rehrnan Amber 1992. Two new N–acetylated spermidine alkaloids from *Capparis decidua*. *Journal of Natural Products*, 55(10): 1509–1512.

Ahmed, V.U, Fizza, K., Amber, A.U.R. and Arif, S., 1987. Cadabicine and cadabicine diacetate from *Crataeva nurvala* and *Cadaba farinosa*. *J. Nat. Prod.*, 50(6):1186.

Anon, 2001. *Ker: A Genetic Wealth of Haryana*. CCS HAU, Research Courier 6(1):4,

Anon., 1992. *The Wealth of India*, Vol. 3. Council of Scientific and Industrial Research Publication, New Delhi, India, p. 210–212.

Arora, G.L and Tewari, P.K., 1984. Incidence and distribution of *Bruchidius obscurus* Arora (Bruchidae: Coleoptera). Research Bulletin of the Punjab University Science 35(1–2): 97–98.

Baumer, M., 1983. Notes on trees and shrubs in arid and semi–arid regions. Rome FAO. Forestry Division.

Bhansali, R., 1994. In: *Advances in Plant Tissue Culture in India*, (Ed.) P. Tendon. Pragati Prakashan, Meerut (UP).

Bhansali, R.R., 2002. Capparis. In: *Fruits Tropical and Subtropical, Vol II*. Naya Udyog, Bidhan Sarni, Calcutta.

Bhargava, R. and Saroj, P.L., 2003. Standardization of protocol for micropropagation of *Capparis decidua*. *CIAH Annual Report* 2002–2003, p. 125–126.

Bhargava, R., Saroj, P.L. Neelam Chauhan and Verma, Pyush, 2002. Development of propagation techniques for *Capparis decidua*. In: *CIAH, Annual Report* 2001–2002.

Bhargava, Rakesh, Nath, Vishal, Pareek, O.P., Shargava. R. and Nath, V., 2000. Note on the role of plant growth inhibitors in sprouting of *Capparis decidua* cuttings. *Current Agriculture*, 24(1–2): 131–133.

Booth, F.E.M. and Wickens, G.E., 1988. Non-timber uses of selected arid zone trees and shrubs in Africa. *FAO Conservation Guide* No. 19, Rome.

CABI, 2000. *Global Forestry Compendium*, CD–ROM. CABI.

Chandra, A., Chandra, Anju and Gupta, I.C., 1994. In. *Arid Fruit Research*. Scientific Publisher, New Pali Road, Jodhpur, p. 299.

Chandra, H., 1988. Some observations on *Belenois aurota* Fabricius (Lepidotera: Pieridae). *Plant Protection Bulletin, India*, 40(2): 22.

Chaturvedi, V. and Nagar, R., 2001. Levels of β-carotene and effects of processing on selected fruits and vegetables of the arid zone of India. *Plant Foods for Human Nutrition*, 56(2): 127–132.

Chauhan, S.N., Duhan, A. and Bhat, C.N., 1986 Nutritional value of ker (*Capparis decidua*) fruit. *Journal of Food Science and Technology*, 23(2): 106–108.

Chopra, R.N., Nayar, S.L. and Chopra, I.C., 1999. *Glossary of Indian Medicinal Plants*. CSIR, New Delhi, p. 49.

Chundawat, B.S., 1990. *Arid Fruit Culture*. Oxford & IBH Publishing Co. Pvt. Ltd., New Delhi, pp. 187–189.

Dafni, A., Eisikowitch, O. and lvn, V., 1987. Nectar flow and pollinators efficiency in two co-occurring species of *Capparis* (Capparaceae) in Israel. *Plant Systematics and Evolution*, 157: 181–186.

Dahot, M.U., 1993. Chemical evaluation of the nutritive value of flowers and fruits of *Capparis decidua*. *J. Chem. Soc. Pak.*, 15(1): 78–81.

Deora, N.S. and Shekhawat, N.S., 1995. Micropropagation of *Capparis decidua* (Forsk.) Edgew.: A tree of arid horticulture. *Plant Cell Reports*, 15(3–4): 276–281.

Duhan, A., Chauhan, B.N. and Punia, D., 1992 Nutritional value of some non-conventional plant foods of India. *Plant Foods for Human Nutrition*, 42(3): 193–200

Eisikowitch, O, lvri, Y. and Dafni, A., 1986. Reward partitioning in *Capparis spp*. Along ecological gradient. *Oecologia*, 71(4): 47–50.

Gaind, K. N., 1969. Antibacterial and antifungal studies. *Indian J. Pharm.*, 31: 124–125.

Gaind, K. N., Juneja, T.R. and Jain, P.C., 1969. Anthelmintic and Purgative Activity of *Capparis decidua* Edgew. *Indian J. Hosp.Pharm.*, July–August, 153–155.

Gaind, K.N., Juneja, T.R. and Bhandarkar, P.N., 1966, 1972. *Ind. J. Pharm.*, 28:343, 34: 86.

Garg, S.P., Rai, Sushila and Rai, B., 1993. Chemical examination of medicinal desert plant, *Capparis decidua*. Part I. Isolation and mass spectral studies of N-nonacosane and N-hentriacontane from its flowers. *Current Agriculture*, 17(1–2): 97–99.

Gautam, D.D. and Bishnol, S., 1991. Influence of root and shoot extracts of some desert plants on seed germination of *Callsonurn polygonoideg* Linn. and *Lasiuruss indicus* Henr. *Journal of Phytological Research*, 4(2): 177–181

Gehlot, Praveen, Purohit, O.K. and Gehlot, P., 2001. A new fruit rot disease of *Capparis*. *Indian Phytopathology*, 54(1): 140.

Ghazanfar, S.A., 1994. *CRC Handbook of Arabian Medicinal Plants*. CRC Press Inc., Boca Raton, FL. 265 p.

Goyal, Madhu and Sharma, S.K., 2009. Traditional wisdom and value added prospects of arid foods of desert of North West region. *Indian Journal of Traditional Knowledge*, 8(4): 581–584.

Goyal, R. and Grewal R.B., 2003. The influence of Teent (*C. decidua*) on human plasma triglycerides, total lipids and phospholipids. *Nutr. Health.*, 17(1): 71–76.

Gupta, I.C., Harsh, L.N., Sharikarnarayan, K.A. and Sharma, B.D., 1989. Wealth from wastelands. *Indian Farming*, 38(11): 18–19, 24.

Gupta, J. and Ali, M., 1997. Oxygenated heterocyclic constituents from *Capparis decidua* root bark. *Indian J. Heterocycl. Chem.*, 6(4): 295–302.

Harsh, L.N. and Tiwari, J.C., 1998. Biodiversity of vegetational complex in arid regions of India. In: *Biodiversity of Forest Species*, (Eds.) R. Bawa and P.K. Khosla. Bishen Singh Mahender Pal Singh, Dehradun, India, p. 91–10.

Hocking, D., 1993. *Trees for Drylands*. Oxford & IBH Publishing Co. Pvt. Ltd., New Delhi, pp. 141–143.

Hooker Joseph. O., 1960. An enumeration of the genera and species of flowering plants. In: *Index Kewensis*, Vol. 1. Clarendon Press, London, pp. 415–418.

IBPGR–Kew., 1984. *Forage and Browse Plants for Arid and Semi-arid Africa*. IBPGR, Rome.

Jam, R.K., 1994. Fuelwood characteristics of medium tree and shrub species of India. *Bioresource Technology*, 47(1): 81–84.

Juneja, T.R., Gaind, K.N. and Panesar, A.S., 1970. *Res. Bull. Punjab Univ. Sci.*, 21(1–2):67(pt.3–4), 519–21.

Kayani, S.A., Ahmed, Tanweer, Achackzai, A.K. and Qadeer, Abdul, 1985. Some studies on abundance of termites in various plant communities of eastern Baluchistan, Pakistan. *Pakistan Journal of Forestry*, 35(4): 169–174:

Khera, A.P. and Singh, J.P., 1975. *Capparis aphylla*: A nutritive fruit for arid region. *Indian Horticulture*, 20(2): 10.

Kirtikar, K.R. and Basu, B.D., 1993. *Indian Medicinal Plants*, Vol.II, 2nd edn. Lalit Mohan Basu Prakashan, Allahabad, p. 197.

Kumar, N.K. and Bhandari, M.M., 1993. Impact of human activities on the pattern and process of sand dune vegetation in the Rajasthan Desert. *Desertification Bull.*, 22: 45–54.

Mali, R. G., Hundiwale, J. C., Sonawane, R. S., Patil, R. N. and Hatapakki, B.C., 2004. Evaluation of *Capparis decidua* for anthelmintic and antimicrobial activities. *Indian J. Nat. Prod.*, 20(4): 10–13.

Meghwal, P R., 2002. Judging the culinary maturity of ker fruits (Teent) (*Capparis decidua* Forsk). *Current Agriculture*, 26(1–2): 75–76.

Meghwal, P.R. and Vashishtha, B.B., 1998. Effect of time of planting and auxins on rooting of cuttings of ker (*Capparis decidua* Forsk). *Annals of Aid Zone*, 37(4): 401–404.

Pandey A.N. and Rokad, M.V., 1992. Sand dune stabilization: an investigation in the Thar desert of India. *Journal of Arid Environments*, 22(3): 287–292.

Pandey, R.K. and Sharma, R.B., 2003. In Journal of "*Current Bio Science*" 1 (3–4): 256–257.

Pandey, R.P. and Shetty, B.V., 1985. Rare and threatened plants of Rajasthan. In: *Proc. Nat. Symp. on Evaluation of Environment Species*, (Eds.) S.D. Mishra, D.N. Sen and J. Ahmed. Geobios Univ., Jodhpur, India: 238–241,

Pareek, O.P., 1998. In: *50 Years of Horticultural Research*, (Eds.) S.P. Ghosh and R.S. Bhatnagar. Division of Horticulture, ICAR, New Delhi, pp. 95–99.

Pareek, O.P., Sharma, B.D. and Sharma, S., 1998. *Wasteland Horticulture*. Malhotra Publishing House, New Delhi, India, 136 pp.

Rai, S., 1987. Chemical examination of edible plants of Rajasthan desert with special ref. to Capparadaceae. *Curr. Agric.*, 11(2–2):15–23.

Rai, Sushila and Rai, S., 1987. Oils and fats in arid plants with particular reference to *Capparis decidua*. *Transactions of Indian Society of Desert Technology*, 12(2): 99–105.

Rai, Sushila and Shekhawat, N.S., 2000. Biotectinological studies on *Capparis decidua*. *National Symposium on Prospects and Potentials of Plant Biotechnology in India in the 21st Century and 23rd Annual Meeting of PTCA(1)*: 18–21 October, 56 p.

Rao, B.H. and Rao, P.S.P., 1977. Anther in *Capparris decidua* (Forsk.) Pax. *Current Science*, 46:19.

Rashid, S., Lodhi, F., Ahmad, M. and Usmanghani, K., 1989. Preliminary cardiovascular activity evaluation of capparidisine, a spermidine alkaloid from *Capparis decidua*. *Pak. J. Pharmacol.*, 6(1–2): 6–16.

Saroj, P.L., and Awasthi, O.P., 2006. *Ker* (*Capparis decidua*). In: *Advances in Arid Horticulture*, Vol. 11, pp. 257–266.

Satyanarayana, T., Anjana, A., Mathews and Vijetha, P., 2008. Phytochemical and pharmacological review of some Indian *Capparis* species. *Pharmacognosy Reviews [Phcog Rev.]*–Supplement Vol 2, Issue 4, Jul–Dec, p. 36–45.

Shankarnaryan, K.A., Harsh, L.N. and Kathju, S., 1987. Agroforestry in the arid zones of India. *Agroforestry Systems*, 5: 69–88.

Sharma, I.K., 1981. The carpenter bee (*Xylocopa fenestra*) in the Indian Thar desert. *Journal of the Bombay Natural History Society*, 78(2): 408–409.

Sharma, R., 2003. *Medicinal Plants of India: An Encyclopedia*. Daya Publishing House, New Delhi, p. 42–43.

Shekhawat, J.S., 1999. Flower and fruit development in *kair* (*Capparis decidua* [Forsk.] Edgew). In: *Recent Advances in Management of Arid Ecosystem. Proceedings of a Symposium held in India*, (Eds.) A.S. Faroda, N.L. Joshi, S. Kathju and Amal Kar. Arid Zone Research Association of India, Jodhpur, pp. 383–386.

Shekhwat, J S., 1994 Studies on growth, fruit development and quality of *ker* (*Capparis decidua* (Forsk) Edgew). *Ph.D. Thesis*, Rajasthan Agricultural University, Bikaner.

Singh, Manjit, Jindal, S.K. and Sivadasan, Rekha, 2005. *Capparis decidua*: A multipurpose shrub. In: *Shrubs of Indian Arid Zone*. Arid Agroecosystem Directorate, CAZRI, Jodhpur, India, pp. 97–104.

Singh, Kalyan and Singh, K., 1994. Site suitability and tolerance limits of trees, shrubs and grasses on sodic soils of Ganga-Yamuna Doab. *Indian Forester*, 120(3): 225–235.

Vashishtha, B.B., 1987. Vegetative propagation of *Capparis decidua Annals of Arid Zone*, 26(1&2): 123–124.

Vogt, K., 1995. A field guide to the identification, propagation and uses of common trees and shrubs of dryland Sudan. SOS, Sahel International (UK).

Von Maydell, H.J., 1986. *Trees and Shrubs of the Sahel: Their Characteristics and Uses*. GTZ 6 MBH, Eschborn.

Vyas, Govind K., R., Sharma, Vinod Kumar, Sharma, T B. and Khandelwal, V., 2009. Diversity analysis of *Capparis decidua* (Forsk.) Edgew. using biochemical and molecular parameters. *Genetic Resources and Crop Evolution*, 56 (7): 905–911

Yadav, P., Sarkar, S. and Bhatnagar, D., 1997. Lipid peroxidation and antioxidant enzymes in erythrocytes and tissues in aged diabetic rats. *Indian J. Exp. Biol.*, 35(4): 389–392.

2011, Future Crops Vol. 1 ***Pages 79–91***
***Editor:* Professor K.V. Peter**
***Published by:* DAYA PUBLISHING HOUSE, NEW DELHI**

Chapter 6
LETTUCE
(*Lactuca sativa* L.)

Anant Bahadur and K.P. Singh
Indian Institute of Vegetable Research,
Varanasi – 221 305, U.P.
E-mail: singhab98@gmail.com

Lettuce (*Lactuca sativa*L. Family: Asteraceae, 2n = 2x = 18) is a temperate annual or biennial herb. Both the English and the Latin name of the genus are derived from *lac*, the Latin word for 'milk'. It is often grown as a leaf vegetable. It is eaten raw in salads, sandwiches, hamburgers, tacos and in many other dishes. In eastern countries like China and Egypt, stems instead of leaves of lettuce are consumed, either cooked, raw, pickled, dried, or as a sauce. Lactucarium, the dried latex from a wild lettuce relative, *Lactuca virosa* L., is used to make a sleep-inducing medicine (Ryder, 1986).

Area and Distribution

As a cool-season crop, lettuce is extensively grown on all the continents, particularly in temperate and subtropical. World lettuce production was more than 22 million tonnes from about 1 million ha in 2005. About two-thirds of the total production area in the world is in Asia. China is the largest lettuce producer and accounts for about half of the world's total production and area, primarily for stems. The United States has the largest production of lettuce as a salad crop, and produced 22 per cent of the world's lettuce supply from only 13 per cent of the production area due to its higher yield. Western Europe claims about 13 per cent of the total lettuce production and area in the world. Lettuce is also grown in large areas in India, Japan, Mexico, and Turkey. India produces 7.9-lakh tonnes lettuce from about 1.2 lakh hectare area with a productivity of 6.58 tonnes/ha.

Currently, 62 per cent of the lettuce production in the United States is of the crisp head type, 23 per cent is romaine, and 15 per cent is leaf and butterhead types (USDA-NASS, 2006). In UK, about 75 per cent of the production is crisp head, 15 per cent is butterhead, and 10 per cent is romaine (Ryder, 1999). Lettuce production in Germany consists of about 2/3 butterhead and 1/3 crisphead. A large portion of lettuce in the Netherlands, mostly butterhead type, is in greenhouses. France produces mostly

Butterhead and Batavia lettuces for domestic consumption and iceberg-type crisphead for export. In Australia, production is of the iceberg type, and to a lesser extent, of romaine and leaf lettuces. Japan produces mostly iceberg, leaf and butterhead lettuces.

Botany

Lettuce belongs to family Asteraceae (Compositae), tribe Cichoreae, and genus *Lactuca*. Of about 100 species of *Lactuca*, only four can be crossed to each other by conventional hybridization methods and thus form the most important breeding group. They include *L. sativa* L., *L. serriola* L., *L. saligna* L., and *L. virosa* L.

Lettuce is a self-pollinated annual plant. It forms a deep taproot with largely horizontal lateral roots, most densely near the soil surface. Leaves are nearly sessile, spirally arranged in a dense rosette on the shortened stem. Leaf margins may be entire, lobed, incised, indented, or undulating. Leaf surface can be smooth, savoy, or crinkled. The colour of lettuce leaves ranges yellow to dark green with different degrees of shininess; anthocyanin may cover all or part of the leaves, or in a spotting pattern or just along the margin. Stem elongation signals the end of vegetative stage and reproductive phase begins. A single stem is usually formed to bear the inflorescence, which is a dense corymbose panicle composed of many capitula, each consisting of many florets. The number of florets usually ranges 12 to 20, but can be as few as seven and as many as 35 (Feráková, 1977). Each floret produces a single-seeded achene, which is ribbed and topped with a pappus hair. Seed colour varies, including white, yellow, brown, gray, and black. Newly harvested seeds usually have a short period of dormancy, and most cultivars exhibit varying levels of thermodormancy.

Origin and Domestication

It is still not exactly clear which species were involved in the evolution that led to modern-day lettuce. However, it is certain that *L. serriola* is one of or the only direct ancestor(s) (de Vries, 1990, 1997; Kesseli *et al.*, 1991). The chromosomes of *L. sativa* and *L. serriola* are similar morphologically and they cross freely with each other. It is the most likely that changes in *L. serriola* caused by mutations led to the appearance of forms favoured by humans, particularly forms without spines on stems and leaves and with large seeds. They were then selected for use and further modified to fit human needs. Several primitive forms still exist and are used for these purposes in Egypt today (Harlan, 1986: Ryder, 1999). The existence of these primitive forms in the Middle East provides strong support for the idea that lettuce probably originated in the eastern Mediterranean basin.

From Egypt, cultivated lettuce spread to Greece and Rome and throughout the Mediterranean region (Lindqvist, 1960). Around the Mediterranean basin, the romaine type of lettuce, also known as cos type predominated and still is the most common lettuce type. The first written records of lettuce cultivation are credited to Herodotus who mentioned that about 550 B C a cos-like lettuce was eaten at the Persian Court (de Vries, 1997). Later, lettuce-growing areas further expanded from the Mediterranean region to the rest of Europe. Lettuce was brought to the New World by Christopher Columbus. In the next 400 years following its introduction, an assortment of leaf types was grown in America, including the loosely packed, softheaded crisp lettuce, known as Batavia types. At the beginning of the 20th century, more popular lettuces were butterhead types. In the 1940s, T.W. Whitaker crossed a Batavia cultivar 'Imperial' with an heirloom variety 'Brittle Ice' and developed the first true iceberg lettuce, 'Great Lakes', which is a compact firm-headed crisp head lettuce. Cultivated lettuce was introduced into China between 600 and 900 A.D. (de Vries, 1997). The Chinese selected lettuce for the bulky succulent non-bitter stem and long narrow leaves for consumption of the stem mainly as a cooked vegetable.

Domestication of the wild types of lettuce resulted in the loss of prickles from leaves and stems, less latex and tissue bitterness, reduced suckering, slow bolting except for stem lettuce, an increase in seed size, and non-reflexed involucres to prevent seed shattering. Human selection and later breeding efforts also led to changes in size, shape, colour, texture, and taste of leaves and plants, heading habits, resistance to diseases and insects, and adaptation to different geographic areas and environments.

Horticultural Types

There is a great diversity of shape, size, and colour among lettuce cultivars, which are classified into six horticultural types mainly based on leaf shape, size, texture, head formation and stem type. These types are crisphead, butterhead, romaine, leaf, stem, and Latin.

Crisphead

Commonly called "iceberg" or head lettuce, the crisphead type produces a spherical firm head that resembles cabbage and weighs 0.7 to 1 kg. The tightly folded inner leaves are rugose, brittle, and crispy with a mild taste. A crisp subtype called Batavia included in this type that forms a less dense and softer head and weighs about 500 g when mature. A cultivar named 'Iceberg' is actually a Batavia type of lettuce.

In United States, four crisphead lettuce groups were developed with different colour, leaf texture and shape, head size and shape and butt appearance (Ryder, 1986). These include: (1) *Imperial* (light- or medium-green leaves serrated or wavy, relatively soft texture, and varied in butt colour and ribbiness), (2) *Great Lakes* (bright-green leaf colour in various shades, serrated very crisp leaves, and whitish butt with prominent ribs), (3) *Empire* (deeply serrated light-green leaves, very crisp, heads often conical in shape, and whitish butt with flat ribs), and (4) *Vanguard* (dull-green leaves, softer texture than Great Lakes, wavy leaf margins, and green butt with flat ribs).

Butterhead

Sometimes called cabbage lettuce; butterhead lettuce produces a smaller and less compact head than the crisphead type. Leaves are broad, crumpled, relatively thin and tender with a soft buttery texture. In Europe, there are two subtypes based on day-length sensitivity and growing season. Day-neutral types are grown outdoors in summer and weigh about 350 g. They have firmer heads at maturity and are slower bolting than the short-day types which are commonly grown in protective shelters in winter and may weigh 150-200 g. There are also two subtypes in the U.S. based on head appearance and size. The Boston type is larger, lighter in colour, and has a closed head. The Bibb type is smaller and darker green, and the head is relatively open at the top.

Romaine

The romaine type is also known as roman or cos lettuce. It has elongated, coarse, and relatively crispy textured leaves with prominent broad midribs. Plants tend to have an upright stature and form a loaf-shaped head after the rosette stage. The heads are either closed or relatively open at the top. The closed type is used for "romaine heart" production, in which only the closed inner leaves are harvested. Romaine heads may weigh as much as 750g. They have a sweeter and stronger taste than that of crisphead lettuces.

Leaf

Leaf or cutting, lettuces have considerable variation in leaf size, shape, margin, colour and texture. They form a bunch or rosette of leaves which may have broad, elongated, or lobed shape like oak leaves, smooth or frilled margins, and yellow, green, or red colours in varying shades. Texture ranges from crispy to soft. Its relatively open growth habit gives it a fewer bleached leaves and a stronger taste than the crisp head types. Leaf lettuce may weigh up to 0.5 kg when all leaves are harvested.

Stem

Stem lettuce is also called celtuce, stalk, or asparagus lettuce. Plants are grown mainly for the erect thickened stem (4-10 cm in diameter) that may be as long as 50-60 cm. The leaves are usually long and narrow, but may be as broad as romaine leaves. Except for the young foliage, leaves are not palatable due to their high latex content and bitterness, and are removed before consumption. Stems are peeled and the soft translucent green core is cooked or eaten raw. Stem lettuce is grown mainly in China and Egypt.

Latin

Latin lettuce has an upright stature and somewhat resembles romaine lettuce, but the leaves are shorter and less crispy. Heads are sometimes closed and leaf texture resembles Bibb-type butterhead. Latin lettuce is popular in Mediterranean regions and to some extent in Argentina and Chile.

Table 6.1: Important lettuce cultivars in different groups.

Groups		*Important Cultivars*
Crisphead	:	Great Lakes, Iceberg, Pennlake, New York, Imperial, Imperial 44, Imperial 859, New York 515, Calmar, Climax, Sea Green, Red Coach 74, Vanguard 74, Punjab Lettuce No. 1
Butterhead	:	White Boston, Big Boston, Bibb, May King, Butter-crunch, Butter King, Tom Thumb
Cos or Romaine	:	Paris Island, Dark Green, Paris White (White Paris), Valmaine
Leaf	:	Grand Rapids, Early Curled Simpson, Slobolt, Salad Bowl, Walsmann's Green
Stem	:	Celtuce

Great Lakes, Chinese Yellow, Imperial 859 and Slobolt are the important cultivars of lettuce and suitable for growing in India.

Nutritional Importance

Presence of organic acids, especially citric acid soothes the nervous system and improves the sleep. It also strengthens walls of blood vessels. Consumption of lettuce reduces blood cholesterol, obesity and diabetes, and improves blood structure. Fresh lettuce juice can be taken in chronic gastritis ulcers. Lettuce is also known to be sedative, diuretic and expectorant. Lettuce contains vitamins C, B_1, B_2, K and A. They also contain organic acids, calcium, iron, manganese, potash salts, cobalt, copper, iodine, zinc, molybdenum and boron.

Table 6.2: Nutritive values of different lettuce types (per 100 g of edible portion).

Nutrients	*Crisphead*	*Butterhead*	*Green Leaf*	*Romaine*	*Stem*
Water per cent	95.6	95.6	95.1	94.6	
Energy, kcal	14	13	15	17	
Protein, g	0.90	1.35	1.36	1.23	0.60
Carbohydrate, g	2.97	2.23	2.79	3.28	1.90

Contd...

Table 6.2–Contd...

Nutrients	*Crisphead*	*Butterhead*	*Green leaf*	*Romaine*	*Stem*
Fat, g	0.14	0.22	0.15	0.30	0.10
Dietary fiber, g	1.2	1.1	1.3	2.1	
Total sugar, g	1.76	0.94	0.78	1.19	
Calcium, mg	18	35	36	33	7
Iron, mg	0.41	1.24	0.86	0.97	2
Magnesium, mg	7	13	13	14	
Phosphorus, mg	20	33	29	30	31
Potassium, mg	141	238	194	247	
Sodium, mg	10	5	28	8	
Zinc, mg	0.15	0.20	0.18	0.23	
Copper, mg	0.025	0.016	0.029	0.048	
Manganese, mg	0.125	0.179	0.250	0.155	
Vitamin A, IU	502	3312	7405	5807	33
Vitamin B_6, mg	0.042	0.082	0.090	0.074	
Vitamin C, mg	2.8	3.7	18.0	24.0	1.0
Vitamin E, mg	0.18	0.18	0.29	0.13	
Vitamin K, μg	24.1	102.3	173.6	102.5	
Folate, μg	29	73	38	136	
Niacin, mg	0.123	0.357	0.375	0.313	0.500
Pantothenic acid, mg	0.091	0.150	0.134	0.142	
Riboflavin, mg	0.025	0.062	0.080	0.067	0.020
Thiamin, mg	0.041	0.057	0.070	0.072	0.030

Source: USDA-ARS, 2005 except for stem lettuce (Lu, 2000).

Cultural Practices

Soil and Climatic Requirements

A well-drained sandy loam or loam soil, rich in organic matter is the best. Lettuce is very sensitive to high soil acidity, and fairly tolerant to salt. Excess salinity results in poor seed germination and small heads. Soil having pH 6-7 is the best for lettuce growing.

Lettuce is a cool season vegetable crop. The temperature range for seed germination is 10° to 27°C with 24°C as optimum. At lower temperature, seed germination is delayed; however, germination(per cent) is not affected. Lettuce is sensitive to frost; it may cause blistering of epidermal tissues of the outermost leaves. The germination of lettuce seed can be inhibited in dark above 27°C due to high temperature induced seed dormancy (thermo-inhibition). As temperatures rises 2° or 3°C above the optimum (usually 15-22°C), lettuce germination sharply declines from nearly 100 per cent to near zero (Reynolds and Thomson 1971). The thermo-dormancy may be broken by exposure of seed to chilling at 4° to 6°C for 3-5 days. The optimum temperature range for growth is 19° to 23°C during the day and 10 to 12°C at night. Heading is prevented and shooting to seed induced at temperatures between 21°

to 26°C. High temperatures (above 22°C) also hampers the edible quality and gives a strong bitter flavour. Crisp-head lettuce is more sensitive to heat stress (35°C) causing rib discolouration when stress is imposed two weeks after heading (Jenni 2005). Leafy type lettuce is adapted to a wider range of temperature than heading type.

Sowing and Transplanting Requirement

Lettuce seeds may directly be sown in field or in nursery to produce transplants. In most of the developed countries, it is directly sown in field, while in India it is mostly grown through transplants. Lettuce seeds have thermodormancy, and can be overcome by priming in aerated 1 per cent solution of $K_3 PO_4$ for 6-8 h. Lettuce seeds are small and irregular in shape, and are difficult to be singularized for uniform sowing. Seed coating with *Trichoderma viride* and growth regulators increases the percentage and seedling emergence speed index (Diniz *et al.,* 2006). For direct sowing, the seed requirement is 1.2-1.5 kg/ha. Seeds are sown at 0.5-1 cm deep. In field lettuce, seed can directly be sown in 3rd week of October (Rehman *et al.,* 2001). Three weeks after sowing, the plants are thinned to 30-40 cm between plants and 45 cm between rows in head lettuce. Under direct sowing, the maximum leaf yield in crisphead lettuce can be achieved at a sowing distance of 20 x 20 cm (da-Silva *et al.,* 2000).

The seed rate in lettuce to raise transplants for one-hectare area varies from 150g under precision nursery production to 450 g for commercial nursery growing. For transplanting in one-hectare area, about 70-80 m^2 seedbed area is required. One-gram lettuce seed contains 700-1000 seeds. The plant population/hectare varies from about 50 thousands in head lettuce to around 300 thousands in leaf lettuce. Seeds are sown in nursery beds in October-November. The soil mixture comprising 40 per cent of garden soil, 40 per cent peat and 20 per cent sand are the optimum for growing nursery (Asaftei and Berar, 2006). Seedlings become ready for transplanting in 20-25 days. Before transplanting, seedlings are hardened by withholding irrigation for 5-6 days. Seedlings are generally transplanted in flat beds at a spacing of 45 x 45 cm or 45 x 30 cm. Seedlings may be transplanted on raised beds at similar distances. Soon after transplanting, a light irrigation is required or individual seedling is watered with rose can for 3-4 days.

Manure and Fertilizer Requirements

Lettuce is responsive to fertilizer application. It is advantageous to apply 10-15 tonnes of well rotten FYM during land preparation. Soil application of digested sludge @ 10-20 tonnes/ha can be also used as a partial substitute of chemical fertilizer and as a soil conditioner without harmful accumulation of heavy metals (Dolgen *et al.,* 2004; Bahadur *et al.,* 2009). Besides, 75-80 kg nitrogen, 50 kg P_2O_5 and 50 kg K_2O are also given (Rehman *et al.,* 2001). Application of more than 100 kg N/ha has an adverse effect on colour, flavour, and compactness in lettuce (D'-Antuono and Neri, 2001). Full P_2O_5, K_2O and 1/3rd N are given as basal. Remaining N is applied 30 and 45 days after transplanting in equal doses as top dressing. In direct seeded crop, N is applied just after thinning and during later growth. In fertigation, the better head yield and quality may be obtained with application of N at 60-90 mg/l at every second or third day (Soundy *et al.,* 2005). In zinc deficient soil, 20-25 kg/ha zinc sulphate may be incorporated to the soil. Soil application of FYM and digested sludge (10 tones/ha) and seedling inoculation in biofertilizers, particularly P-solubilizers or VAM increase head production, photosynthesis and vitamins contents in lettuce, besides considerable savings of chemical fertilizers (Bahadur *et al.,* 2009). Seed inoculation in *Azospirillum* enhanced the germination and vegetative growth, and also alleviated the salt stress to some extent (Barassi *et al.,* 2006). It is also observed that the lettuce produced with organic manures has lower nitrates accumulation in leaves than produced with chemical fertilizers (Stopes *et al.,* 1989).

Irrigation

Lettuce requires uniform supply of soil moisture (60 per cent available soil moisture) throughout the growth period. Light irrigation is provided at frequent intervals as per need of the crop. Lettuce root zone soil moisture should be constantly maintained at field capacity through drip system (Sutton and Merit,1993). Water requirements may vary from 450 to 700 mm/ha depending upon soil type, variety and growing temperature. Lack of soil moisture particularly during head enlargement favours early bolting and affects quality of head/leaf. Over watering should be avoided as it encourages bottom rot disease.

Intercultural Operations

The critical period of crop-weed competition is between 2-5 weeks. The crop should be kept weed free during this period. One or two manual weedings and hoeings may take care of weeds. Herbicides can also be used to control weeds. Pre-emergence application of pendimethalin or alachlor (1 kg/ha) before transplanting can manage weeds up to 30-35 days after transplanting. If there is weed problem later in the season, one hand weeding is advised.

Soil mulching is beneficial for reducing weed growth, nitrate content in leaves and increasing head/leaf yield. The black paper/polythene offers complete weed control, although weeds under the other mulches do not develop beyond the cotyledonary or two-true-leaf stage (Brault 2002; Siwek *et al.*, 2007). Kocar (2001) noticed higher early yield with transparent mulch while black-white film had higher total yield and head weight than other mulch materials. White or transparent polythene mulch has reduced the nitrate contents in lettuce leaves (Wojciechowska *et al.*, 2007).

Harvesting

Harvesting is done by cutting the plant just below the soil surface or pulling them from the ground and trimming off the roots, outer leaves and any necrotic, diseased or damaged leaves. Generally, the harvest maturity in crisp head type is 70-120 days, butter head and cos is 55-70 days and for loose-leaf types 40-45 days. In lettuce,there is a shift in diffuse reflectance (reflected light just below the leaf surface) from 640–660 nm to 700–750 nm during maturation, which could be used to determine harvest maturity (Brach *et al.* 1982).

Crisp head lettuce should be harvested when the heads are well developed and firm, but before it begins to bolt. At initiation of flower stalk, they lose their crisp texture, become tough and may have a bitter taste. Immature heads are spongy and are not fit for postharvest handling. Butter heads and cos types can be picked after formation of head or in the semi-heading stage. Leaf lettuce is harvested as soon as the leaves are large enough, but before becoming tough and bitter.

Yield

The average yield of lettuce varies from 15-20 tonnes/hectare. By adopting good agro-techniques, a yield of around 30 tonnes/ha can be obtained in head lettuce (Bahadur *et al.*, 2009).

Postharvest Management

Lettuce is highly perishable. Pre-cooling is very important, and vacuum cooling is the best for lowering product temperature quickly. Forced-air cooling with high humidity was also suitable, but takes more time. Hydro cooling is effective for non-heading lettuce, but it is not used for head lettuce. After 25-30 minutes of vacuum cooling at 4.0-4.6 mmHg with a condenser temperature of–1.7 to 0°C the lettuce temperature falls down to 1°C (Barger 1963). Perforated polyethylene film is used as head wraps or plastic film box liners to reduce weight loss. If the crop is to be vacuum cooled, the film must be perforated. The minimally processed lettuce when packed in multi-layer film made by laminating an aluminum foil with a polyethylene film and stored at 4 °C shows minimum decay in lettuce quality (Del Nobile *et al.*, 2008).

At room temperature of 20°C and 60 per cent RH, lettuce could be kept for about 2 days (Mercantilia 1989). At 0°C and 90–95 per cent RH, butterhead can be kept for 8–12 days (Mercantilia 1989, SeaLand 1991). In CA storage at 0°C with 2 per cent CO_2 and 3 per cent O_2, lettuce can be stored for 1 month.

Physiological Disorders

Tip Burn

Tipburn is an abiotic disorder characterized by marginal collapse and necrosis at or near the margins of rapidly expanding inner leaves. It is associated with low tissue concentration of calcium, and an imbalance of other nutrients (Sonneveld and Mook 1983). It is associated with rapid growth and restricted transpiration during sudden rise in temperature and reduced calcium transport to rapidly growing new tissues (Saure 1998). Plant tissues affected with tip burn have low level of calcium even though calcium level in the soil may be quite high. Tipburn predisposing factors include: warm temperatures, excess fertility (particularly nitrogen), and an increase in light intensity. Hartz *et al.* (2007) reported that soil available Ca plays no substantive role in tipburn severity, and Ca fertigation does not improve lettuce Ca uptake or reduce tipburn.

Management strategies for tipburn include production practices which avoid rapid and excessive growth. Regular irrigation and foliar spray of calcium may provide some reduction of tipburn incidence in open-head lettuce types but not crisphead lettuce. Calcium sources added to the soil are usually ineffective due to slow translocation.

Bolting

Lettuce plants are very prone to premature flowering, if exposed to warm weather (usually above 25°C). Such heads become unmarketable. Sometimes, the heads may appear normal, but they have elongated core and stem inside. Long days may induce seed stalk formation and flowering.

Brown Stain

It is caused by exposure of crisphead lettuce to elevated levels of CO_2. Brown stain lesions are oval, brown and slightly sunken. Chlorophyll-free midribs at the base of lettuce leaves are the most susceptible. Brown stain severity increases as CO_2 concentrations in storage atmospheres increase from 1 to 5 per cent, and as O_2 concentrations decrease from 21 to 1 per cent (Ke and Saltveit 1989). Brown stain can be avoided by maintaining CO_2 levels below 2 per cent and O_2 levels above 10 per cent. Lettuce harvested at the proper maturity and selected varieties are more resistant to brown stain.

Insect Pests

Lettuce is attacked by several kinds of insects. The most destructive ones are six-spotted leafhoppers, wireworms, plant bugs, aphids, whiteflies, and several types of caterpillars. In addition, corn earworms will occasionally attack lettuce during the early stages of growth.

Aphids are the most serious insect pest and can cause physical damage as well as serving as vectors for lettuce mosaic virus. The caterpillars include loopers, variegated cutworms, and various species of armyworms. The cabbage looper (*Trichoplusia sp*) feeds on the underside of the foliage and eventually develops into a green caterpillar. The emerging larvae of beet armyworm (*Spodoptera exigua*) and the yellow-striped armyworm (*S. ornithogalli*) can cause severe seedling damage at the ground. Later generations or older larvae of caterpillars can enter the head, making control difficult. Insecticides give good control if applied early. The bacterial insecticide *Bacillus thuringiensis* is frequently used to control certain caterpillars as a substitute for synthetic insecticides.

Diseases

Bottom Rot

This common fungal disease is caused by the same organism that causes damping-off (*Rhizoctonia solani*). It is the most prevalent during wet weather. The plants may be attacked at nearly all stages of growth, and the disease may be present throughout the season. Infection starts at the lower leaves which are in contact with the soil and later progresses into the head. Tissues deteriorate and become brown and slimy.

Optimum control of bottom rot is achieved by using various cultural practices in combination with fungicidal control. Ploughing before planting effectively buries sclerotia. Crop rotation and effective weed control also reduce inoculum potential. Growing lettuce on raised beds improves aeration and minimizes foliar contact with the soil, the point at which infection typically occurs. Standard bottom rot fungicides such as iprodione and vinclozolin have also found effective (Mahr *et al.* 1986).

Lettuce Mosaic Virus

Lettuce mosaic is one of the most common and potentially devastating diseases of lettuce caused by the *Lettuce mosaic potyvirus* (LMV) The disease is widespread worldwide (Dinant and Lot 1992). The initial leaves of plants originating from LMV-infected seed are irregularly shaped, exhibiting a light green mottle or mosaic. Mottling and mosaic are usually more pronounced on leaf lettuce than on crisphead cultivars. Leaf vein-clearing and necrotic flecking may also appear. Growth of infected plants is stunted and frequently, distorted. Infected heads may appear dull green, with outer leaves developing a characteristic downward curling. The downward curling symptom helps to distinguish LMV infections from CMV infections in the field. The virus is transmitted in a non-persistent manner by a number of aphid species, including the green peach (*Myzus persicae* Sulzer) and the cotton aphid (*Aphis gossypii* Glover). LMV may also be mechanically transmitted in sap, and is the only economically important virus of lettuce that is seed transmitted.

Management of LMV is successfully accomplished through the use of virus-indexed seed and/or host-plant resistance. Indexing is commonly done by certified labs using ELISA. To complement the planting of indexed seed, a lettuce-free period of at least two weeks is recommended, along with adequate weed management. LMV resistant cultivars are more heavily relied upon for control of LMV in Europe, South America, and lettuce-growing regions other than the U.S.

Downy Mildew

Downy mildew is caused by the fungus *Bremia lactucae*. Although fairly widespread, it is more common in western production areas. The first symptoms appear on older leaves as pale, yellowish-green, blotchy areas on the upper surfaces. On the lower sides, a distinct visible downy or velvety growth occurs on the affected surface. The affected areas enlarge and coalesce. Wild lettuce is a host for the disease organism and should be eradicated from nearby fields. Fungicide sprays and dusts are effective in controlling the disease, if applied as soon as the first sign of infection appears.

Lettuce Drop

Drop is one of the most destructive diseases of lettuce worldwide, characterized by rapid collapse of infected plants (Patterson *et al.*, 1986). It is a fungal disease caused by *Sclerotinia sclerotiorum*. Initial symptom is wilting of the lower, outermost leaves. This gives the plants a faded stressed appearance. As the infection advances, the entire plant wilts. Collapsed outer leaf layers flatten, turn chlorotic and then necrotic.

Deep-ploughing to bury inoculums deep in the soil has proven effective under low inoculum densities (Subbarao *et al.* 1996). Irrigation methods, which avoid wetting the top 5-8 cm of soil, such as subsurface drip or seepage irrigation, assist in controlling the disease. Fungicidal control of drop is achieved using dicarboximide fungicides and several new fungicides look promising (Matheron and Porchas 2000). Roguing of plants exhibiting drop, if feasible, may be successful in limiting sclerotial densities of *S. minor*, thereby reducing disease incidence (Patterson and Grogan 1985). *Trichoderma hamatum*, *T. virens* and *Coniothyrium minitans* reduced *Sclerotinia* lettuce drop disease by 30–50 per cent (Rabeendran *et al.*, 2006).

Lettuce Big-vein

Lettuce big-vein (LBV) is caused by virus (Vetten *et al.* 1987). LBV is also associated with the soilborne fungus, *Olpidium brassicae* (Campbell and Grogan 1963). LBV may be the most ubiquitous lettuce disease for which there is presently no control (Falk 1997). The most obvious symptom is clearing along the veins, making them appear larger than normal. Affected leaves frequently pucker and leaf margins, appear ruffled. Leaves of infected head lettuce often have an upright growth habit, making it more closely resemble leaf lettuce. Lettuce infected at an early growth stage with LBV may be severely stunted or delayed in their maturity, while plants infected in the late rosette stage may possess conspicuous vein- clearing symptoms yet remain marketable. LBV is favoured by cool temperatures and moist soils.

References

Asaftei, V.I. and Berar, V., 2006. Research concerning influence of the nutritive mixture on seedling vigour in some ornamental vegetable species. *Seria Horticulturae*, 63: 32–36.

Bahadur, Anant, Singh, J., Singh, K.P., Upadhyay, A.K. and Rai, M., 2009. Morpho-physiological, yield and quality traits in lettuce (*Lactuca sativa*) as influenced by use of organic manures and biofertilizers. *Indian Journal of Agricultural Sciences*, 79: 282–285.

Barassi, C.A., Ayrault, G., Creus, C.M., Sueldo, R.J. and Sobrero, M.T., 2006. Seed inoculation with *Azospirillum* mitigates NaCl effects on lettuce. *Scientia Horticulturae*, 109: 8–14.

Barger, W R., 1963. Vacuum pre-cooling: A comparison of cooling of different vegetable. *United States of America Department of Agriculture Market Research Report*, 600, 12 pp.

Brach, E.J., Phan, C.T., Poushinsky, B., Jasmin, J.J. and Aube, C.B., 1982. Lettuce maturity detection in the visible 380–720 nm, far red 680–750 nm and near infrared 800–1850 nm wavelength band, *Lactuca sativa.Agronomique Science Prod. Vég. et de l'Environ.*, 2: 685–694.

Brault, D., Stewart, K.A. and Jenni, S., 2002. Optical properties of paper and polyethylene mulches used for weed control in lettuce. *Hort. Science*, 37: 87–91.

Campbell, R.N. and Grogan, R.G., 1963. Big-vein virus of lettuce and its transmission by *Olpidium brassicae*. *Phytopathology*, 53: 252–259.

D' Antuono and Neri, R., 2001. Lettuce, the quality. *Colture Protette*, 30: 73–74.

da-Silva, V.F., Bezerra, N.F., de-Negreiros, M.Z. and Pedrosa, J.F., 2000. Effects of lettuce cultivars and spacings on lettuce leaf yield under high temperature and sunlight. *Horticultura Brasileira*, 18: 183–287.

de Vries, I.M., 1990. Crossing experiments of lettuce cultivars and species (*Lactuca* sect. *Lactuca*, Compositae). *Pl. Syst. Evol.*, 171: 233–248.

de Vries, I.M., 1997. Origin and domestication of *Lactuca sativa* L. Genet. *Resour. Crop Evol.*, 44: 165–174.

Del Nobile, M.A., Conte, A., Cannarsi, M. and Sinigaglia, M., 2008. Use of biodegradable films for prolonging the shelf–life of minimally processed lettuce. *Journal of Food Engineering*, 85: 317–325.

Dinant, S. and Lot, H., 1992. Lettuce mosaic virus: A review. *Plant Pathology*, 41:528–542.

Diniz, K.A., Oleivera, J.A., Guimares, R.M., Carvalho, M.L.M.De and Machado, J.D.A.C., 2006. Incorporation of microorganism, amino-acids, micronutrients and growth regulators in lettuce seed through coating technique. *Revista Brasileira de Sementes*, 28: 37–43.

Dolgen, D., Alpaslan, N.M. and Delen, N., 2004. Use of an agro-industry treatment plant sludge on iceberg lettuce growth. *Ecological Engineering*, 23: 117–125.

Falk, B.W., 1997. Lettuce big-vein. In: *Compendium of Lettuce Diseases*, (Eds.) Davis, R.M., Subbarao, K.V., Raid, R.N. and Kurtz, E.A. APS Press, St. Paul, MN, pp. 41–42.

Harlan, J.R., 1986. Lettuce and the sycomore: Sex and romance in ancient Egypt. *Economic Botany*, 40: 4–15.

Hartz, T.K., Johnstone, P.R., Smith, R.F. and Cahn, M.D., 2007. Soil calcium status unrelated to tipburn of romaine lettuce. *HortScience*, 42: 1681–1684.

Jenni, S., 2005. Rib discoloration: A physiological disorder induced by heat stress in crisphead lettuce. *HortScience*, 40: 2031–2035.

Ke, D. and Saltveit, M.E., 1989. Carbon dioxide-induced brown stain development as related to phenolic metabolism in iceberg lettuce. *Journal of the American Society of Horticultural Science*, 114: 789–794.

Lu, G., 2000. *Lettuce Cultivation Techniques*. Jindun Publishing, Beijing (in Chinese).

Kesseli, R., Ochoa, O. and Michelmore, R., 1991. Variation at RFLP loci in *Lactuca* spp. and origin of cultivated lettuce (*L. sativa*). *Genome*, 34: 430–436.

Kocar, G., 2001. Effects of mulching with different coloured polyethylene on lettuce production in greenhouse. *Anadolu*, 11: 47–55.

Lindqvist, K., 1960. On the origin of cultivated lettuce. *Hereditas*, 46: 319–350.

Mahr, S.E.R., Stevenson, W.R. and Sequeira, L., 1986. Control of bottom rot of head lettuce with iprodione. *Plant Disease*, 70: 506–509.

Matheron, M.E. and Porchas, M., 2000. Evaluation of new fungicides for management of *Sclerotinia* leaf drop of lettuce, 1999. *American Phytopathological Society, Fungicide and Nematicide Tests*, 54:167.

Mercantilia, 1989. *Guide to Food Transport Fruit and Vegetables.* Mercantilia Publishers, Copenhagen, 247 pp.

Patterson, C.L. and Grogan, R.G., 1985. Differences in epidemiology and control of lettuce drop caused by *Sclerotinia minor* and *S. sclerotiorum. Plant Disease*, 69: 766–770.

Patterson, C.L., Grogan, R.G. and Campbell, R.N., 1986. Economically important diseases of lettuce. *Plant Disease*, 70: 982–987.

Rabeendran, N., Jones, E.E., Moot, D.J. and Stewart, A., 2006. Biocontrol of *Sclerotinia* lettuce drop by *Coniothyrium minitans* and *Trichoderma hamatum. Biological Control*, 39: 352–362.

Rehman, Z., Hussain, S.A., Abdur, R. and Muhammad, S., 2001. Effect of nitrogen levels and sowing dates on yield of lettuce. *Sarhad Journal of Agriculture*, 17: 359–362.

Reynolds, T. and Thomson, P.A., 1971. Characterization of high temperature inhibition of germination of lettuce (*L. sativa*). *Physiology Plantarum*, 24: 544–547.

Ryder, E.J., 1986. Lettuce breeding. In: *Breeding Vegetable Crops*, (Ed.) M.J. Bassett. AVI Publishing, Westport, pp. 433–474.

Ryder, E.J., 1999. *Lettuce, Endive and Chicory*. CABI Publishing, New York.

Saure, M.C., 1998. Causes of the tipburn disorder in leaves of vegetables. *Scientia Horticulturae*, 76: 131–147.

SeaLand, 1991. *Shipping Guide to Perishables*. SeaLand Services, Iselim, NJ.

Siwek, P., Kalisz, A. and Wojciechowska, R., 2007. Effect of mulching with film of different colours made from original and recycled polythene on the yield of butterhead lettuce and celery. *Folia Horticulturae*, 19: 25–35.

Sonneveld, C. and Mook, E., 1983. Lettuce tipburn as related to the cation contents of different plant parts. *Plant and Soil*, 75: 29–40.

Soundy, P., Cantliffe, D.J., Hochmuth, G.J. and Stoffella, P.J., 2005. Management of nitrogen and irrigation in lettuce transplant production affects transplant root and shoot development and subsequent crop yields. *HortScience*, 40: 607–610.

Stopes, C., Woodward, L., Forde, G. and Vogtmann, H., 1989. Effects of composted FYM and a compound fertiliser on yield and nitrate accumulation in three summer lettuce cultivars grown in an organic system. *Agriculture, Ecosystems and Environment*, 27: 555–559.

Subbarao, K.V., Koike, S.T. and Hubbard, J.C., 1996. Effects of deep plowing on the distribution and density of *Sclerotina minor* sclerotia and lettuce drop incidence. *Plant Disease*, 80: 28–33.

Sutton, B.G. and Merit, N., 1993. Maintenance of lettuce root zone at field capacity gives best yields with drip irrigation. *Scientia Horticulturae*, 56: 1–11.

U.S. Dept. of Agriculture, Agricultural Research Service (USDA–ARS)., 2005. USDA National Nutrient Database for Standard Reference, Release 18, Nutrient Data Laboratory Home Page, (April 11 2006), http://www.nal.usda.gov/fnic/foodcomp/Data.

U.S. Dept. of Agriculture, National Agricultural Statistics Services (USDA–NASS)., 2006.

Vegetables 2005 Summary, (March 3, 2006), http://usda.mannlib.cornell.edu/reports/nassr/fruit/pvg–bban/vgan0106.txt.

Vetten, H.J., Lesemann, D.E. and Dalchow, J., 1987. Electron microscopical and serological detection of virus-like particles associated with lettuce big-vein disease. *Journal of Phytopathology*, 120: 53–59.

Wojciechowska, R., Siwek, P. and Libik, A., 2007. Effect of mulching with various films on the yield and quality of butterhead lettuce and celery stalk with special reference to nitrate metabolism. *Folia Horticulturae*, 19: 37–44.

2011, Future Crops Vol. 1 ***Pages 92–108***
***Editor:* Professor K.V. Peter**
***Published by:* DAYA PUBLISHING HOUSE, NEW DELHI**

Chapter 7
Lichens and Future Food

*Trupti Mohaptra, R. Valarmathi and G.N. Hariharan**
Lichen Ecology and Bioprospecting Laboratory,
M.S. Swaminathan Research Foundation,
3rd Cross Street, Taramani Institutional Area, Taramani, Chennai – 600 113
**E-mail: gnhariharan@mssrf.res.in*

Today, there is much wider acceptance of the urgent need to diversify our food basket with new sources of food since changing climate is a major threat to the existing food production scenario. Declining land and water resources, increasing pests and pathogens and exploding human population necessitate greater demand for food, fiber, medicine and fuel. We also need to produce more crops per unit area achievable through crop improvement, disease resistance, abiotic stress tolerance and with enhanced nutritional status. The major question in front of us today is to find the alternate resources as our future food and feed source to meet the challenges ahead. Today the content of the dining plate is mainly from limited crop and animal resources. It consists of three major cereals (rice, wheat and maize),a few legume species, vegetables and animal products. The current agricultural practices emphasize production of this limited food producing species. They fail to pay equal attention to the large scale production of new sources of food (*e.g.* mushrooms) and converting the wildly harvested food yielding plants and fungi as food crops. It is important to explore the wild biota to identify suitable organisms (uncultivated edible mushrooms, tubers, vascular and non-vascular epiphytes) as future food stuff. Several species are used as food stuffs by the tribal communities in different parts of the world.

Lichens as Untapped Resources

Lichens are fungi which live in symbiotic association with one or more green algae or cyanobacteria, constituting 47 per cent of ascomycetes and one fifth of all known fungi (Crittenden *et al.*, 1995) approximately numbering 18, 500 species (Stocker-Wörgötter, 2008). There are records of lichens being used as food by many different human cultures across the world. Lichens are eaten by people in North America, Europe, Asia and Africa, and perhaps elsewhere. Often lichens are merely famine foods eaten in times of dire needs, but in some cultures, lichens are a staple food or even a delicacy. Several lichen species are used as food stuffs in a very limited way in various parts of the world.They are *Laconera esculenta, Umbellicaria esculenta* and *Cladina* sp. in bread making, soups, desert and as an additive to vegetable preparations in Scandinavian, Mediterranean and in Japan. Considering the diverse lichen species as our food stuff may be a wild thinking today but it is expected that certain biomass producing, secondary compound rich lichens/their genes/their compounds will soon enter human food chain. Already many fungal species including (macro fungi — yeasts) are contributing to food and nutritional security of humans. Many of these fungal strains are still wild in nature. The key steps involved in bringing this potential wild biota into the human food chain need to be based on extensive research.

Traditional Uses of Lichens as Food

Manna lichen (*Leconora esculenta*) is known as Biblical manna found in Western Asia and North Africa. It is mentioned that the food was miraculously sent to the Israelites during their forty years of stay in the desert. It was described as small white flakes of grains falling on the ground in the night like hoar frost. The manna had to be gathered in the morning. Though it was probably eatable in the natural state, it was usually ground in a mill or beaten in a mortar and then boiled and made into cakes.

As a food stuff, lichens have often been directly consumed by man, but they have also been used indirectly to make alcohol or molasses, or to feed to livestock. Lichens have frequently been used as food by people of different regions of world. They are often used as famine food, but there are also many people who used lichens for food on a more regular basis. Lichens are sometimes used as a delicacy (like *Umbilicaria esculenta* in Japan) or a dessert (like *Cladina, Cetraria islandica* in Scandinavia). *Bryoria* used by native people of North America, was the most widely used edible lichen -a famine food for many groups. Generally most taxa are bitter tasting and provide little nutritional value *e.g., Bryoria fremontii. Umbilicaria esculenta* (Japanese: Iwatake; Korean: Seogi) is collected from cliffs for use in a variety of traditional Korean and Japanese foods. Lichen flour is used by indigenous people of Northern Mexico as an ingredient of Tesguino. *Bryoria fremontii* was the most widely used edible lichen in North America--a famine food for many groups, and a delicacy for some. *Cladina rangiferina*, or reindeer lichen, is a staple food of reindeer and caribou in the Arctic. Peoples in North America and Siberia traditionally eat the partially digested lichen after they remove it from the rumen of caribou that is killed. *Cladina rangiferina*, or reindeer lichen, is a staple food of reindeer and caribou in the Arctic. Northern peoples in North America and Siberia traditionally eat the partially digested lichen after they remove it from the rumen of caribou that is killed. Species mixture of *Parmelia, Heterodermia, Pyxine* and *Physcia* are extensively used to increase flavour of the Indian spicy dish 'Briyani' from Kashmir to Kanyakumari.

The following are the concerns about lichens as food to humans.

- ✰ Many consider that the secondary compounds of lichens cannot be digested in human digestive tract.
- ✰ The complex carbohydrates of lichens are not easily broken down in the human digestive tract *e.g.*, lichenin (soluble in hot water) and/or isolichenin (soluble in cold water, turns iodine purple), and often also contain other lichen polysaccharides like evernin and usnin (Swartz, 1911).

On the other hand, lichens are autotrophs and occupy the producer role in the ecosystems and are consumed by a variety of organisms including protozoans to mammals. Even though, some of the lichen compounds have anti-herbivorous properties in its natural state, they are available in very low

Table 7.1: Preparation methods of lichens used as food by people.

People	*Lichen Species*	*Preparation*
Thompson, Okanagan, Lillooet, Secwepemc, and Kootenay of British Columbia; Flathead Montana	*Bryoria fremontii*	Harvested, rinsed in fresh water for several hours to overnight, sometimes worked with hands, steam cooked in a covered pit with plant material (often onion) for several days, dried, then stored.
Okanagan	*Bryoria fremontii*	Usually cooked as above, but sometimes roasted over fire until crumbly then boiled in water to form molasses
Secwepemc	*Bryoria fremontii*	Usually cooked as above, but sometimes eaten raw in times of famine
Nez Perce and Blackfoot of Montana; Klamath of Oregon; and Wailaki of California	*Bryoria fremontii*	Eaten as famine food
Yuqpik	*Nephroma arcticum*	Collected and stored until winter, then boiled with crushed fish eggs and eaten
Yuqpik	*Cetraria ericetorum, Flavocetraria cucullata*	Chopped and boiled in soup
Yup'ik Inuit	*Lobaria scrobiculata*	Eaten raw
Ojibwa	*Lobaria amplissima*	Boiled until looked like scrambled eggs, then eaten; or Dried, boiled, then used in broth
Iroquois	*Lobaria amplissima*	Cooked and made into porridge
Potawatomi of Wisconsin	*Hypogymnia physodes*	Boiled in soup
Menomini	*Lobaria amplissima, Sticta glomulerifera*	Gathered in any season, stored dry, then boiled in soup
Algonkin, Naskapi, Chipewyan, Cree, Inuit, and European travelers in boreal North America	*Umbilicaria* spp. and *Lasallia* spp.	Boiled in several changes of water and eaten in soup or alone
Huron	*Umbilicaria* spp. and *Lasallia* spp.	Boiled then eaten
Nihitahawak of Saskatchewan	*Umbilicaria muehlenbergii*	Added to fish broth to make a thick soup
Japan	*Umbilicaria esculenta*	Eaten in soups and salads
Scandinavia, Iceland, Northern Europe and Britain, Northern Russia	*Cetraria islandica*	Lower parts removed, soaked in ash water for several days, boiled, rinsed, dried, ground, mixed with flour, then cooked as bread, or jelly
Turkey, Egypt	*Evernia prunastri, Pseudevernia furfuracea*	Made into bread or jelly, or fermented

Contd...

Table 7.1–Contd...

People	*Lichen Species*	*Preparation*
Lepchas and Nepalis: India	*Everniastrum cirrhata*	Boiled, liquid removed, then fried and eaten as a vegetable
West Central Asia and North Africa	*Aspicilia esculenta*	Ground, mixed with meal to one-third its weight, then made into bread or has been fermented with honey
Western Saudi Arabia, Kuwait, Oman	*Parmotrema tinctorum*	Used as food spice
India	*Parmelia abessinica*	Used as curry powder
Himalayas	*Peltigera canina*	Use as an efficacious medicine in hydrophobia and jaundice
Siberia	*Lobaria pulmonaria*	Used for making beer
Northern Europe and Northern Russia	*Anaptychia ciliaris*, *Bryoria* spp., *Cetraria islandica*, *Cladina rangiferina*, *Ramalina farinacea*, *Ramalina fastigiata*, *Ramalina fraxinea*, and *Usnea florida*	Made brandy: steamed lichen under pressure for several hours, added weak acid, steamed again, neutralized with alkali, then fermented
Northern Russia	*Alectoria ochroleuca*, *Cetraria islandica*, and *Cladina* spp.	Made molasses: soaked in weak alkali, hydrolyzed with dilute sulphuric acid, neutralized with chalk, then purified with activated carbon

quantity in natural thallus. So far there is no report of large-scale lethal effects of lichen compounds on any animal system, with an exception of vulpinine produced by *Letharia vulpine*, found used as a poison against wolves. People have traditionally used many preparation methods to make lichens edible and for long-term storage by removing the secondary compounds of lichens. Various methods of removal of secondary compounds from lichen thallus are outlined elsewhere.

The most frequently used preparation technique is boiling or steaming. This is used by groups of people from North America, Europe and India. Boiling would hydrolyze the lichen polysaccharides into digestible forms. It would also help to remove many lichen compounds. It is recorded that people would boil the lichen and discard the water, which indicated that the boiling water was being used to remove the lichen compounds. The lichen was also often soaked or rinsed with water. This could have removed some lichen compounds as well, but they are generally not very soluble in pure water. Both the Iroquois and Northern Europeans soaked the lichens in ash water. Wood ash is alkaline, and so it would be more effective in removing the acidic lichen compounds. Alkali could also hydrolyze lichen polysaccharides. The addition of dilute acid, or acidic things like onion, is common when cooking lichen. Acids could possibly have helped to hydrolyze lichen polysaccharides, or they might make some lichen compounds more water soluble.

Nutritive Value

As a biological system, lichens also contain carbohydrates, proteins, fats, fibres, amino acids, vitamins and little quantity of micro nutrients. Lichens used as food is documented by Turner (1977, 1978). *Bryoria fremontii* and *Cladonia* sp. are used as important items in the diet of people of California,USA. The nutritive value of *Bryoria* is higher than in *Cladonia*, and consists 5 per cent proteins, 3 per cent fats and 4 per cent minerals. Crude fraction of fibre includes sugars, starch, pectins and lignins which together total 40 per cent (Pulliainen, 1971).

Lichens like *Parmelia cirrhata, Parmelia quercina, Parmelia tinctorum, Usnea orientalis, Roccella montagnei, Usnea stirtoniana, Ramalina sinensis, Peltigera canina* and *Usnea longissima* also provide some other micro-nutrients like iron, calcium, phosphorus; vitamins like B and C and found to be more than in cereals and more comparable to green leaf materials (Lal and Ranganatha Rao, 1956). The calcium to phosphorus ratio was from 2 to 14, showing that lichens could serve as a good source of calcium. *Peltigera canina* is relatively high in protein and essential amino acids. More research need to be done with respect to micro-nutrient and vitamin content of these lichens. The various findings are not consistent due to variation in nutrient composition between and within species. The value of lichens as a food stuff is possibly as a source of carbohydrate. The nutrient composition of lichens varies widely between different species of lichens but they are generally high in carbohydrates and low in most other nutrients.

Lichens as Medicines

Lichens are used in folk medicines for centuries. Indians (Saklani and Upreti, 1992), Americans (Turner, 1977) and Chinese employed lichens to treat a variety of ailments in their traditional medicines, primarily as expectorants (Elix, 1996). *Peltigera canina*, a foliose lichen containing a cyanobacterial photobiont, is used in India as a remedy for liver ailments. High levels of amino acid methionine in this lichen are attributed for its use in the treatment of liver ailment. Several species having therapeutic activity are listed in various pharmacopoeias.They include *Cetraria islandica, C. nivalis, Cladonia coccifera, C. pyxidata, Usnea plicata, Peltigera canina, P. horizontalis, P. polydactyla, Lobaria pulmonaria, Xanthoria parietina* and *Evernia prunastri* (Saklani and Upreti, 1992).

Table 7.2: A summary of studies on the nutritional composition of lichens

Lichen Species	Ash %	Fat %	Ether Extractives %	Protein g/100g	Crude Fiber %	Carbohydrates %	Type of Carbohydrate	Vit. C mg/100g	Vit. B mg/100g	Ca mg/100g	P mg/100g	Fe mg/100g	Other	Reference
Parmelia cirrhata	12.3		2.6	8.2[a]	11.0	22.5[b]	Lichenin and/or Isolichenin	4.4	210[c]	398	156	52		Lal and Ranganatha Rao, 1956
Parmelia quercina	2.4		4.5	12.8[a]	10.7	25.0[b]	Lichenin and/or Isolichenin	2.7	530[c]	1250	195	272		Lal and Ranganatha Rao, 1956
Parmelia tinctorum	12.6		6.0	13.8[a]	13.4	25.0[b]	Lichenin and/or Isolichenin	5.0	100[c]	1728	118	36		Lal and Ranganatha Rao, 1956
Usnea orientalis	8.9		1.6	5.1[a]	6.1	50.0[b]	Lichenin and/or Isolichenin	9.4	80[c]	902	110	314		Lal and Ranganatha Rao, 1956
Usnea longissima	3.5		2.2	5.4[a]	3.8	46.3[b]	Lichenin and/or Isolichenin	3.4	310[c]	1028	80	90		Lal and Ranganatha Rao, 1956
Usnea stirtoniana	2.9		2.0	4.6[a]	3.7	36.0[b]	Lichenin and/or Isolichenin	4.5	160[c]	773	90	30		Lal and Ranganatha Rao, 1956
Ramalina sinensis	4.6		0.2	6.8[a]	4.1	52.0[b]	Lichenin and/or Isolichenin	6.3	140[c]	1022	150	81		Lal and Ranganatha Rao, 1956
Peltigera canina	8.7		0.1	21.3[a]	17.5		Lichenin and/or Isolichenin	4.4	430[c]	486	235	49		Lal and Ranganatha Rao 1956
Roccella montagnei	4.4		1.0	14.2[a]	12.9	15.0[b]	Isolichenin	3.7	750[c]	394	145	46		Lal and Ranganatha Rao, 1956
Aspicilia esculenta						23.0[b]	Lichenin						Calcium oxalate[h]	Crum, 1993
Bryoria spp.				7.77[a]										Perez-Llano, 1944
Umbilicaria pennsylvanica				6.27[a]										Perez-Llano, 1944

Contd...

Table 7.2–Contd...

Lichen Species	Ash %	Fat %	Ether Extractives %	Protein g/100g	Crude Fiber %	Carbohydrates %	Type of Carbohydrate	Vit. C mg/100g	Vit. B mg/100g	Ca mg/100g	P mg/100g	Fe mg/100g	Other	Reference
Cetraria islandica		2.6		3.0[a]		80.0[b]	Isolichenin and Lichenin	0.75		4.8	50.0	53.0	Pb, Hg, Cd[f]	Airaksinen *et al.*, 1986
Cetraria islandica	2.2	1.2		0.32[a]		48.6								Swartz, 1911
Cladina spp.										5.15	610.0	79.0	Pb, Hg, Cd[g]	Airaksinen *et al.*, 1986
Cetraria islandica						66.0								Perez-Llano, 1944
Cladina rangiferina	1.4	2.1		5.4[a]	32.9				501[d]					Barberie, 1946
Peltigera canina				21[e]										Subramanian and Ramakrishnan, 1964

a: Kjeldahl (N x 6.25); b: Only lichenin and isolichenin measured; c: Riboflavin in mg/100g; d: Niacin in mg/100g; e: Contains 13 amino acids, including 8 out of 9 essential amino acids; f: In units of mg/kg of lichen: Pb=30, Hg=0.075, Cd=0.30; g: In units of mg/kg of lichen: Pb=30, Hg=0.12, Cd=0.20; h: Per cent calcium oxalate: 60 to 66.

A lichen's usefulness as medicine probably comes from the secondary compounds abundant in most lichen thalli. Use of lichens for medical purposes is of great antiquity and reviewed previously (Richardson, 1988). Lichens produce a wide variety of these compounds, most of which are unique to lichens. The exact use of these lichen compounds is still being debated, but some lichen compounds act as antibiotics, fungicides and herbivore deterrents (Lawrey, 1986). This undoubtedly gives the lichen some protection and probably endows the lichen with some medicinal characters as well.

The 50 per cent of all lichen species have antibiotic properties (Sharnoff, 1997). Research on antibiotics in lichens started in 1944 when Burkholder found that extracts from 27 out of 42 different species of lichen inhibit growth of certain bacteria. Lichen compounds act as anti-tumour agents (Takai *et al.*, 1979), antibiotics (Burkholder *et al.*, 1944 and Vartia, 1973) and anti-inflammatories (Handa *et al.*, 1992).

Some of the most widely studied lichen compounds are usnic acid, vulpinic acid, atranorin and protolichesterinic acid. Usnic acid is found in large quantities in *Usnea* spp., as well as in several other lichen genera. It is a fairly wide spectrum antibiotic and is the most active antibiotic to be characterized from lichens (Abo-Khatwa *et al.*, 1996, Rowe *et al.*, 1991 and Dobrescu *et al.*, 1993). Usnic acid and diffractaic acid (a derivative of usnic acid) have both been demonstrated to be analgesic when tested on mice (Okuyama *et al.*, 1995). A mixture of usnic acid and isolichenin is demonstrated to have moderate activity against sarcoma 180 and Ehrlich tumor cells (Periera *et al.*, 1994).

There is research to indicate that protolicheresterinic acid may be valuable in the treatment of ulcers and cancers, and in AIDS prevention. It is documented that protolicheresterinic acid has *in vitro* activity against *Helicobacter pylori* (Ingolfsdottir *et al.*, 1997) and DNA polymerase activity of human immunodeficiency virus-1 reverse transcriptase (Pengsuparp *et al.*, 1995). Protolicheresterinic acid was found antiproliferative and cytotoxic to T-47D and ZR-75-1 cell lines cultured from breast carcinomas and to K-562 from erythro-leukemia (Ogmundsdottir *et al.*, 1998). Protolichesterinic acid may perform these functions by inhibiting 5-lipoxygenase, and this would also contribute to protolichesterinic acids with anti-inflammatory actions (Ogmundsdottir *et al.*, 1998). Vulpinic acid is also reported to have mild antibiotic properties, but it is not as strong as an antibiotic as usnic acid. It is a significant herbivore deterrent and is toxic to animals in large doses (Lawrey, 1986). Atranorin is less biologically active than the above mentioned compounds (Lawrey, 1986 and Abo-Khatwa *et al.*, 1996).

Another property of lichens, used for medicines, is their cool little shapes. According to the 'Doctrine of Signatures' of the 15th century Europe, a plant could be used to treat whatever ailments it most looked like. Some lichens commonly used according to the Doctrine of Signatures include species of *Cladonia, Evernia, Lobaria, Parmelia, Peltigera, Pertusaria, Physcia, Roccella, Usnea,* and *Xanthoria.* Importance of this use is evident when one looks at the origin of the word 'lichen'. 'Lichen' comes from the Greek word 'Leprous' and refers to the use of some lichens for treating cutaneous diseases due to their peeling-skin appearance (Perez-Llano, 1944).

According to the 'Doctrine of Signatures', *Lobaria pulmonaria* and *Parmelia sulcata* were used to cure lung disease, *Peltigera canina* to cure for rabies, *Letharia vulpine* used as a poison against wolves. Other species are used in "folk medicines" around the world, usually prepared as a "tea". Usnic acid inhibits growth of gram '+ve' bacteria and is an active ingredient in some topical ointments sold in Europe and Russia ("USNO, BINAN"). A few lichen fatty acids and carbohydrates have anti-tumour effects in study animals. *Parmotrema chinense, Parmotrema perforatum,* and *Everniastrum cirrhatum* are all called 'Chharila' in India and are used medicinally. They are also used as a snuff and as an aphrodisiac (Chandra and Singh, 1971).

Biology of Lichens

Lichens are organisms between fungus and algae or cyanobacterium (blue-green algae) resulting in a phenotype distinct from either of the symbiont (Goward *et al.*, 1994). Although lichens appear to be single plant-like organisms, under a microscope, the association seems to consist of millions of cells of algae (called the phycobiont) woven into a matrix formed of the filaments of the fungus (called the mycobiont). The majority of mycobionts are placed in a single group of ascomycota.Remaining mycobionts are distributed among different fungal groups. Although there are various types of phycobionts, half the lichen associations contain species of Trebouxia, a single-celled green algae. There are about 15 species of cyanobacteria which act as the photobiont in lichen associations, including some members of the genera *Calothrix*, *Gloeocapsa*, and *Nostoc*. In general terms, the photobiont (cyanobacteria/algae) supplies the association with organic carbon from photosynthesis, and the mycobiont (fungi) ensures protection and supply of minerals and water. Most commonly, photobionts are located in a layer within the fungal tissue. The layer is generally oriented in a manner that maximizes photosynthesis, and is protected from rapid changes in water availability.

Lichens are a biological group lacking formal status in the taxonomic framework of living organisms. Although the mycobiont and phycobiont have Latin names, the product of their interaction, lichen, does not. Earlier names given to lichens as a whole are considered names for the fungus alone. Classification of lichens is difficult and remains controversial due to their changeable morphological pattern according to the environment. Part of the problem is that the taxonomy of lichens was established before their dual nature was recognized, *i.e.*, the association was treated as a single entity.

Some lichens are leafy and form beautiful rosettes on rocks and tree trunks; others are filamentous and drape the branches of trees. At the opposite extreme are those smaller than a pin head and seen only with a magnifying lens. The structure and origin of these plants were a mystery to botanists for years. Now it is found that lichen are a combination of an algae and a fungus. In this mutually beneficial partnership, the algae, since it is green, can utilize sunlight and carbon dioxide to make food; the fungus holds water and provides structure.

Lichen as Extremophile

They grow almost anywhere where a stable and reasonably well-lit surface occurs. This may include soil, rock, or even sides of trees. They are pioneers on bare rock, desert sand, cleared soil, dead wood, animal bones, rusty metal and living bark. Capable of shut down metabolically during periods of unfavorable conditions, they survive extremes of heat, cold and drought. Given appropriate amounts of light and moisture, clean air and freedom from competition, lichens colonize almost any undisturbed surface. As a result of their desiccation-tolerance, lichens are extremophiles and can live in places no higher plant can (Kranner *et al.*, 2008). They are very adaptable to become dormant during dry periods and at low temperatures, and regain their metabolic activity when the conditions improve. A lichen may absorb certain mineral nutrients from any of these substrates on which it grows, but is generally self-reliant in feeding itself through photosynthesis in the algal cells. Most lichens are temperate or arctic, though there are many tropical and desert species. Lichens seem to do better in drier environments, where they are not often left in standing water. What the lichen considers dry, however, may not be what we would consider to be dry. In bayous and cool rainforests, large lichens known as "old man's beard" may often be seen hanging from the branches of trees. Though there is considerable water in these habitats, the air is not saturated, and drying breezes may serve to desiccate arboreal organisms.

Lichens are the most noticeable on the tundra, where they constitute majority of ground cover. This cover insulates the ground and provides forage for grazing animals. The so-called Reindeer "moss" is one such lichen. Lichens grow in the leftover spots of the natural world that are too harsh or limited for most other organisms. They are not parasites, but like to grow on tree trunks. Lichens growing in trees are using the tree as a base/home. Lichens growing on rocks, though, may release chemicals which speed the degradation of the rock into soil and thus promote production of new soils.

Among 800 species of Antarctic plants, 350 are lichens. Although slow-growing organisms, they are particularly well adapted to Antarctic survival. They can endure lengthy high-stress periods in dormancy and almost instantly become photosynthetic when conditions improve. In extensive swampy wetlands, lichens are also generally well distributed on tree trunks and especially in the canopy of older conifers throughout the boreal forest which constitutes a significant part of the ground cover in the lichen woodland. Considering the climatic conditions, tundra vegetation is quite varied. The long daylight periods of spring and summer contribute to sudden and rapid growth. Lichen tundra is found in the drier and better-drained parts of ground usually covered to form a thick carpet.

Lichen Diversity, Ecology and Distribution: Global and the Indian Scenario

Ecology of lichens refers to the place they hold in ecosystems. It is estimated that nearly 8 per cent of the earth's terrestrial surface is covered by lichens. This enables them to make a significant ecological contribution by generating carbohydrates through photosynthesis, reducing carbon dioxide levels and fixing nitrogen so that it is in a form suitable for living organisms.

The estimated number of different lichen species ranges approximately 15,000 to 20,000 world wide, some of which provide forage for reindeer and products for humans (Negi, 2000). In India, around 2000–2200 species are reported (Negi, 2000). Lichens are vital components of ecosystem functions–in food web and food chains, nutrient cycling and are used by many animal groups as brooding/nest building spaces. Squirrels use lichens to line inside of their nests. Birds use lichens to patch outside of their nests (feathers, where ever possible are used on the inside) to camouflage them. *Parmelia* and *Physcia* species are used most frequently. Birds play an important part in dispersing lichen soredia. The soredia are small and light enough to adhere to bird's feet and feathers and consequently are carried to other trees and habitats. Moths and butterflies feed on lichens and use them for camouflage.

Ecology of Lichens

Lichens through their successful evolutionary strategies such as ability to form symbiotic association and having genetic shields to withstand harsh environments have resulted in a rich diversity of fungal species (DePriest, 2004). Lichen symbiosis was supposed to have happened > 900 million years ago. Lichen is considered the early land colonizers of newly exposed surfaces and even in extreme environments like bare soil, rock surfaces in the glaciers, solidified lava flows which are non-vegetated and uninhabitable (Rundel, 1978 and Heckman *et al.*, 2001,). Often lichens are the only available nutrient sources at the primary producer level in extreme habitats and secondary compounds produced by lichens under such habitats provide self defense and an effective survival strategy. Rundel (1978) suggested that lichen compounds may be vital in determining the relative ecological success of an individual lichen species and the structure and diversity of lichen communities are very much determined by the light screening, chemical weathering, allelopathic and antiherbivory-defense properties of secondary compounds of litchens. Rundel (1969) showed increased concentrations of usnic acid in the cortex of *Cladina subtenuis* with higher light intensity indicating the photoprotective role of usnic acid. The hydrophobic properties of lichen compounds help lichen bionts to prevent "drowning" by permitting continuous gaseous exchange with the atmosphere (Armaleo, 1993).The lichen thallus may have been evolved as protective shell that guards newly formed cyanobacterium or green algae (photobiont) in the terrestrial and harsh environments such as water and nutrient deficit, fluctuating temperatures and UV-rays (Heckman *et al.*, 2001). Successful establishment and tolerance of lichens in terrestrial habitat are also related to poikilohydric nature and ability to repeatedly become desiccated and hydrated. Under sufficient hydrated situations, maximum metabolic activities like photosynthesis, respiration, N_2-fixation (cyanolichens) are maintained (Rundel, 1982 and Sundberg *et al.*, 1997).

Huneck (1999) outlined the significant biological role of lichen substances for effective survival under desiccation, high light intensity and protection against UV radiation. Further, he suggested that lichen secondary compounds also protect lichen thalli against viruses, bacteria, algae, bryophytes, plants and animals. Hence lichen secondary metabolites act as allelopathic and antibiotics to combat biotic stresses. This self defense is vital since lichens may be the only available nutrient source at the primary producer level in the extreme environments. They lack an epidermis or any hard tissue for protection and their growth rates *in nature* is one among the slowest. Some secondary compounds are reported to temporarily increase the permeability of cells of algae *Trebouxia erici* (Vainshtein, 1991). Lawrey, (1984) observed that the licheniferous slug *Pallifera varia* preferred to feed on lichens with the lowest diversity of secondary metabolites indicating the anti herbivorous roles of lichen metabolites. Lichen compounds like usnic acid, vulpininc acid and 3-hydroxyphysodic acid reduce the growth of polyphagous insect larvae (*Spodoptera littoralis*) (Emmerich *et al.*, 1993). Barbatic acid impairs the photosystem II function in higher plants indicating its phytotoxic activity (Takahagi *et al.*, 2006). Kathirgamanathar *et al.* (2008) reported the feeding and sequestration of lichen compound of *Leproloma sipmanianum* by the larvae of *Talicada nyseus nyseus* (Lycaenidae) to obtain protection against predators during their caterpillar stage. Snail and some moths (Lepidoptera) of the family Arctiidae accumulate lichen phenolics in their body and use them as defense from predators (Kathirgamanathar *et al.*, 2008).

Alternative Approaches to Harness Lichen Products

Demands for new promising natural pharmacologically active molecules are increasing due to emergence of new dreaded diseases and multi resistant pathogens. Majority of the synthetic drugs are instable at higher temperatures and cause deleterious side effects. Lichen metabolites in large number can be added to the pool of new and potential drugs against various diseases (Oksanen, 2006).

Lichens and lichen products are used in traditional medicines for centuries (Richardson, 1991) and still hold considerable interest as alternative treatments in various parts of the world. Indeed, today a variety of lichen-based tonics, lotions and lozenges can be purchased mainly in Iceland. However, lichens are essentially ignored by the modern pharmaceutical industry, despite the fact that lichens produce a large number of low molecular weight molecules with diverse structures and biological activity. The reasons for this are:

- ☆ Slow growth rate of lichens *i.e.*, 1-2 mm/year in nature.
- ☆ Low biomass availability
- ☆ Many lichens are crusts growing adpressed to the substratum, making it difficult to harvest and extensive lichen collection is expected to jeopardize conservation interest.
- ☆ Though, *in vitro* protocols for lichen secondary compound production are available for a few species, many of the tropical lichens rich in compounds are not brought under culture yet

- ☆ Protocols for production of secondary compounds in desired quantities (even at lab level) for preliminary and secondary bioprospecting screenings and for down stream applications (cytotoxic, genotoxic and pre-clinical studies) are yet to be standardized

In vitro Lichen Cultures for Large Scale Production of Lichen Compounds

The *in vitro* establishment of lichen cultures is basic for experimental lichenology to harvest sufficient lichen biomass and to overcome the above bottle necks (Crittenden *et al.*, 1995). Many authors established lichen cultures to synthesize secondary compounds found in natural thallus (Hamada and Miyagawa, 1995, Behera *et al.*, 2000 and Yamamoto, 1990). Lichen culture studies focus on *in vitro* production of novel secondary compounds with the mycobiont or whole thallus cultures by modulating the culture conditions or media composition (Valarmathi and Hariharan, 2007 and Stocker-Wörgötter, 2008). Excess sucrose concentrations simulate drought conditions to the mycobiont, resulting in high induction of secondary metabolite pathways. Specifically, many studies reported the synthesis of novel depsides and depsidones derived from polyketide pathway in mycobiont cultures maintained under osmotic conditions (created by excess sucrose concentration) (Hamada, 1993, 1996 and Hamada *et al.*, 1996). These are dibenzofurans from *Evernia esorediosa* and *Lecanora iseana* (Miyagawa *et al.*, 1993, Takenaka *et al.*, 2005), graphenone from *Graphis scripta* (Miyagawa *et al.*, 1994), xanthones from *Pyrenula japonica* and *P. pseudobufonia* (Tanahashi *et al.*, 1999), Isocoumarins from *Graphis* sp.(Tanahashi *et al.*, 2000), Isofuranonapthoquinone from *Arthonia cinnabarina* (Yamamoto *et al.*, 2002), physcion, emodin, physcion- bisanthrone, teloschistin monoacetate from *Xanthoria elegans* (Brunauer *et al.*, 2007). For many compound rich species such approaches are lacking. Recent studies showed that ecological factors, including physiological stresses specifically UV radiation and cold (4°c) also induce secondary metabolite pathways (Stocker-Wörgötter, 2001 and Hager *et al.*, 2008). Hager *et al.* (2008) showed production of diffractic acid in UV-C treated samples and barbatic acid in cold temperature stressed (4°C) mycobiont cultures of *Heterodea muelleri*. Ecological conditions (day-night cycle, moisture content and pH) and cell and thallus differentiation lead to functional regulation of genes involved in secondary metabolism (Stocker-Wörgötter, 2008).

Significant improvements in culture conditions, specifically in composed nutrient media, inoculum selection and stress inception contributed to enormous progress in experimental lichenology and *in vitro* secondary compounds production of temperate species (Hager *et al.*, 2008, Stocker-Wörgötter, 2008 and 2001). Such data for tropical lichens are scarce and worth investigating.

Molecular Approaches to Harvest Lichen Products

Molecular biology provides the means to access lichen genomes to reveal and eventually to harvest their potential for production of novel secondary metabolites. As described later, the direct extraction of DNA from lichen thalli and subsequent survey and study of DNA sequences related to secondary metabolite biosynthesis will result in more productive source of novel products (Miao *et al.*, 2001). Advancements in molecular techniques like genetic engineering, cloning, and homologous or heterologous expression provide the means to directly access lichen genomes and to harvest their potential for novel secondary compound production (Miao *et al.*, 2001). Secondary metabolism in non–lichen fungi are extensively studied. Currently the information on lichen PKS genes are very limited (Schümann and Hertweck, 2006). Only three full length PKS gene are available for lichens. Chooi *et al.* (2007) characterized the full length PKS gene from *Xanthoria semivirids* and also achieved heterologous expression in *Aspergillus nidulans*. Brunauer *et al.* (2009) characterized the full length

PKS gene in *Xanthoria elegans* (XsePKS). Still for many lichen species, the nature of PKS gene, the exact mechanisms regulating PKS expression, copy number, location of the PKS genes etc.are still unclear.

Initially the technique involved to identify PKS gene was hybridization with DNA probes of desired PKS from heterologous species or by gene fragments amplified by PCR of genomic DNA using degenerate primers (Santi *et al.*, 2000, Schmitt *et al.*, 2005, 2008 and Grube and Blaha, 2003). Genomic library screening, RACE PCR and cDNA techniques are now used to isolate full-length polyketide genes from lichen species *Xanthoria elegans* and *Xanthoria semiviridis* (Brunauer *et al.*, 2009 and Chooi *et al.*, 2007). The third full length PKS sequence information from the lichen *Dirinaria applanata* (DnPKS) using a genome-walking approach (TAIL-PCR) was reported by Valarmathi *et al.* (2009).

Lichen cultures are used to study the regulation of polyketide genes under different culture/ stress conditions (Stocker-Wörgötter, 2008). High expression of secondary metabolites in mycobiont cultures under optimized culture conditions will help in understanding the functions of PKS (Transcription of PKS gene) and their control by PKS genes (Stocker-Wörgötter, 2008).

References

Abo–Khatwa, A. N., Al–Robia, A. A and Jawhari, D. A., 1996. Isolation and identification of usnic acid and atranorin from some Saudi Arabian lichens. *Journal of Scientific Research.* 15: 15–28.

Airaksinen, M. M., Peura, P., Ala–Fossi–Salokangas, L., Antere, S., Lukkarinen, J., Saikkonen M and StenbŠck, F., 1986. Toxicity of plant material used as emergency food during famines in Finland. *Journal of Ethnopharmacology.* 18: 273–296.

Armaleo, D., 1993. Proceedings of the International Botanical Congress, Tokyo–Yokohama, Abstract 15.

Barberie, M. A., 1946. Edible wild plants of New Brunswick as a source of riboflavin. M. Sc. thesis, University of New Brunswick, Fredericton, NB.

Behera, B. C., Makhija, U and Adawadker, B., 2000. Tissue culture of *Bulbothrix setschwanensis* (lichenized ascomycetes) *in vitro. Current Science.* 78: 781–783.

Brunauer, G., Hager, A., Grube, M., Türk, R and Stocker–Wörgötter, E., 2007. Alterations in secondary metabolism of aposymbiotically grown mycobionts of *Xanthoria elegans* and cultured resynthesis stages. *Plant Physiology and Biochemistry.* 45(2): 146–151.

Brunauer, G., Muggia L., Stocker–Wörgötter, E and Grube, M., 2009. A transcribed polyketide syntahse gene from *Xathoria elegans. Mycological Research.* 113: 82–92.

Burkholder, P. R., Evans, A. W., McVeigh, I and Thornton. H. K., 1944. Antibiotic activity of lichens. *Proceedings of the National Academy of Sciences of the U. S. A.* 30(9): 250–255.

Chandra, S and Singh, A., 1971. A lichen crude drug (chharila) from India. *Journal of Research in Indian Medicine.* 6: 209–215.

Chooi, Y. H., David, M. S., Davis, M. A., Fujii, I., Aelix, J., Louwhoff, S. H. J. J and Clawrie, A., 2007. Cloning and sequence characterization of a non–reducing polyketide synthase gene from the lichen *Xanthoparmelia semiviridis. Mycological Research.* 112: 147–161.

Crittenden, P. D., David, J. C., Hawksworth, D. L and Campbell, F. S., 1995. Attempted isolation and success in the culturing of a broad spectrum of lichen–forming and lichenicolous fungi. *New Phytologist.* 130: 267–297.

Crum, H., 1993. A lichenologist's view of manna. Contributions from the University of Michigan Herbarium., 19: 293–306.

De Priest, P. T., 2004. Early molecular investigations of lichen–forming symbionts: 1986–2001. *Annual Review of Microbiology*. 58: 273–301.

Dobrescu D., Tanasescu M., Mezdrea A., Ivan C., Ordosch E., Neagoe F., Rizeanu A., Trifu L and Enescu V., 1993. Contributions to the complex study of some lichens—*Usnea* genus. Pharmacological studies on *Usnea barbata* and *Usnea hirta* species. *Rom J Physiol*. 30(1–2): 101–107.

Elix, J. A., 1996. Biochemistry and secondary metabolites. In: Nash, TH III (Ed.) Lichen biology. Cambridge University Press, Cambridge, pp 154–180.

Emmerich, R., Giez, I., Lange, O. L and Proksch, P., 1993. Toxicity and antifeedant activity of lichen compounds against the polyphagous herbivorous insect *Spodoptera littoralis*. *Phytochemistry*. 33: 1389–1394.

Goward, T., Bruce McCune and Del Meidinger 1994. The lichens of British Columbia–Illustrated Keys–Part 1–Foliose and Squamulose Species. Province of British Columbia, Ministry of Forests Research Program, pp.181.

Grube, M and Blaha, J., 2003. On the phylogeny of some polyketide synthase genes in the lichenized genus Lecanora. *Mycological Research*. 107(12): 1419–1426.

Hager, A., Brunauer, G., Türk, R and Stocker–Wörgötter, E., 2008. Production and bioactivity of common lichen metabolites as exemplified by *Heterodea muelleri* (Hampe) Nyl. *Journal of Chemical Ecology*. 34(2):113–20.

Hamada, N and Miyagawa, H., 1995. Secondary metabolites from isolated lichen mycobionts cultured under different osmotic conditions. *Lichenologist*. 27: 201–205.

Hamada, N., 1993. Effects of osmotic culture conditions on isolated lichen mycobionts. *The Bryologist*. 96: 569–572.

Hamada, N., 1996. Induction of the Production of Lichen Substances by Non–Metabolites. *The Bryologist*. 99(1): 68–70.

Hamada, N., Miyagawa, H., Miyawaki, H and Inoue, M., 1996. Lichen substances in mycobionts of crustose lichens cultured on media with extra sucrose. *The Bryologist*. 99(1): 71–74.

Handa, S. S., Chawla, A. S and Sharma, A. K., 1992. Plants with anti–inflammatory activity. *Fitoterapia* Vol. LXIII, No. 1.

Heckman, D. S., Geiser, D. M., Eidell, B. R., Stauffer, R. L., Kardos, N. L and Hedges, S. B., 2001. Molecular evidence for the early colonization of land by fungi and plants. *Science*, 293: 1129–1133.

Huneck, S., 1999. The significance of lichens and their metabolites. *Die Naturwissenschaften*. 86: 559–570.

Ingólfsdóttir, K., Hjalmarsdottir, M. A., Sigurdsson, A., Gudjonsdottir, G. A., Brynjolfsdottir, A and Steingrimsson, O., 1997. *In vitro* susceptibility of *Helicobacter pylori* to protolichesterinic acid from the lichen *Cetraria islandica*. *Antimicrobial Agents and Chemotherapy*. 41: 215–217.

Kathirgamanathar, S., Wijesekera, A., Wijesundara, D. S. A and Karunaratne, V., 2008. Insights on the chemical association between he larvae of the butterfly *Talicada nyseus* uses the lichen *Leproloma sipmanianum*, *Plant Interactions*. 3: 25–30.

Kranner I., Beckertt R., Hochman A and Nash Thomas H., 2008. Desiccation–tolerance in lichens: a review. *The Bryologist*. 111: 576–593.

Lal, B. M and Ranganatha Rao K., 1956. The food value of some Indian lichens. *Journal of Scientific and Industrial Research*. 15(C): 71–73.

Lawrey, J. D., 1986. Biological role of lichen substances. *The Bryologist*. 89: 111–122.

Lawrey, J. D., 1984. Biology of lichenized fungi. Praeger Publishers, New York. pp 408.

Miao, V., Coëffet-LeGal, M-F., Brown, D., Sinnemann, S., Donaldson, G and Davies, J., 2001. Genetic approaches to harvesting lichen products. *Trends in Biotechnology*., 19: 349–355.

Miyagawa, H., Hamada, N., Sato, M and Ueno, T., 1994. Pigments from the cultured lichen mycobionts of *Graphis scripta* and *G. desquamescens*. *Phytochemistry*. 36(5): 1319–1322.

Miyagawa, H., Hamada, N., Satoh, M and Ueno, T., 1993. Hypostrepsilic acid, a new dibenzofuran from the cultured lichen mycobiont of *Evernia esorediosa*. *Phytochemistry*. 34: 589–591.

Negi, H, R., 2000. Spatial patterns of biodiversity of lichens. *Journal of Indian Institute of Science*. 80: 571–589.

Ogmundsdottir, H. M., Zoega, G. M., Gissurarson, S. R and Ingolfsdottir, K., 1998. Anti–proliferative effects of lichen–derived inhibitors of 5–lipoxygenase on malignant cell–lines and mitogen–stimulated lymphocytes. *J. Pharm. Pharmacol*., 50(1): 107–115.

Oksanen, I., 2006. Ecological and biotechnological aspects of lichens. *Applied Microbiology and Biotechnology*. 73(4): 723–734.

Okuyama, E., Umeyama, K., Yamazaki, M., Kinoshita, Y and Yamamoto, Y., 1995. Usnic acid and diffractaic acid as analgesic and antipyretic components of *Usnea diffracta*. *Planta Med*. 61(2): 113–115.

Pengsuparp, T., Cai, L., Contstant, H., Fong, H. H. S., Lin, L–Z., Kinghorn, A. D., Pezzuto, J. M., Cordell, G. A., Ingolfsdóttir, K., Wagner, H and Hughes, S. H., 1995. Mechanistic evaluation of new plant–derived compounds that inhibit HIV–1 reverse transcriptase. *Journal of Natural Products*. 58(7): 1024–1031.

Perez–Llano, G. A., 1944. Lichens: their biological and economic significance. *The Botanical Review*. 10(1): 1–65.

Pereira, E. C., Nascimento, S. C., Lima, R. C., Silva, N. H., Oliveira, A. F. M., Bandeira, E., Boitard, M., Beriel, H., Vicente, C and Legaz, M.E., 1994. Analysis of *Usnea fasciata* crude extracts with antineoplastic activity. *Tokai Journal of Experimental and Clinical Medicine*., 19(1): 47–52.

Pulliainen, E., 1971. Nutritive values of some lichens used as food by reindeer in northeastern Lapland. *Ann. Zool. Fenn*. 8: 385–389.

Richardson, D. H. S., 1988. Medicinal and other economic aspects of lichens. Pages 93–108 In M. Galun (ed.), ed. CRC Handbook of Lichenology. Volume III.

Richardson, D. H. S., 1991. Lichens and man. In: Hawksworth, DL (Ed.): Frontiers in Mycology. CAB International, Wallingford, pp. 187–210.

Rowe, J. G., Saenz, M. T., Garcia, M. D and Gil. A. M., 1991. Additional contribution to the study of the antimicrobial activity and identification of lichenic substances in some lichens from Southern Spain. *Ann. Pharm. Fr.* 49(5): 278–285.

Rundel, P. W., 1969. Clinal variation in the production of usnic acid in *Cladonia subtenuis* along light gradients. *The Bryologist.* 72: 40–44.

Rundel, P. W., 1978. The ecological role of secondary lichen substances. *Biochemical and Systematic Ecology.* 6: 157–170.

Rundel, P. W., 1982. The role of morphology in the water relations of desert lichens. *Journal of the Hattori Botanical Laboratory.* 53: 315–320.

Saklani, A and Upreti, D. K., 1992. Folk uses of some lichens in Sikkim. *Journal of Ethnopharmacology.* 37(3): 229–233.

Santi, D.V., Siani, M.A, Julien, B., Kupfer, D. and Roe, B., 2000. An approach for obtaining perfect hybridization probes for unknown polyketide synthase genes: a search for the epothilone gene cluster. *Gene.* 247: 97–102.

Schmitt, I., Kautz, S and Lumbsch, H. T., 2008. 6–MSAS–like polyketide synthase genes occur in lichenized ascomycetes. *Mycological Research.* 112: 289–296.

Schmitt, I., Martin, P., Kautz, S and Lumbsch, H. T., 2005. Diversity of non–reducing polyketide synthase genes in the pertusarials (lichenized ascomycota): a phylogenetic perspective. *Phytochemistry.* 66: 1241–1253.

Schümann, J and Hertweck, C., 2006. Advances in cloning, functional analysis and heterologous expression of fungal polyketide synthase genes. *Journal of Biotechnology.* 43: 12–14.

Sharnoff, S. D 1997. Lichens and people. Online: <http://www.lichen.com/people.html>

Stocker–Wörgötter, E., 2001. Experimental Lichenology and Microbiology of Lichens Culture Experiments, Secondary Chemistry of Cultured Mycobionts, Resynthesis, and Thallus Morphogenesis. *The Bryologist.* 104: 576–581.

Stocker–Wörgötter, E., 2008. Metabolic diversity of lichen–forming ascomycetous fungi: culturing, polyketide and shikimate metabolite production, and PKS genes. *Natural Product Reports. 25:* 188–200.

Subramanian, S. S and Ramakrishnan, S., 1964. Amino–acids of *Peltigera canina. Current Sci.* 33(17): 522.

Sundberg, B., Campbell, D and Palmqvist, K., 1997. Predicting CO_2 gain and photosynthetic light acclimation from fluorescence yield and quenching in cyano–lichens. *Planta.* 201(2): 138–145.

Swartz, M. D., 1911. Nutritional investigations on the carbohydrates of lichens, algae, and related substances. *Transactions of the Connecticut Academy of Arts and Science.* 16: 249–382.

Takahagi, T., Ikezawa, N., Endo, T., Ifuku, K., Yamamoto, Y., Kinoshita, Y., Takeshita, S and Sato, F., 2006. Inhibition of PSII in atrazine–tolerant tobacco cells by barbatic acid, a lichen–derived depside. *Bioscience, Biotechnology, and Biochemistry.* 70(1): 266–268.

Takai, M., Uehara, Y and Beisler. J. A 1979. Usnic acid derivatives as potential antineoplastic agents. *Journal of Medicinal Chemistry.* 22: 1380–1384.

Takenaka, Y., Hamada, N and Tanahashi, T., 2005. Monomeric and dimeric dibenzofurans from cultured mycobionts of *Lecanora iseana*. *Phytochemistry*. 66: 665–668.

Tanahashi, T., Takenaka, Y., Ikuta, Y., Tani, K., Nagakura, N and Hamada, N., 1999. Xanthones from the cultured lichen mycobionts of *Pyrenula japonica* and *Pyrenula pseudobufonia*. *Phytochemistry*. 52(3): 401–405.

Tanahashi, T., Takenaka, Y., Nagakura, N., Hamada, N. and Miyawaki, H., 2000. Two isocoumarins from the cultured lichen mycobiont of *Graphis* sp. *Heterocycles*. 53(3): 723–728.

Turner, N. J., 1977. Economic importance of black tree lichen (*Bryoria fremontii*) to the Indians of western North America. *Economic Botany*. 31: 461–470.

Turner, N. J., 1978. Food plants of the British Columbia Indians. Part 2. Interior peoples. British Columbia Provincial Museum Handbook No. 36, Victoria, British Columbia.

Vainshtein, E. A., 1991. The role of lichen substances in symbiosis. *Botanicheskii Zhurnal [Moscow & Leningrad]*. 76(6): 793–800.

Valarmathi, R and Hariharan, G.N., 2007. Soredial Culture of *Dirinaria applanata* (Fée) Awasthi: Observations on developmental stages and compound production. *Symbiosis*. 43(3):137–142.

Valarmathi, R., Hariharan, G.N., Gayatri Venkataraman and Ajay Parida., 2009. Characterization of a non–reducing polyketide synthase gene from lichen *Dirinaria applanata*. *Phytochemistry*. 70: 721–729.

Vartia, K. O., 1973. Antibiotics in lichens. *In:* The Lichens (eds. V. Ahmadjian and M. E. Hale), Academic Press, New York, pp. 547–561.

Yamamoto, Y., 1990. Studies of cell aggregates and the production of natural pigments in plant cell culture. Nippon Paint Publications, Osaka.

Yamamoto, Y., Kinoshita, Y., Thor, G. R., Hasumi, M., Kinoshita, K., Koyama, K., Takahasi, K and Yoshimura, I., 2002. Isofuranonaphthoquinone derivatives from cultures of the lichen *Arthonia cinnabarina* (DC.) Wallr. *Phytochemistry*. 60(7): 741–745.

2011, Future Crops Vol. 1 ***Pages 109–142***
***Editor:* Professor K.V. Peter**
***Published by:* DAYA PUBLISHING HOUSE, NEW DELHI**

Chapter 8

LONGAN

(*Dimocarpous longan* Lour)

Rajesh Kumar
National Research Centre for Litchi,
Muzaffarpur – 842 002, Bihar
E-mail: rajeshkr_5@yahoo.com

Longan (*Dimocarpous longan* Lour or syn. *Euphoria longan*) fruit is a kind of very special food enriched with nutritive values for human consumption. It is closely allied to the glamorous litchi, literally meaning as dragon's eye or eyeball, and known locally as *Kathlitchi* in India, which is referred to as the "little brother of the litchi" or "cousin". It belongs to litchi family *i.e.*, Sapindaceae (which is also called Soapberry family) and shares many similarities in growth and fruiting habit, environmental requirements, cultivation aspects and production trends. Longan in particular is cultivated for more than 2000 years in China, which is an originative and producing country with rich germplasm resources, large planting areas and high gross yield accounting for 70 per cent in house and more than 50 per cent in the world respectively (Chen, 2003; Gu *et al.*, 2008). At present the commercial cultivation of longan is done in Thailand, China, Vietnam, Malayasia, with sporadic cultivation in Mayanmar, Laos, Indonesia, Cambodia, Israel, Australia and USA. Potential of this fruit still remains to be explored and utilized. This situation exists largely because of very little research and developmental efforts made in the difficult and exotic subtropical habitat. Longan is an underutilized and underexploited edible fruit cultivated in China where the more popular litchi originated. Recent years witnessed considerable improvement and attention on production of underutilized fruits. As alternate crops, these fruit crops help in diversification besides meeting multifarious human needs. According to Prof. G. Weidman Groff, longan is less important to the Chinese as an edible fruit but more widely used than litchi in Oriental medicine. The longan is a highly esteemed arilloid fruit species in Asia. Despite this importance, there is a general lack of published literature on the statistics and production of longan. The crop is currently gaining more and more acceptance in its commercially growing countries, sometimes in preference to litchi. The fruit is easily distinguished from litchi in that it is smaller and the shell is tan and smooth. The fruit flavour is more musky and sweet. The fruit is esteemed by many

Chinese, particularly in subtropical areas. The fruits of longan have received more attention and consumption of longan is expected to increase further when more people outside Asia develop a taste for this fruit. In Thailand, longan's contribution towards local economy far exceeds that of litchi. In China, the longan industry is running almost parallel to litchi and on the whole, this is an economically important crop in a number of countries of the Asia-Pacific Region. Its role in the rural economy is increasing, especially with respect to employment and income generation.

Origin and Spread

The origin of longan is not authentically documented, as some authors limit the area of origin of longan to the mountain chain from Myanmar through Southern China, others extend it to southwest India and Sri Lanka, including the lowlands. In China, it is suggested that the primary centre of origin of longan is Yunnan, and the secondary centres are Guangdong, Guangxi and Hainan provinces (Ke *et al.*, 2000). The longan is native to Southern China, in the provinces of Kwangtung, Kwangsi, Schezwan and Fukien, between elevations of 150-450 m and China has a history of over 2,000 years of longan cultivation. There are a fewer ancient monographs recording the history of this Sapindaceae crop, eventhough it has almost similar facts and figures compared to litchi for origin and spread.

Longan has spread from China across the subtropical countries since 17th century. Extensive international exchange of information on production and research of longan and other Sapindaceae fruits did not occur till recently. The longan is seldom grown under orchard conditions compared to litchi. There is only increasing demand for fruits since a few decades back. It is believed that trees were brought to Thailand from China in the late 1800s, although longans were also found growing wild in Northern Thailand over 200 years ago (Angsith *et al.*, 1999). According to Groff (1921), the longan was introduced into India in 1798 but in Indian literature, it is averred that the longan is native not only to China but also to southwestern India and the forests of upper Assam and the Garo hills, and is cultivated in Bengal and elsewhere as an ornamental and shade tree. It is commonly grown in former Indo-china (Thailand, Cambodia, Laos and Vietnam and in Taiwan). The tree grows but does not fruit in Malaya and the Philippines. There are many of the trees in Reúnion and Mauritius. The crop was taken to Australia in the mid -1800s and to Hawaii and Florida by the early to mid-1900s (Menzel *et al.*, 1989). In the western hemisphere, specimen trees are found in some Central American countries and Hawaii and Florida in the United States. There are commercial orchards in Florida and it also merits wider planting in home gardens for its fruit and as an ornamental. The longan was introduced into Florida from Southern China by the United States Department of Agriculture in 1903 and flourished in a few locations. A longan tree flourished in the Atkins Garden in Cuba and seedlings were distributed but found to fruit irregularly and came to be valued mostly for their shade and ornamental quality. In Hawaii, the longan grew faster and more vigorously compared to litchi. The longan is cultivated in much of southeast Asia, including southern China, Taiwan, Thailand, Cambodia, Vietnam, Laos, Burma and northern India.

Botany and Taxonomy

The soapberry family, Sapindaceae, includes about 2,000 tropical, subtropical and temperate species. Under the family Sapindaceae, the genus *Dimocarpus* contains six species of trees and shrubs (Leenhouts, 1971, 1973). Five of the species (*Dimocarpus longan, Dimocarpus dentatus, Dimocarpus gardneri, Dimocarpus foveolatus*, and *Dimocarpus fumatus*) are found in Asia from Sri Lanka and India to eastern Malaysia; one (*Dimocarpus australianus*) exists in Queensland, Australia.

The longan, (*Dimocarpus longan*, sometimes known as *Euphoria* or *Nephelium longana*), is the more temperate zone relative comes under subtropical group. Botanically, it is placed in a separate genus, and is currently designated *Dimocarpus longan* Lour. (syns. *Euphoria longan* Steud.; *E. longana* Lam.; *Nephelium longana* Cambess.). Many other scientific names are given to the longan, which include *Nephelium longana* (Lam.) Cam. and *Euphoria longana* Lam. The species *Dimocarpus longan* contains two subspecies *i.e.,* subspecies *longan* and subspecies *malesianus*, each with several varieties. Within the subspecies *longan,* the most commonly cultivated taxon is *Dimocarpus longan* ssp. *longan* var. *longan,* which is the commercial longan. Three edible longan types can be distinguished within the variety *longan* in Thailand (Subhadrabandhu, 1990). The first type is a large forest tree with small fruits and a very thin aril, possibly of interest for breeding purposes. The common longan is from Southern China to Myanmar. The second one, is the native longan, growing in the northern part of the country as an erect tree producing small fruits with large seeds and is recommended as a rootstock for commercial cultivars. The third type is the commercial longan which produces large fruits and small seeds. The *Dimocarpus longan* ssp.*longan* contains four varieties (i) ssp. *longan* var. *longan,* the common longan from southern China to Myanmar. (ii ssp. *longan* var. *longepetiolulatus,* from southern Vietnam. (iii) ssp. *malecianus* var. *malecianus,* from Malaysia and Indochina. (iv) *malecianus* var. *echinatus,* from Borneo and Philippines. (Choo and Kesta,1992). Besides the variety *longan,* there exists at least three other varieties, *viz.* variety *longepetiolulatus,* variety *obtusus* and variety *magnifolius,* which all are growing wild in China (Huang, 1999).

Among these species, the most commonly cultivated species is *Dimocarpus longan* where the taxon *Dimocarpus longan* spp. *longan* var. *longan* is commonly known as the commercial longan. Botanically, it is placed in a separate genus, and is currently designated *Dimocarpus longan* Lour. (syn. *Euphoria longan* Steud.; *E. longana* Lam.; *Nephelium longana* Cambess.). The taxonomical classification of longan is as follows:

- ✰ Kingdom–Plantae
- ✰ Phylum–Magnoliophyta
- ✰ Division–Dicoteledonae
- ✰ Class–Eudicots
- ✰ Sub-class–Rosids
- ✰ Order–Sapindales
- ✰ Family–Sapindaceae
- ✰ Sub-family–Sapindoideae
- ✰ Genus–*Dimocarpus*
- ✰ Species–*Dimocarpus longan*

Morphological Attributes of Longan

Tree

The longan tree is handsome evergreen tree which grows up to 20 m and possesses a spreading or erect habit, depending on the cultivars. In general, it measures upto 9-12 m in height and to 1 -4 m in width, with rough-barked trunk to 76.2 cm thick and long, spreading, slightly drooping, heavily foliaged branches. The brittle trunk and branches have a corky bark which gives a split and peel appearance, unlike those of litchi which are smooth.

Leaves

The evergreen, alternate, paripinnate leaves have 4 to 10 opposite leaflets, elliptic, ovate-oblong or lanceolate, blunt-tipped; 10-20 cm long and 3.5-5 cm wide; leathery, wavy, glossy-green on the upper surface, minutely hairy and greyish-green beneath. New growth is wine-colored and showy. The compound leaves are alternate and paripinnate with 6-9 pairs of leaflets which are dark glossy green on the upper surface and paler green on the lower surface. The young leaf flushes are reddish brown in colour and with maturity turn to light green.

Flowers

The pale-yellow, 5- to 6- sepalled, 5- to 6-petalled, hairy-stalked flowers which are borne in upright terminal panicles, male and female mingled. Inflorescences are terminal, 8-40 cm long, densely tufted-tomentose, leafless and greatly branched. They are borne on new growth produced during midsummer or autumn, although sometimes they are borne on shoots produced in spring on terminals which have not set panicles (Cymules) The flowers of longan within a panicle are made up of staminate (pistil non-functional), pistillate (stamens non-functional) and hermophrodite flowers. The staminate flower has 8 or less hairy stamens arranged in a single row on a light brown disc. The pistillate flower has anthers which are sterile and non-functional. Hermaphrodite flower contains bicarpellated and densely-hairy ovary with bilobed stigma. Normally, only one carpel develops into fruit. The stamens of the hermaphrodite flower consist of 8 sessile filaments with anthers producing viable pollens.

Fruits

The fruits in drooping clusters, are globose, 1.25-2.5 cm in diameter, with thin, brittle, yellow-brown to light reddish-brown rind, more or less rough (pebbled) without protuberances. In Thailand, a panicle may carry up to 80 individual fruits which vary in weight from 5 to 20 g. The premium commercial grades of longan fruits are in the range of 14 to 18 g. The fruit rind is thin but tough and leathery and changes colour from greenish yellow to yellowish brown with advancing maturity. Tubercles are typically flattened or indistinct. The aril has total soluble solid values ranging 15 to 25 per cent. It is translucent white to off-white and may constitute 60 to 75 per cent of the total fruit weight. Texture of the aril ranges from juicy to very crispy and flavour ranges from bland to sweet and aromatic, but seldom acidic.

Seeds

The seed is small, round to ovoid in shape and glossy reddish brown to black in colour and easily detached from the aril. Only one seed is present in each fruit and in some cultivars, there are a certain percentage of small-seeded fruits. Longan seeds are recalcitrant and, therefore, short-lived and the best sown fresh. Germination takes 7-10 days.

Climate and Soil

The commercial longan is a subtropical tree that grows well in the tropics but requires a prominent change of seasons for satisfactory flowering. A short (2-3 months) but cool (mean temperature 15–22°C) winter season brings out a prolific bloom and in this respect, longan is less demanding and more predictable than litchi. Longan is sensitive to frost. High night temperature beyond 25°C is detrimental for fruit development and temperature above 40°C causes fruit damage and fruit drop. In Thailand, the ideal temperatures for flowering and fruit set are 20–25°C (Morton, 1987).

Ample soil moisture is needed from fruit set until maturity. Suitable annual precipitation is about 1,500 mm. Drought during flowering and fruit set period can reduce fruit productivity. Excessive rainfall during flowering results in poor pollination and increase in flower drop, while overcast weather prior to harvest leads to fruit shedding, possibly due to poor production of photosynthate. Longan is sensitive to wind damage and strong cyclonic winds can cause branch splitting and fruit shedding. Trees in China and Thailand are sometimes mounded after planting and/or branches supported by posts and bamboo pole fences to reduce wind damage, especially when the tree is carrying a heavy crop. The mounds are built up gradually over time around the trunk of the tree to about 1 m high (Menzel *et al.*, 1990).

Longan thrives on rich sandy loams, it does well on oolitic limestone; moderately acid sandy soils are more marginal and on organic muck soils flowering is deficient, probably because shoot growth continues for too long. In Thailand, the soils yielding high fruit production are the heavy alluvial soils with access to water table. The roots grow down 2-4 m to the water table. In eastern Australia, the preferred soils for longan growing are heavy, fine textured soils and red loams with high fertility and good water holding capacity. In general wet lowlands or heavy clay soils are the best avoided.

Production and Productivity

China has the largest longan industry in terms of cultivated area and production in the world. The extensive international exchanges in information on the production and research of longan and other Sapindaceae fruits got momentum only from last decade in a systematic manner. In 2000, there were 465,000 ha in mainland China producing 608,000 tonne. There was setback in acerage to 444,000 ha in 2001 due to frost in late 1999. Then statistical figures remain changed like this as, during the year, 1997 the cultivated area of longan was 4,32,400 ha with a production of 2,32,000 metric tons (Liu and Ma, 2000), while it was recorded as 11,13,800 metric tons from 4,03,260 ha during 2008. Major provinces growing longan are Guangxi, Guangdong, Fujian, Sichuan, Yunnan and Hainan. A large increase in area under longan cultivation during 1987 to 1997 was recorded in the three main longan growing provinces of Guangxi, Guangdong and Fujian but it remained almost static during 1997 to 2007 but production increased dramatically. There are over 400 cultivars of longan in China (Liu and Ma, 2000). Among these cultivars, 14 per cent are early-maturing 68 per cent are mid-maturing, and 18 per cent are late-maturing. Fruit yield ranges from 5 tons/ha in off season to 10 tons/ha in "on" year.

Longan production areas are concentrated in central and southern Taiwan Province of China. The area planted in 1998 was 11,808 ha and production was 53,385 tons. There is minimal increase in planted area for the period 1998 to 2007 and production pattern is evidence of the alternate bearing phenomenon (Yen, 2000). There are over 52 longan cultivars existing in the country. Over 98 per cent of the longan area are planted with the cultivar 'Fengko'. This is followed by the cultivars 'Hongko' and 'Chingko'. A great variation exists in physicochemical characteristics of fruits among cultivars grown at different locations and years of production (Yen and Chang, 1991). Most cultivars flower from March to early April and harvesting of fruits is from August to early September.

Longan production in Vietnam is concentrated mainly along the Mekong delta, the northern region and the southeast region. Among these longan areas, the Mekong delta accounted for 70–80 per cent of the total planted area. There was increase in the longan planted area and production trend (Nguyen, 2000).

Longan production in Thailand is confined mainly to the northern provinces of Lamphun, Chiang Mai and Chiang Rai where the planted area makes up 37.6, 24.1 and 8.0 per cent of the total planted area, respectively. Though longan cultivation has extended to the eastern and central regions of the country like Nan, Phra Yao, Lampang, Phare,Chanthaburi etc, but they constitute only a small percentage of total production. In this country, the production of longan during 2001 was about 1,86,8,000 tons with a planted area of 57,261 ha (Anupunt and Sukhvibul, 2005). The productivity of longan in northern provinces of Thailand ranged from 3,000 to 3,400 kg/ha. Although there was annual increase in planted area, the production has not increased at the same pace as the planted area. Such a scenario is attributed to the sensitivity of longan to the changing climatic conditions from year to year. Variable field management has also accounted for the varying production in different years. The productivity of longan is being raised by managing the off season and enhancing the production with the chemical potassium chlorate treatment. The longan industry in Thailand is based on seedlings developed from cultivars brought in from Southern China. The present Thai cultivars are generally more superior in fruit quality to the Chinese cultivars.

In Australia, the total planted area in 1995 was estimated 200 ha. Over the last five years about 72,000 trees were planted (Singh *et al.*, 2000) using cultivars as 'Kohala', 'Biew Khiew', 'Chompoo,' 'Haew,' 'Dang', 'Kay Sweeney', and 'Fuhko2' (Chacko and Downton, 1995; Singh *et al.*, 2000). Current production of longan is between 300 to 1,000 tons of fruit each season.

The current planted area of longan is between 140 to 180 ha in Southern Florida (Campbell and Campbell, 2000). Elsewhere longan is grown on a smaller scale, for example Cambodia, Laos, Myanmar, Indonesia and Malaysia.

Countries like India and South Africa grow longan on smaller area since priority is given to litchi cultivation. There is also some growing of longan, mainly cultivars from Thailand and Israel (Blumenfeld *et al.*, 2000). Elsewhere longan is grown on a smaller scale, for example Cambodia, Laos, Myanmar, Indonesia and Malaysia.

The production and productivity in all growing countries depend on many bio-physical factors like method of planting, planting density, pruning and training for canopy management, nutrition and water management, pests and disease management, location and weather condtions. In China, if the longan is raised on the lowlands it is always put on the edges of raised beds. On high ground, the trees are placed in pre-enriched holes on the surface. The trees are fertilized after the fruit harvest and during the blooming season, at which time proportion of nitrogen is reduced. Fresh, rich soil is added around the base of the trees year after year. The longan needs an adequate supply of water at definite period to give enhanced quality fruit production and particularly irrigation is necessary in dry periods. An important operation is pruning and training which can affect the bearing potential of the tree. Later, the fruit clusters are also thinned to increase the size and quality of the fruits. Generally, the trees are planted too close together, seriously inhibiting productivity when they become overcrowded. In China, full-grown trees given sufficient room–at least 12 m apart–may yield 180-225 kg in good years. Crops in Florida from trees 6 m tall and broad, have varied from light–22.5-45 kg–to medium–68-113 kg, and heavy–135-225 kg. Rarely such trees may produce 272-317 kg. Larger trees have larger crops but if the trees become too tall harvesting is too difficult, and they should be topped. Harvesters, working manually from ladders, or using pruning poles cut the entire cluster of fruit with leaves attached. A serious problem with the longan is its irregular bearing–often one good year followed by 1 or 2 poor years. Another handicap is the ripening season–early to mid-August in China, which is the time of typhoons; August and September in Florida which is during the hurricane season. Rain is a major nuisance in harvesting and in conveying the fruit to market or to drying sheds or processing plants.

Composition and Uses

The longan fruit is normally eaten fresh like litchi. It is particularly popular among Asians, particularly the Chinese. There are claims among some people in China and Thailand that the taste of longan equals or is superior to that of litchi. The edible portion of export quality fruit ranges from 67 to 78 per cent of the whole fruit. The energy value averages 458 kJ/100 g. According to taste and physicochemical properties, longan fruit is very sweet and juicy, and it can be eaten either raw, dry, or cooked. Longans are much eaten fresh, out-of-hand, but some have maintained that the fruit is improved by cooking. Raw longan fruit (100g) contains many nutrients like Vitamin B_1 (Thiamin), Vitamin B2 (Riboflavin), Vitamin B_3 (Niacin), Vitamin C (Extreme amount), Protein, Iron, Magnesium, Phosphorus, Potassium, Copper, Manganese, Natural Fats, Dietary Fiber, Calories, Carbohydrates, acids and more particularly the energy value averages 458 kJ/100 g. The studies on aril composition of longan revealed that Total Soluble Solids, total sugars and ascorbic acids contents increase and total acid decreases during longan fruit ripening (Chattopadhay and Ghosh, 1991; Ke *et al.*, 1992 and Lin, 2002). An analytical study of longan fruits showed that the Vitamin C content was 63.3 and 55.3 mg/100 g for the longan cultivars, Biew Kiew and Sri Chompoo, respectively. Longans were a good source of K (324.9 mg/100 g) and Cu (0.26 mg/100 g) (Wall, 2006). The main sugars present in the pulp are sucrose, fructose and glucose (Paull and Chen, 1987b) with higher concentration of Sucrose in some cultivars (Lin *et al.*, 2002e). The relative ratio of these sugars varies according to invertase activity, stage of maturity and cultivar (Huang, 1995; Lin, 2002). The main organic acids in the pulp are succinic acid, malic acid and citric acid (10:5:1) (Paull and Chen, 1987b). The pulp tends to have a relatively high sugar and low acid content as titrable acidity in pulp usually decreases rapidly after harvest (Han *et al.*, 2001; Lin *et al.*, 2002d).

Table 8.1: Nutritional composition of longan fruit.

Calorie Unit	*Moisture (%)*	*Protein (g)*	*Fat (g)*	*Carbohydrate (g)*	*Fibre (g)*	*Ca (mg)*	*P (mg)*	*Fe (mg)*	*Vit. A (I.U.)*	*Vit. B_1 (mg)*	*Vit. B_2 (mg)*	*Niacin (mg)*	*Vit. C (mg)*
109.0	72.4	1.0	0.5	25.2	0.4	2.0	6.0	0.3	28.0	0.04	0.07	0.6	8.0

Source: Wong and Saichol (1991).

Table 8.2: Nutritional composition of longan fruit.

	Food Value Per 100 g of Edible Portion			*Food Value Per 100 g of Edible Portion*	
	Fresh	*Dried*		*Fresh*	*Dried*
Calories	61	286	Moisture	82.4 g	17.6 g
Protein	1.0 g	4.9 g	Fat	0.1 g	0.4 g
Carbohydrates	15.8 g	74.0 g	Fiber	0.4 g	2.0 g
Ash	0.7 g	3.1 g	Calcium	10 mg	45 mg
Phosphorus	42 mg	196 mg	Iron	1.2 mg	5.4 mg
Thiamine	-	0.04 mg	Ascorbic Acid	6 mg (possibly)	28 mg

Source: http://www.google.co.in/hort.purdue.edu/longan+fruit+culinary+uses&aq/html#Food Uses.

Longan fruit contains a significant amount of polyphenols. A study of longan's antioxidant activity found that 4-*O*-methylgallic acid had higher reducing power and 2,2-diphenyl-1-picrylhydrazyl-(DPPH), hydroxyl radical- and superoxide radical-scavenging activities than (-)-epicathechi (Jain *et al.*, 2007). An other study of longan extraction by 70 per cent methanol from peel, pulp and seed revealed that the major components were identified as gallic acid, corilagin (an ellagitannin) and ellagic acid (Rangkadilok *et al.*, 2005). The analyses showed that there was a large variation in the contents of gallic acid, corilagin and ellagic acid in different plant tissues and from different cultivars. The seed contained the highest level of the three phenolics, while the pulp contained the lowest. Amongst commercial cultivars, Biewkiew and Edor contained the highest levels of gallic and ellagic acid, while Srichompoo contained the highest content of corilagin. The researchers suggested that these three cultivars might be used in directed breeding and cultivation programs and to develop concentrated longan seed extracts to promote good health. Utilization of this by-product material would solve a major problem of thousands of tons of waste longan seeds from the canned longan pulp production industry.

Culinary Uses

The fruit is edible, and is often used in East Asian soups, snacks, desserts and sweet and sour foods, either fresh or dried, sometimes canned with syrup in supermarkets. The seeds of fresh longan can be boiled and eaten, with a distinctive nutty flavor. Dried longan (round meat) is often used in Chinese cuisine and Chinese sweet dessert soups. In Chinese food therapy and herbal medicine, it is believed to have an effect on relaxation. In contrast with the fresh fruit, which is juicy and white, the flesh of dried longans is dark brown to almost black. In Chinese medicine the longan, much like the litchi, is considered a "warm" fruit. For drying, the fruits are first heated to shrink the flesh and facilitate peeling of the rind. Then the seeds are removed and the flesh dried over a slow fire. The dried product is black, leathery and smoky in flavor and is mainly used to prepare an infusion drink for refreshment. A liqueur is made by macerating the longan flesh in alcohol.

Medicinal Uses

The flesh of the fruit is administered as a stomachic, febrifuge and vermifuge, and is an antidote for poison. A decoction of the dried flesh is taken as a tonic and treatment for insomnia and neurasthenic neurosis. In both North and South Vietnam, the "eye" of the longan seed is pressed against a snakebite in the belief that it will absorb the venom. Dried leaves, which contain quercetin and quercitrin, and flowers of longan are also sold as ingredients in Chinese herbal medicine. The nutraceutical properties and benefits which longan fruits could provide are found as it improves wound healing, prevents cells from damages, improves gums health, improves teeth health, improves Immune System, protects from free radicals, reduces Ageing, lowers risks of Cancers, improves iron absorption, improves lungs health, prevents from frequent colds, protects from frequent infections, reduces sensitivity to light, improves skin around eyes, reduces oral sores and cracks, reduces skin peeling around nose, supports good vision, supports healthy skin, supports cells production, improves nerve function, improves muscle function, reduces confusion, reduces irritability, reduces fatigue, lowers chance of heart problems, lowers chance of high blood pressure, protects from chronic diarrhoea and improves exercise ability. The health benefits of fruits acting against chronic diseases are ascribed to their antioxidant activities which are mainly responsible due to presence of phenolic compounds, which showed higher antioxidant and antityrosinase activities. Leaves and flowers are sold in Chinese herb markets but are not a part of ancient traditional medicine. The leaves contain quercetin and quercitrin. Burkill says that the dried flowers are exported to Malaysia for medicinal purposes. The seeds are administered to counteract heavy sweating and the pulverized kernel, which contains saponin, tannin and fat, serves as a styptic.

Other Uses

The seeds, because of their saponin content, can be used for shampooing the hair. The seeds and the rind are burned for fuel and are part of the payment of the Chinese women who attend to the drying operation. While the tree is not often cut for timber, the wood is used for posts, agricultural implements, furniture and construction. The heartwood is red, hard, and takes a fine polish. It is not highly valued for fuel.

Breeding and Genetic improvement

Breeding Objectives

The major breeding objectives of longan are high and stable yield, but now it is focussed on the cultivation of longan cultivars with early or late ripening periods, special qualities, storability (increasing shelf life), processability, high and stable yields, resolving problem of alternate/irregular bearing and stress (biotic and abiotic) resistance.

Longan Genetic Resources

Among the existing different types of longan in China, the first type is a large forest tree with small fruits and a very thin aril, possibly of interest for breeding purposes, while the second one is the native longan ('lamyai kradook' or small 'lamyai'), growing in the northern part of the country as an erect tree producing small fruits with large seeds and recommended as a rootstock for commercial cultivars and the third type is the commercial longan ('lamyai kraloke') which produces large fruits and small seeds. Beside these, there also exist at least three other varieties, *viz.* variety *longepetiolulatus*, variety *obtusus* and variety *magnifolius.* All three varieties grow wild in China (Huang, 1999). The other subspecies *malesianus*, contains mainly unexploited genetic materials which may have great potential to be developed into commercial fruits in their own right or as breeding materials for the commercial longan. The diversity of this subspecies in Sarawak in the Borneo island was documented by Wong and Gan (1992) and Wong (2000). The variety *echinatus* differs from the variety *malesianus* in that the fruits have rather long spines resembling the 'rambutan' (*Nephelium lappaceum*). This variety is found in Sabah where the 'kakus' also exists (Welzen *et al.*, 1988). These include variety *malesianus* and variety *echinatus.*, while the commercial longan is adapted only to the subtropics and will not flower when grown in the true tropics. The subspecies *malesianus* is fully adapted to the unchanging heat, humidity, daylength and other conditions of the equatorial zone. The subspecies *malesianus* var. *malesianus* is native to Southeast Asia with the greatest variation in Borneo where it might be possible to distinguish between 30 to 40 local races (Leenhouts, 1971; van Welzen *et al.*, 1988). The diversity of this subspecies in Sarawak in the Borneo island was documented by Wong and Gan (1992) and Wong (2000). The fruits are globular to slightly oblong and smooth to warty. In Peninsular Malaysia, the most common form of this taxon is the one with globose smooth fruits which turn brown when ripe. This is the true 'mata kuching' or 'cat's eye' which is often identified as *Euphoria malaiensis.* It has a very thin aril and is hardly worth eating. This form also exists in Borneo and Sumatra. The more superior forms are found in Sarawak, all with densely thick warty fruits and greater aril recovery (per cent). The true tropical longan offers the greatest opportunity for selecting superior material and thus offers an attractive possibility of longan becoming a new fruit crop for the humid lowlands throughout the tropics. Thus variable types of inter and intraspecies of longan germplasm are available in its original place China and to some extent, in other longan growing countries of the world.

For germplasm collections, various programmes were established in longan growing centres worldwide. The national longan germplasm repository of China was established in the Fujian Academy of Agricultural Sciences in 1981 and so far 202 cultivars were collected and preserved which included 3 wild, I dwarf, 15 aborted-seed and 1 pink flesh species and also about more than 300 types were identified in agronomic characters and fruit quality and cold hardiness determined (Zheng, 2001). There are also collections of longan in Fujian, Guangdong and Taiwan Province of China. In Malaysia, the wild longan and related species were collected from South East Asia (Wong, 2001). Collections of accessions are also reported from Hawaii, Florida, Australia, Indonesia, Israel and Thailand. The basic research under germplasm resources of longan are the establishment of germplasm bank and conservation *in vitro* of longan alongwith research morphology and palynology of germplasm resources of longan which are in progress in China (Pan *et al.*, 2010). The genetic resources of longan collected and preserved in bank provided a novel way for research and utilization of germplasm. (Ye, 2006). Thailand has also considerable number of collection of longan germplasm in its central and northern parts (Yapwattanaphun and Subhadrabandhu, 2001). Guo *et al.* (2006) successfully developed the method for cryopreservation of embryonic cultures of different genotypes of longan by vitrification. Lot of work in genetic relationship, origin, evolution and taxonomy of longan were done using morphological characteristics (Lin *et al.*, 2003; Zhuang, 2006).

Breeding Methods

Globally, very little work is being undertaken to breed specifically for longan genetic improvement. A few concerted efforts were made in China and Thailand to breed new varieties which produce high yield with quality fruits. To achieve this, introduction and selection of cultivars of longan are still the leading way.

Introduction

Introduction is one of the ways to improve cultivars of fruit trees and that too exploited in longan. The spread of longan from its original abode to various growing regions of the world only happened through introduction. Later, it became a potential step in improvement aspects. Since 1980s, some novel cultivars of longan were introduced from Thailand by Agricultural Institutions and Academies by various channels. In China, Maioqiao was introduced from Thailand first in 1981 and later many more. These are providing scientific foundations for efficient utilizations in China.

Selection

Progeny Selection

Seedling progenies are extremely variable and small fruit size appears to be a dominant characteristic. Therefore, through the centuries, improved cultivars have resulted merely from selection amongst existing cultivars or selections amongst open-pollinated seedlings from such cultivars. The true tropical longan offers the greatest opportunity for selecting superior material and thus offers an attractive possibility of longan becoming a new fruit crop for the humid lowlands throughout the tropics. Seedling progenies are extremely variable and small fruit size appears to be a dominant characteristic. Therefore, through the centuries improved cultivars resulted merely from selection amongst existing cultivars or selections amongst open-pollinated seedlings from such cultivars. Selection is based particularly on large fruit size, high edible portion, crisp flesh, good flavour and high sugar content. In so doing, heavy and regular yields appear to have been sacrificed. Future selection has to include marketing characteristics, such as early or late harvest, a long shelf-life and a pure white aril for the canned product. In Fujian province, China, a selection programme was initiated over 30 years ago to select abort-seeded longan cultivars. Five promising strains with abort-seeded fruits, namely, 'Minjiao No.1', 'Minjiao No.2', 'Minjiao No.3', 'Minjiao No,4' and 'Minjiao No.5' were successfully selected (Huang *et al.*, 2000). Among them 'Minjiao No,4' is the most promising due to its stable and higher rate of seed abortion, large fruit size, thick and good quality aril and high yield. Seedling selection is still a main method in the breeding of longan in China and Thailand. A variety of seedlings were selected from cultivars like Wulonglin, Youtanben, Caiuo etc. In recent years new cultivars and lines or single plants were selected such as Shuinan 1, Lidongben, Songfengben, Quanlong 104, Guiming 1 and Linglong. These were selected from seedling variations (Chen, 2001; Peng *et al.*, 2002; Liao *et al.*, 2006). In Fujian province, China, a selection programme was initiated over 30 years ago to select abort-seeded longan cultivars. Five promising strains with abort-seeded fruits, namely, 'Minjiao No.1', 'Minjiao No.2', 'Minjiao No.3', 'Minjiao No,4' and 'Minjiao No.5' were successfully selected (Huang *et al.*, 2000). Among them 'Minjiao No,4' is the most promising due to its stable and higher rate of seed abortion, large fruit size, thick and good quality aril and high yield.

Bud Sport Selection

Layering is used in propagation of longan for a long time to obtain some sports, like Baike, Fuyan, Qiugpi Fuyan, Huake Fuyan, Chike Fuyan, Chike Fuyan etc., which all belong to FuYan series. Guihua cultivar was selected from the sport line of Shixia cultivar in Guangdong. Many cultivars/varieties were selected from sports emerging in process of long cultivation in China and Thailand. Sport selection in longan has a great potential in exploring germplasm and breeding.

General Characteristics of Selection

Selection is based particularly on large fruit size, high edible portion, crisp flesh, good flavour and high sugar content. In doing so, heavy and regular yields are sacrificed many a times. Future selection has to include marketing characteristics like early or late harvest, a long shelf-life and a pure white aril for the canned product.

Hybridization

In case of longan, the cultivars with excellent quality are relatively a few. To fasten the progress in germplasm innovation and crop improvement in longan, cross breeding should be applied. Breeding targets which mainly pursued high and stable yield before were changed into cultivars with early or late ripening, special qualities (high sugar, large fruit, aborted seed, and unique flavours, etc.). An example of cross breeding between maternal parent Lidongben and parental parent Qingkebaoyuan was made and a new cultivar Dongbao No. 9 obtained (Zheng *et al.*, 2001). According to characteristics of the cultivars/varieties various cross combinations were made and attempted in various research institutes of China and Thailand but success is not high and still selection and investigation of traits in hybrid population are going on.

Biotechnology

Breeding process can be geared up by means of biotechnology because longan has long juvenile phase, high hybrid genotype and on the other side the traditional breeding is a long term venture or cumbersome to be applied in longan cultivation. In China, biotechnology research has achieved great progress in terms of its contribution to longan genetic improvement and production, especially genetic improvement through modern high technology (Lai *et al.*, 2000). Some of the biotechnology research includes tissue and organ culture, *i.e.*, studies on plant regeneration obtained from explants like cotyledons, immature embryos, anthers, leaves of bearing trees, stem apices and stem axes. The biotechnological research for longan improvement got momentum only two decades back, mainly emphasizing on cell engineering based on tissue culture, including: 1. Saving cultures with young and small embryos; 2. Anther or pollen culture to obtain haploid plants; and 3. Basic research focused on protoplast culture. Subsequently, with the rapid development of biotechnology, research was re-oriented towards genetic engineering, including mainly molecular marker techniques and new germplasm construction by protoplast fusion and transformation (Lai and Chen, 1997; Lai, 1997). Induction of embryonic cultures from microspores of cultured longan anthers was reported with some cultivars (Yang and Wei, 1984; Litz, 1988; Wei, 1990). Plant regeneration was successfully carried out via somatic embryogenesis from calli derived from the cotyledon of an immature embryo. Research studies on biochemical markers of germplasm of longan of China and Thailand were successfully made by Chen (1988) using SOD Isozymes. Molecular markers applied to longan classification are mainly RAPD, ISSR and AFLP (Lin, 1998a, b; Chen *et al.*, 2001; Zhong *et al.*, 2001, 2006, 2007; Yonemoto *et al.*, 2006). A breakthrough was achieved in the induction and long-term maintenance of friable embryogenic calli, and somatic embryogenesis at high frequency. Single cell cloning technology and a highly effective protoplast regeneration system were established from embryogenic suspension cell lines by isolating single cells and protoplasts. *Agrobacterium-mediated* genetic transformation was established and transgenic plantlets of longan obtained (Zeng *et al.*, 2000). Culture of shoot tips and shoot segments was successfully carried out to obtain virus-free plantlets for overcoming witches' broom disease. The basic *in vitro* protocol for manipulating elite longan selections is available, and could be utilized to address various plant breeding objectives using biotechnology procedures. Genetic transformation of longan is possible (Zheng *et al.*, 2001), because it can be regenerated from embryonic cultures of elite selections with high rate of conservation, could be genetically transformed with genes which encode for resistance to fungal diseases (anthracnose). It should also be possible to apply *in vitro* mutation induction and selection procedures to address certain fungal disease problems of known longan cultivars. Genetic transformation could be utilized to develop the preferred chicken tongue seeds using the *picta* gene from *Arabidopsis thaliana* that mediates seedlessness. Protoplast technology could be harnessed to produce somatic hybrids between haploid and diploid as a way of producing seedless triploid. McConchie *et al.* (1994) reported that longan and litchi can be sexually hybridized and intergeneric hybrids can be recovered. The resolution of problems associated with rapid deterioration of fruit during postharvest are more difficult to be addressed due to the fact that the longan fruits are non-climacteric. There is need to go a long way to exploit the untapped potential by applying biotechnology for improving longan.

Varieties

Longan will be a promising fruit crop for the New World as new superior cultivars/varieties are coming foreward under plantation and area expansion. The main reason of its low popularity and little interest is the type of longan brought was not one of the best. The earlier mention of the four types of varieties by Groff, but none of the other 4 varieties described by him have any great merit. There were no "chicken- tongue" (aborted seed) varieties in China.

The characteristics of the more popular cultivars in longan growing countries are as follows:

China

Chuliang

It is characterized by large fruit size (average fruit weight of 12.0–16.5 g), thick and firm aril, high aril recovery rate (edible part of 69–74 per cent), sweet with Total Soluble Solids (TSS) of 20–23 per cent, and fragrant flavour. The dry fruit rate ranges from 35-38 per cent and the dry flesh rate from 13-16 per cent. The colour of processed dried flesh is golden yellow. The cultivar produces high and

stable yield. Its clones show the features of early bearing, high and stable yielding and hereditary stability. (Huang *et al.*, 2000).

Table 8.3: Cultivars popular in provinces.

Sl.No.	*Countries*	*Province*	*Name of Varieties/Cultivars*
1.	China	Guangxi province	Dawuyuan, Guangyan
		Guangdong province	Chuliang Shixia, Wuyuan
		Fujian province	Tuyan, Wulongling Chike, Dongbi Honghezi
2.	Taiwan		Fongko, Hongko, Chingko
3.	Vietnam	Mekong delta	Longnhan, Tieuhue Xuongcomvang
		Northern region	Longhungyen
		Southeast region	Tieuhue, Xuongcomvang
4.	Thailand		Kohala, Biew Kiew, Chom Poo, Homestead, Haew, Dang and Seedling lines
5.	India		Seedlings
6.	Australia		Kohala, Biew Kiew, Chom Poo, Homestead, Haew, Dang and Seedling lines

Shixia

The fruit is small, but has crisp, thick and sweet aril with TSS of 19-20 per cent and with good flavour. It is the best cultivar for consuming as fresh fruit and produces heavy crops regularly on unusually large panicles (Menzel *et al.*, 1990). This cultivar is planted under a high density planting system coupled with constant pruning to produce dwarf trees. This system is reported early yielding and produces high and stable yield of quality fruits.

Wuyuan

Wuyuan' (Black Round) is another important longan cultivar. The fruit is medium in size (15 g) with large seed and soft and juicy aril of average quality. Yield is high and TSS value is 14-15 per cent. The fruit is suitable for eating fresh and dried. Seedlings are vigorous and consequently useful as rootstock (Menzel *et al.*, 1990).

Fuyan

This is an important longan cultivar in Fujian province and accounted for over 90 per cent of plantings in the province. The fruit is large (18 g) with thin skin, small seed and thick crisp aril. Yield is high. It is the best used for canning since the fruits have a low TSS value of 15-16 per cent (Menzel *et al.*, 1990).

Wulongling

This is also a major longan cultivar in Fujian province. It is in existence for more than 150 years. The fruit is medium in size (15 g) with thick skin, good aril recovery and sweet flavour (TSS of 21-23 per cent) (Menzel *et al.*, 1990). 'Wulongling' has a distinct alternate bearing cropping.

Taiwan Province of China

Fengko

This is the most popular longan cultivar in Taiwan province of China and over 98 per cent of the longan area are planted with this cultivar. The cultivar is a good yielder. Fruit is evenly large in size.

Colour of fruit rind is yellowish brown which becomes brighter at low temperature. The aril is very sweet with a TSS of 20 per cent. Fruits are not easily detached from the stalk, considered as a good quality. Propagation by grafting usually yields a high percentage of success.

Hongko

This cultivar is grown mainly in the southern region of Taiwan province of China. It is a good yielder. The fruit is of reasonably large size with fruit rind dark brown in colour. The sweetness of the aril is less than that of 'Fengko'.

Chingko

The fruit rind is light brownish green when ripe. The quality of the fruit is easily affected by environmental conditions. The aril is less sweet when compared to that of 'Fengko'. Fruits are easily detached from the stalk and is considered of a poor quality.

Vietnam

Longnha

This is a truly tropical longan which can yield two crops/year.

Tieuhue

This is also a true tropical longan which can yield three crops in two years.

Longhungyen

This longan thrives only in subtropical conditions and produces only one crop/year.

Thailand

Daw

This is the most popular cultivar in Thailand and about 73 per cent of the total longan area in the country are planted with this cultivar. It is an early-maturing cultivar as indicated by the name 'Daw' meaning early. The fruits fetch high premium in foreign markets due to the early harvest. This cultivar is the most consistent bearer and has no alternative bearing problem. In addition, the cultivar is relatively free from infection by witches' broom disease (Subhadrabandhu, 2000). Although the fruit is large, it has a big seed and, therefore, low aril recovery rate. The aril, which is sweet with good flavour, is rather tough and not so crispy. The fruits can be consumed fresh or processed.

Chompoo

It is a mid-maturing cultivar which flowers in December to early January and the fruits can be harvested from middle July to early August. The fruit is of medium size, oval in shape and with greenish light brown skin. Its small seed leads to high aril recovery percentage. The aril is slightly pink and thus the name 'Chompoo' meaning pink. It is very sweet (TSS 21-22 per cent) and has a pleasant aroma. The weakness of this cultivar is its irregular cropping habit as a result of poor flowering. For high production, the tree requires high fertilizer application and good management.

Biew Khiew

It is a late-maturing cultivar with flowering in late January and harvesting of fruits in late August to September. Mature fruit is round, large, brownish green in colour and has good aril recovery percentage. The aril is crispy, pleasant scented and sweet (TSS 22 per cent) and of excellent quality. The fruit skin is rather thick and, therefore, advantageous to longer shelf life. This cultivar exhibits irregular bearing and is susceptible to witches' broom disorder.

Haew

This is a late-maturing cultivar.The fruit is medium to large in size with rather small seed and has an average recovery percentage. The aril is firm and of good eating quality. The fruit rind is rather rough and thick and, therefore, advantageous for longer shelf-life. This cultivar flowers easily and produces heavily. Its weakness is its alternate bearing habit. The fruits are suitable for canning.

Dang

This is a mid-maturing cultivar whose fruits can be harvested in mid-July to early August. Fruit is large with reddish brown rind. However, it also has quite a large seed leading to rather poor recovery percentage. With maturity, the fruit's quality declines. Its productivity and quality are similar to those of 'Daw'. The trees are susceptible to waterlogging.

Baidum

This is a mid-maturing cultivar which flowers in late December to early January and harvesting of fruits is around mid-July to early August. The fruit is medium in size with rough rind, small seed and an average aril recovery percentage. The aril is of acceptable flavour, crispy, very sweet and is bright white in colour. This cultivar is a regular bearer and can withstand drought quite well.

Talub Nak

This is an early-maturing cultivar whose fruits are harvested in mid to late July. Fruit is medium in size, with small seed and a high aril recovery percentage. The aril, which is bright white in colour, is less sweet when compared to those of other cultivars. This cultivar can produce well when given good management.

Phetsakon

It is a lowland cultivar which does not require a cool climate for induction of flowering. 'Phetsakon' is an early-maturing cultivar and it flowers readily. It is grown in the central region of the country in Samut Sakhon and Ratchaburi provinces.

Indonesia

There are three cultivars of longan namely Batu, Kopyor and Lumut (Poerwanto, 2005). The characteristics of these popular cultivars are as follows:

Batu

The cultivar produces big fruits with small seeds. The flesh is thick, rather bitter and sprouting. It has sweet refreshing taste and nice aroma. It has pale brownish yellow skin.

Kopyor

The cultivar produces medium sized fruits with big seeds. The flesh is not very thick, sticky and sprouting. It has sweet taste with not so strong aroma. It has pale brownish yellow skin.

Lumut

The cultivar produces small fruits with medium sized seeds. The flesh is thin and sticky and not so sweet with unpleasant aroma. It has dark brown skin.

Florida

Kohala: It is only reliable in its blooming and fruiting. The trees are also easily propagated by air-layering. Fruit of 'Kohala' is light and thin-skinned. It is large in size, and has a sweet aroma with a spicy flavour. Fruit generally has a high aril recovery percentage. Air-layered trees will begin bearing-very early, and a three-year old tree may carry 10 kg of fruit. The flower panicle is normally pruned to 2/3 its size to obtain large fruit size.

Australia

Kohala, Biew Kiew, Chom Poo, Homestead, Haew, Dang and Seedling lines are commonly grown in Australia. These varieties performed differently in different regions and even though Kohala is an early season variety but has poor quality. The varieties Biew Kiew and Chompoo are somehow dominating in the market.

Propagation

Longan can be successfully propagated by employing both sexual and asexual methods of propagation (*i.e.* from seed, air-layering, budding, grafting, cutting and inarching).

Seed/Sexual Propagation

Propagation by seed is not advisable since the seedling takes a long time to bear fruit (7-8 years) and the planting material is not true to type. Most longan trees are grown from seed. The seeds lose viability quickly. After drying in the shade for 4 days, they should be planted without delay, but no more than 2 cm deep, otherwise they may send up more than one sprout. Germination takes place within a week or 10 days. The seedlings are transplanted to shaded nursery rows the following spring and set in the field 2-3 years later during winter dormancy.

Asexual/Vegetative Propagation

Vegetative propagations are, therefore, recommended means of propagating the planting material. Among the vegetative propagations, air layering (marcottage) is the most popular method and widely used for a long time in China, Taiwan Province of China and Thailand. This method yields a high percentage of successful planting material as the marcotted branches produce roots readily. Claims of success rates of 80 to 90 per cent with air-layering are very common. Air-layering is frequently done in Fukien Province and is a feasible means of distributing the 'Kohala' from Hawaii. Air-layers bear in 2 to 3 years after planting. Marcotting is usually carried out during the wet season. A strong healthy branch with matured leaves is chosen. The branch is first girdled or a strip of bark (2 to 2.5 cm wide) is completely removed to the cambium layer. The exposed surface is then scraped to remove the phloem and cambial tissue to prevent premature healing. IBA or any other rooting hormone is usually applied onto the cut surfaces of the cincture (towards the tip of the branch) to encourage rooting. A rootball made up of a suitable medium which holds moisture and is well aerated (*e.g.* wet peat, sphagnum moss or suitable soil mixture) is then wrapped around the cincture. This is followed by wrapping a polythene film, 20 to 25 cm square, around the rootball making sure that it is completely covered and the two ends of the wrapping are tied securely to prevent water from seeping inside. The marcot is removed after two to four months, when the roots have turned from white to creamy brown. In Thailand, the marcotted branches produce roots in about a month when propagation is carried out during the wet season. Pruning of the rooted marcot to reduce the top in proportion to the roots is usually necessary. The rooted marcot is potted in a suitable container and placed under warm, humid and partially shaded conditions in the nursery to allow the plant to acclimatize prior to planting out in the field. Under normal conditions, the marcotted tree is planted out in the field after 6 to 12 months in the nursery. Trees obtained by marcottage are more susceptible to wind damage when compared to grafted trees. This is because tap roots are absent in the marcotted trees. To prevent wind damage in the field, the marcotted trees are either supported by permanent bamboo props, by soil mounded around the trunk, or by rooted seedlings planted close by for later inarching.

Longan can also be propagated by grafting. Grafted trees are used in some orchards. The industry generally feels that the high cost of production of grafted trees without any obvious gain in production does not warrant its continued use for mass multiplication.

In Kwangtung Province, where vegetative propagation is undertaken, it is mostly by means of inarching, nearly always onto 'Wu Yuan' trees 3-5 years old and 1.5-1.8 m high. The union is made no less than 1.2 m from the ground because it is the most convenient.

In China, whip-and-tongue graft is practiced as far back as the late 1970s. Approach grafting is also carried out using seedlings of the same cultivar as the rootstock. Eight to twelve months old seedlings (commonly cultivar 'Wuyuan') are used as rootstocks. They are approach-grafted to a similar sized branch of the scion cultivar. After a union has occurred within 40–60 days, the top of the rootstock plant is removed above the graft and the base of the scion plant removed below the graft. The grafted plant is nursed in a pot or polythene bag for 2-4 weeks before planting out in the field. Sometimes, double or triple graftings (successive grafting operations on top of the previous graft) are carried out by the Chinese growers as they believe that such graftings can give added strength to the tree and produce larger and sweeter fruit. Currently, propagation by budding (patch budding with a single bud) has become popular and the success rate is over 85 per cent.

Grafting is uncommon and when it is done, it is a sandwich graft on longan rootstock, 3 or 4 grafts being made successively, one onto the beheaded top of the preceeding one, in the belief that it makes the graft wind-resistant and that it induces better size and quality in the fruit. Conventional modes of grafting have not been successful in Florida, but whip-grafting has given 80 per cent success in Taiwan. A tree can be converted to a preferred cultivar by cutting it drastically back and veneer-grafting the new shoots.

In Thailand, grafting is commonly done in winter when vegetative growth has ceased. One year old longan seedlings which show signs of good growth are chosen as rootstocks. In this respect fast growing longan such as the native longan (or 'Kradook') can be used as rootstock. After the grafting operation, the grafted plants are covered by plastic bags to provide a high humidity condition around the graft union. The plants are well watered for about one month before the plastic covers are removed. The newly grafted plants are usually nursed in the nursery for about 2 months before planting out in the field. Work from Thailand and Queensland suggest that some scions/rootstocks are incompatible. In China, graft incompatibility has recently been reported to occur in Guangdong. The rate of incompatibility for cultivar 'Chuliang' has attained 36 per cent (Liu and Ma, 2000). Researchers recommend grafting a cultivar onto seedlings of the same cultivar. In Australia, grafting is preferred onto seedlings of the same cultivar or at least not Thai cultivars on Chinese rootstocks or vice versa (Menzel *et al.*, 1990).

Forket budding and cuttings are successful with longan cultivars introduced into Australia (Menzel *et al.*, 1990).

Cultivation of Longan

Planting

Generally, the trees are planted too close together, seriously inhibiting productivity when they become overcrowded. In China, full-grown trees given sufficient room–at least 12 m apart while crops in Florida spaced at 6 m. Larger trees have larger crops but if the trees become too tall harvesting is too difficult, and they should be topped and recent trend is high density plantation (HDP).

Trees are the best planted at the beginning of the wet (monsoon) season. For planting pits of size 60 x 60 x 60 cm are dug and allowed to weather for about two weeks prior to planting. If the planting materials are planted in polybags, make sure that the bags are removed at the time of planting. Organic manure and phosphate fertilizer can be incorporated into the soil to be used to fill the planting hole. Compact the soil after planting to ensure that there is firm contact between the root system and the soil mass. The plant should be firmly staked and tied to avoid been blown over by strong wind. Tree spacings range from 6 x 6 m to 12 x 12 m; giving tree densities ranging from 70 to 300 trees/ha. Mature orchards in Thailand have a tree density of 50 trees/ha. In Taiwan, tree spacings of 4 m to 6 m with a density of 300 to 600 trees/ha are common in mature orchards. Tree densities in Fujian province in China range from 195 to 300 trees/ha with an average of around 270 trees/ha. Under normal spacings, well-grown trees are able to fill their spacing in 6 to 7 years. At the Yehai Litchi Science and Technology Garden in Dongguang, Guangdong province, dwarf 'Shixia' longan (through heavy pruning) was planted under a high density of about 1,500 to 1,800 trees/ha. A serious problem with the longan is its irregular bearing–often one good year followed by 1 to 2 poor years. Another handicap is the variable ripening season–early to mid-August in China. In order to spread the workload in a commercial orchard, it is essential to plant a range of cultivars having different maturity times in any one orchard.

Weed Control

Control of weeds has to be done several times during the rainy season. Both mechanical and chemical control measures are almost equally employed. Two main herbicides commonly used are glyphosate and paraquat. Weed free and well managed orchards always have enhanced fruit yield with quality production.

Mulching

Mulching is normally practiced during the dry season, which coincides with fruit setting and fruit growth period. Mulching prevents fruit cracking due to fluctuation in soil moisture content. Rice straw and dried grass are the two main sources of mulching materials. No plastic film mulching is used.

Plant Growth Regulators and Horticultural Treatments

Growth promoting substances like ethephon when used as a foliar spray is beneficial for floral initiation and inflorescence development (Qiu *et al.*, 2000). A single foliar spray with 400 µl/L ethephon on Shixia longan cultivar increased cytokinin, ABA and cytokinin/gibberellins (GA_{1+3}) ratio in the flower buds, while inhibiting gibberellins activity. The increase in cytokinin level led to the promotion of flower differentiation and morphogenesis. The higher level of ABA exerted a positive effect on bud dormancy and morphogenesis. Also, higher cytokinin/GA_{1+3} ratio favoured flower initiation during the period of floral initiation and inflorescence development. In addition, higher starch content in the ethephon-sprayed trees is beneficial for floral initiation and inflorescence development.

The technique of producing off-season longan is practiced by Thailand growers (Subhadrabandhu and Yapwattanaphun, 2000b). Under normal conditions, the commercial longan requires a period of cool and dry climatic conditions for induction of flowering. The Thai longan growers, however, have discovered an alternative method of floral induction in the absence of suitable climatic conditions. This is by the use of potassium chlorate which is applied to produce off-season longan. With off-season production, the growers are able to obtain a better price for their products. There are many methods of applying potassium chlorate to induce flowering in longan. Many Thai growers apply the chemicals as a soil drench and the rate of application depends on the tree size. Other factors which need to be taken into consideration are soil type, availability of water supply, general health of the tree and management of the orchard. Trees grown on a sandy soil respond better to potassium chlorate than those on a heavy soil type. Watering is essential in an area with a long dry spell, and the tree must be healthy and dormant in vegetative growth at the time of chemical application. Different cultivars require different rates of potassium chlorate for induction of 100 per cent of flowering. Cultivar 'Daw' requires 8 g/m^2 while Chompoo needs 1–4 g/m^2 of potassium chlorate as soil drench. The suitable time for good flower induction by potassium chlorate is from October to February where the average temperature is rather cool with relatively dry condition. Besides soil drenching, the chemical can also be applied as a foliar spray at about 1,000 ppm or as trunk injection at 0.25 g/1 cm diameter of stem size less than 10–15 cm in diameter. Sodium chlorate may be able to substitute potassium chlorate for the induction of flowering in longan. The combination of potassium chlorate application and use of easy to flower cultivars have led to the expansion of longan growing areas in Thailand. These include the central, warm tropical region and also the eastern provinces, where early cultivars are being grown.

Cinturing (of branches and stem) can induce dormancy and gives rise to better flowering, fruiting and production as reported in Thailand. The results are too inconsistent to justify recommendation for commercial application. The easy to flower cultivar, it can be induced to produce early and uniform flowering by cinturing of branches or stems (Subhadrabandhu and Yapwattanaphun, 2000a).

Pruning and Training for Canopy Management

In newly established orchards, young trees should be grown as vigorously as possible for the first four years to attain the greatest tree size and bearing surface. Vigorous growth is achieved by strategic pruning, proper water and nutrient management and protection from weeds, pests and diseases. When the trees are mature and ready to bear fruits, appropriate management in regulating vegetative and flowering cycles is required. It is important to know the phenology of each cultivar so that the vegetative, flowering and fruiting patterns for the cultivar can be monitored. For each year, the tree should be limited to one or two significant vegetative flushes between harvest and the next flowering. Once flowering commences, there should be minimum flower and fruit shedding. This annual cycle can be achieved by strategic applications of water and fertilizers and appropriate timing of pruning of shoots, flowers and fruit.

In China, researchers achieved three successive years of high and stable yields in 'Chuliang' cultivar in Gaozhou, Guangdong province, by effective cultural techniques (Huang *et al.*, 2000). These include mainly fostering two autumn flushes after harvest to form high quality fruiting shoots alongwith integrated measures for manipulation of flushing cycles to promote flower retention and prevention of change of reproductive flush to vegetative flush, and thus increasing flowering rate and fruiting (per cent).

Proper Thinning of Fruit Panicles to Obtain Good Fruit Quality

Immature young longan trees (first 3-4 years from planting) are subjected to formative pruning to obtain 'open' canopy which allows good light penetration. Only limited number of main branches are retained to obtain the desired structure. Surplus branches and all water shoots are removed and the skirt is maintained about 1-2 m above the ground level. In China, one strong branch is retained after every growth flush to form a natural round-shaped crown of 6 to 10 main branches. In Thailand, trees are cut to a 1.2 m trunk and 3-4 vigorous young shoots selected from the re growth to form the main framework of the tree. These laterals are forced into wide angles from the trunk with the aid of sticks.

Two branches are left on each lateral shoot. Similarly, two sub-branches are left on each branch and so on. Finally, the canopy consists of 24-32 sub-branches, which may be achieved in the third or fourth year. Pruning at the immature stage produces well-formed tree canopy, strengthens fruit bearing branches, ensures annual cropping and limits insect pests and diseases. It also reduces the height of the tree to a manageable level. The flower spikes of young trees less than five years of age are normally removed in China and Thailand to encourage growth and crown expansion during summer. Since the productivity on a per plant basis is comparatively low, the higher production on a per hectare basis can be explained by the higher tree density.

In bearing trees,pruning is an essential cultural practice. Harvesting itself is a form of pruning, since the entire panicle is cut. Soon after harvest, this should be followed by pruning any remaining panicle or cutting out some of the subtending twigs. Cutting out these twigs completely simplifies the canopy structure and admits more light to the interior of the tree; it also removes twigs which are least likely to fruit next year, since they fruited this year. If this is not done, side shoots emerge below the cuts of the harvested panicles. These shoots make the canopy denser and come too late to initiate inflorescences for the next crop. Besides pruning, which is carried out during and immediately after fruit harvest, maintenance pruning is also practiced during the period between harvests. This involves pruning water shoots and branches which are dead or infested by pests and diseases. Weak branches, which have lost their vitality, are also pruned. Too much pruning or removal of too much leaf and wood with the fruit panicles at harvest reduce flowering the next season and aggravates biennial bearing.

In China, Taiwan province of China and Thailand, prunings of flower and fruit panicles are usually practiced to overcome alternate bearing phenomenon which is common especially with trees which yield heavily and are older than ten years. Production in an 'off year' is usually about 20 to 40 per cent of the 'on' year (Menzel *et al.*, 1990). In China, pruning of flower panicles involves removal of about 40 per cent of the flower spikes when they are about 10-12 cm long. For pruning of fruit panicles about 30 per cent of the young fruits are removed. The degree of pruning of flower and fruit panicles depends on the crop load and tree vigour. Thinning of fruit in China not only reduces biennial bearing but also increases fruit size. Only large fruits (2.5 cm in diameter or 18 g in weight) attract a premium price. Fruits are thinned about four to six weeks after fruit set when they are of the size of a pea. Fruit thinning is essential after flower thinning because of the high rate of fruit set and greater competition for developing fruit. In Thailand, longan growers reduce the number of flowers by half (each flower spike is retained) before fruit set in an 'on' year. After fruit set, they remove 10 per cent of the fruit.

Physiological Disorders

Chilling Injury

This is a serious postharvest physiological disorder in longan with typical symptoms including darkening of the skin colour, brown water soaked areas with increased susceptibility to decay while under cold storage but it varies with cultivars and production regions (Wang, 1998). Also, some loss of flavour of the arils may occur and ultimately affects the eating quality.

Pericarp Browning/Skin Darkening

Pericarp browning is one of the important physiological disorders in longan (Jiang *et al.*, 2002). This particular browning is attributed to desiccation (Lin *et al.*, 2002), chilling (McGuire, 1998), disease and heat stress (Jiang *et al.*, 2002) and senescence (Huang and Wang, 1990). In general, skin changes from light brown to dark brown as a result of water stress (desiccation) of longan. The desiccation seems to be more correlated with this problem that leads to a rapid loss of bright yellowish skin, which turns to a dull brown colour (Jiang *et al.*, 2002). Fungi associated with pathological skin browning and darkening of the skin along with mycelium include *Lasiodiplodia theobromae, Pestalotiopsis* sp., *Cladosporium* sp., *Fusarium* sp. and *Aspergillus niger* (Sardsud *et al.*, 1992).

Fruit Cracking

The most important pre-harvest disorder in longan *i.e.* fruit skin-cracking (splitting) occurs in almost all the important longan growing countries of the world (mainly throughout Asia and the Pacific). It is the most common in China, Thailand, India, and also occurs in Australia. The incidence and severity vary with season, cultivar and tree morphology. The cracking injury in litchi may vary from short cracks, generally skin deep to larger ruptures extending to most of fruit length. As the underlying flesh (aril) gets exposed to dry air, it withers rapidly. The cracks on the skin may be longitudinal or horizontal. The cracked fruits deteriorate rapidly and thereafter suffer secondary infestation by disease causing organisms and insect-pests become non-marketable and ultimately cause great economic loss to the growers.

Insects Pests and Diseases

Insect Pests

Numerous pests are found on longan. The important insect pests include Stink bugs, Erinose mite, gall mite, scales, fruit flies, aphids, stem borers, leaf eating caterpillars, flower eating caterpillars, mealy bug, fruit spotting bug, elephant beetles and fruit piercing moth.

Stink Bugs (*Tessaratoma papillosa* Drury)

This insect affects all the plant parts like fruit, inflorescence, flowers, stem and leaves of longan (Waite and Hwang, 1999).It is widely distributed in China, India, Indonesia, Malaysia, Pakistan, Philippines, Sri Lanka, Taiwan, Thailand, Vietnam. Of particular importance is the longan stink bug (*Tessaratoma javanica*) which can ruin bloom in a year with light flowering. The adult bug feeds on flower panicles, young fruits and newly developed shoots of longan, causing withering and later abscission of those infested organs. The infested fruit ceases to develop. In Thailand, this stink bug can be of epidemic level during March to April, which coincides with flowering and fruit setting times of longan. At the same time,the population of the bug's natural enemies in the orchard should be maintained at a satisfactory level. The natural predators of this stink bug are identified as *Anastatus* sp., *Micropanurus* sp. and *Eupelmid* sp. Mealybugs also damage the longan fruits.

Control

Studies showed that spraying solution of 20 per cent Sumicidin (fenvalerate) gave good chemical control of (Li *et al.*, 2001). One method of chemically controlling *T. papillosa* involves two applications of trichlorfon. The application of oil soap and/or insecticide reduced the number of scale insects and mealybugs in longan.

Litchi Fruit Borer (*Conopomorpha sinensis* Bradley)

The larvae of this insect penetrate and feed on longan fruit (Waite and Hwang, 2002). Adults feed externally on the fruit (He, 2001; Waite and Hwang, 2002). In longan, damaged fruit often falls from the tree (Waite and Hwang, 2002). It is predominantly found in China, India (Waite and Hwang, 2002); Taiwan (Hwang and Hung, 1996); Thailand (Schuetz *et al.*, 2002; Waite and Hwang, 2002). They prefer orchards with shady, humid and closed conditions (Zhang *et al.*, 1997). During off-season, when fruit is not available, the larvae survive by feeding on young leaves and shoots

(Waite and Hwang, 2002). Adults feed externally on longan fruit and their lifespan is from 5-8 days (Waite and Hwang, 2002). The moths are attracted to leaf flushes which emerge during the rainy season from June to October in Thailand (Menzel, 2000). Affected shoots often wilt.

Control

In Thailand, fruits are inspected weekly from fruit set to detect eggs of *C. sinensis*, which are very small and almost invisible to the naked eye. When the pest becomes more active, permethrin is applied weekly, up to two weeks before harvest. In Taiwan, cypermethrin, deltamethrin, carbofuran or fenthion during early fruit set is recommended to prevent damage later in the season. Moths can be excluded by enclosing the fruit panicles in nylon mesh bags, but this method is uneconomic in areas with high labour costs (Waite and Hwang, 2002).

Leaf Rollers (*Adoxophyes cyrtosema* Meyrick)

This insect affects plant parts like fruit, flower, leaf and new growth (Menzel, 2002) and found mainly in China (Liu *et al.*, 2001), India and Thailand. The larvae web and roll leaves together to form a shelter in which they feed (Waite and Hwang, 2002). This species feeds externally on longan fruit (Waite and Hwang, 2002).

Control

In field trials, chlorpyrifos at various dilutions and abamectin at various dilutions could effectively kill the larvae. Control effectiveness was still 100 per cent at 48 hours after application (Liu *et al.*, 2001).

Other Pests

Of particular importance is fruit bats or flying foxes, which can devour the longan fruits during fruiting seasons. This pest causes serious damage in Thailand and Australia. The only effective means of reducing damage is to erect a protective net around the perimeter of the orchard or over each tree However, this method can be very expensive. Alternatively, a draconian control method is electrocution by a high screen of thin, parallel electric wires in the orchard.

Disease

Fruit Blotch (*Phomopsis longanae*)

Plant parts affected by this insects are mainly stem; twig; bark; fruit and found predominantly occurring in China (Lin & Chi, 1992). Coates *et al.* (2003) listed *Phomopsis* sp. causing stem-end rot of longan. Stem-end rots begin as brown discolouration of the rind at the stem end and expand rapidly and are visually indistinguishable without isolation.

Control

Coates *et al.* (2003) suggested that inoculums from the group of fungi causing stem end of longan, can be reduced by pruning out dead leaves and twigs in the canopy, thus increasing ventilation and providing a less favourable environment. Fruit should be cool-stored after harvest to suppress the development of disease and fungicide treatment can provide some control.

Phytophthora Leaf Blight and Fruit Rot (*Phytophthora palmivora*)

The disease infests young shoots, panicle, root, leaf, fruit (Coates *et al.*, 2003). The disease occurs in Australia; China and Thailand. The symptoms include a dark necrosis of young shoots, brown blight on the panicles, flower drop, irregular brown lesions on fruit and pre-mature fruit drop. During wet weather, the fruit cracks and lesions covered by white sporangia and sporangiophores (Coates *et al.*, 2003). *Phytophthora* leaf blight and fruit rot are serious diseases of longan where fruits are induced to set during the off-season (Visitpanich, *et al.*, 2000).

Control

Control is achieved through good orchard sanitation, including destruction of diseased plant material to reduce the source of inoculum. Fungicide applications may be necessary during disease conducive periods. The soil at the base of the tree trunk can be drenched with metalaxyl as an effective chemical control of this disease.

Longan Witches' Broom

A serious disease in longan is the rosette shoot or witches' broom. This disease is found predominantly occurring in Brazil, China, Taiwan and Thailand (Menzel *et al.*, 1989; Zhu *et al.*, 1994; Koizumi *et al.* 1995). The earliest description of this disease on longans is by Qui (1941). Young leaves of infected shoots are small and light green in colour, with curved margins. They appear stunted and deformed, and tend to roll up rather than expand (Zhang and Zhang, 1999). Adult leaves are light yellow-green with marbled yellow spots, and brown veins. Leaves form blisters and become distorted and dry before falling off (Zhang and Zhang, 1999; Menzel *et al.*, 1989). Shoots on infected branches become compacted clusters and the inflorescences are unable to extend. The flower organs develop abnormally, and consequently, the flowers either fail to produce fruits or develop into small and empty fruits. A characteristic symptom of the disease is the loss of flowers from panicles, resulting in a 'broom-like' appearance of inflorescences (Menzel *et al.*, 1989). Different cultivars of longan vary in their sensitivity to damage by this disease (So and Zee, 1972). However, there is disagreement among the available literature as to whether longan witches' broom is caused by a virus (So and Zee, 1972; Ye *et al.*, 1990; Chen *et al.*, 1996; Chen *et al.*, 2001), a mycoplasma (MLO) (Menzel *et al.*, 1989), or a mite (He *et al.*, 2001).

Control

Integrated management of pruning and spraying with a mitecide on diseased shoots restored blossoming and reduced the average incidence of diseased spikes. No cure is known for witches' broom disease. There are cultivars free from this disease. In Thailand, the cultivar 'Daw' is free from this disease (Subhadrabandhu and Yapwattanaphun, 2000a). In China, cultivars like 'Lidongben' and 'Shuinan No.1' in Fujian province are highly resistant to the disease. An integrated management programme against the disease, *viz.*, strict quarantine inspection; selection and use of disease-resistant cultivars; establishment of virus-free nurseries; timely (chemical) control of insect vectors; removal of infected branches, inflorescence and infected seedlings in the nurseries and orchards may control the spread of pathogen and improving tree vigour by judicious fertilization, irrigation and soil management. Shoot tip culture is successfully employed to produce virus-free plantlets of longan (Wang, 2000; Coates *et al.*, 2003). The best strategy for disease management appears to be controlling the vectors (Zhang and Zhang, 1999; Chen *et al.*, 2001).

Cylindrocladiella Disease (*Cylindrocladiella peruviana*)

Causes root rot, leaf spot and also fruit rot affecting root, leaf and fruit widely distributed in Australia; Brazil; China; India; Jamaica, Mauritius; New Zealand; South Africa; Sri Lanka; Thailand; USA (Florida) (Crous and Wingfield, 1993; Farr *et al.* 1989). The fungus appears to cause root rot and a leaf spot but is recorded on the fruit of longan in both China and Thailand. According to Zhang (1997), infected fruits of longan are visible on the tree by the white mycelium and obvious decay.

Control

There is no specific information or recommendation on control of this pathogen in longan. Good orchard management is necessary particularly intercultural operations and clean cultivation.

Harvesting and Fruit Yield

Longan fruits are non-climacteric and have to be harvested when ripe. Maturity can be determined on the basis of fruit weight, skin colour, flesh sugar concentration, flesh acid concentration, sugar:acid ratio, flavour and/or days from anthesis. Immature fruits are tasteless. The mature longan fruit has a dark, smooth skin, the inside of which is netted and tastes sweet. Most cultivars lose flavour if harvesting is delayed, when fruits are mature. The whole panicle of fruits with one or two leaves is cut with a knife or scissors. Removal of too much leaves and wood with the fruit panicles at harvest reduces flowering in the next season. Panicles should not be dropped. It is recommended that fruit should not be picked during rainy weather as this increases the risk of breakdown after harvest. Longans are picked in the cool of the morning on the panicle and placed in large plastic crates. Harvesting seasons of longan in different growing regions vary from late June to August in Thailand, late July to September in China and India, July to October in Taiwan province, Mid-January to Mid-March and late February to early April in Australia. Longan tree generally bears in the fourth year after planting. The commercial yield after 10th year onwards is obtained in the range of 60 kg/tree to increase upto about 100–200 kg/tree in the fifteenth season. After fifteen years, the yield is in the range of 150–200 kg/tree and this figure lasts for the next ten years. The yield declines when the tree is over thirty five years old. There is considerable variation in yield from year to year and biennial bearing occurs in most cultivars. Some small fruited, inferior types bear more heavily (300–400 kg) and consistently. Productivity of longan is affected both by years and site of production in Thailand (Subhadrabandhu and Yapwattanaphun, 2000a) and in China too longan is grown on poor sites with poor management practices and hence productivity of longan is affected by years and locations (Liu and Ma, 2000, Yen, 2000). In Taiwan Province of China, the average yield/tree also varies with years of production. Longan crops in Florida are medium range yielders with alternate bearing habit. Fruit yields of 40-50 kg/tree are obtained in six to eight years old trees in Australia.

Postharvest Maintenance of Quality Fruit

Longan fruits are non-climacteric with little change in soluble solids concentration (SSC) or titratable acidity (TA) after harvest. It deteriorates rapidly unless proper handling techniques are employed. The major factors reducing storage life and marketability of longan fruit are microbial decay and pericarp browning. Moreover, the quality deteriorates rapidly when removed from cold storage. Under refrigeration, longan fruits have a storage life of approximately 30 days. Pulp quality and disease development are generally stable during cold storage until such time as fruit becomes visually unacceptable from pericarp browning.

Sulfur dioxide fumigation is the most effective postharvest treatment for control of pericarp browning in longan fruit, and is used extensively in commercial situations at present. There is increasing consumer and regulatory resistance to the use of this chemical. Insect disinfestation has become increasingly important with the expanding export market. Irradiation and heat treatments for insect dis-infestation of longan fruit are alternatives to treatment with insecticides. Recent research has focused on reducing these major postharvest problems to produce light-coloured, chemical-free fruit without disease or insect infestation.

The treatment of fresh longan fruit with SO_2 fumigation combined with suitable storage condition improved the overall longan fruit quality, especially on inner and outer peel tissue and aril color than no SO_2 treatment. Treatment stabilizes peel color with no subsequent loss of color during storage (fruit color was bright-yellowish color). The activity of PPO enzyme in control fruit (no SO_2 treatment) gradually lowers more than SO_2 treatments. Fruit exposed to cool storage temperature (2°C) exhibited a lower PPO enzymatic activity compared to those kept in high storage temperature (7 degrees C). Moreover, PPO enzymatic activity significantly increased over the storage durations The additional SO_2 treatment caused no subsequent loss of weight of longan fruit during storage. However, the sulphite residues could detect immediately after SO_2 treatment in all parts of longan fruit, especially on aril tissue. The SEM evaluation found that the surface cracking also impairs the physiological function of the cuticle and increasing water permeability, which may cause water soaking at the inner side of the peel. The injured cell would accelerate the oxidation of phenolic substances and the oxidative products resulting in dark colour of inner and outer peel

Sorting, Grading and Packaging

Harvested fruits are moved rapidly back to packing shed where panicles are trimmed to small stalks with 5-10 fruits. Fruits graded with blemishes, stung or undersized fruit are removed. Fruits larger than 30 mm in diameter achieve grade I status. Thailand national standard for longan was documented by the Ministry of Agriculture and Cooperative in 1998. It is applied for longan that is produced for fresh consumption and this standard indicates provision concerning quality, classification of quality and sizing. At present, the Thai farmers have established grading of longan fruits into Grades A, B and C. The Grade A fruits consist of 55-75 fruits/kg (14-18 g/fruit), Grade B consists of 76-80 fruits/kg (12.5-13.2 g/fruit) and Grade C consists of more than 80 fruits/kg which encompasses the most small-fruited seedling types (Subhadrabandhu, 1990). After harvesting, sorting and grading, fruit should be packed as quickly as possible, as their quality deteriorates markedly, if they are exposed to sun even for a few hours. Longan fruit cannot be kept for more than a few days after harvest, at room temperature. Israel had an innovative technology to keep perishable fruits like litchi fresh for at least 28 days. Access to this technology will increase earnings manifold. Cool chain is essential during the transport of export quality commodity all the way from farm to the customer. This helps in maintaining the temperature inside the box at the same low level as in pack-houses.

Fruits destined for SO_2 fumigation are generally packed into an 8 kg plastic mesh lidded tray, whereas untreated fruits are generally packed into a 5.0 kg cardboard lidded carton. During fumigation SO_2 is applied at the rate of 540 mg/kg of fruit for duration of approximately 20 minutes. Treated fruits must have a withholding period (WHP) of 2 days before retail sale. The WHP generally coincides with the time it takes road freighted fruit to reach the distant whole sale markets. Treated fruits can be safely stored at 4-5°C without suffering chilling injury, where as untreated fruits are generally stored above 10°C (Drinnan, 2004). Alternatives to SO_2 treatment, in particular HCl deep treatment are studied (Drinnan, 2004), but are not commercialized. They are sorted for size, insect damage and skin blemishes and bunched in a bulk tray. In Thailand, longan fruits with stalk intact (about 21-22 kg) are packed in 35 cm x 50 cm round woven bamboo baskets lined with longan leaves. Stalked fruits remain fresher than separated fruits after harvest. Fruits for export, often detached from the panicles, may be packed in corrugated boxes or plastic baskets. At room temperature, longans remain in good condition for only 2-3 days, though the firmer rind, the fruit is less perishable than the litchi. Preliminary tests in Florida indicate that the fruit can be frozen and will not break down as quickly as the litchi when thawed.

Transportation and Marketing

There is always a need for special transport system maintaining the cool chain either by road, ship or air for this highly perishable fruit. Maintenance of cool chain right from farm gate to the consumer's point assures the supply of garden fresh produce to the consumer. The longan has potential domestic and export market. There will be not much difficulty in extending better market at domestic level. There should be proper road linkages and cold chain facilities, no administrative hurdles, considerable freight charges to encourage investors and streamlined procedures will act as steps for expansion of longan market in growing regions and away from the production site in domestic market. The market chain is made up of growers, marketing groups, agents and Asian Wholesalers, boutique fruit retailers and supermarkets. There are potential export markets of United States and European Union. Buyers are clearly demarcated, the bulk of sales are to the ethnic Vietnamese and Chinese communities in the major states capital cities of Sydney, Melbourne and Brisbane. Smaller volumes are sold into Adelaide and Perth. This market is served by both ethnic and nonethnic retailers (Noller, 2001). The perishability of the fruits limits marketability of this crop and especially of export to countries far away from the main areas of production. The main reasons behind inefficient marketing are (*i*) High perishability, (*ii*) Consumer unawareness about the economic value of this crop at the global level. (*iii*) System of pre-harvest, (*iv*) Lack of skilled or trained persons engaged in this fruit and (*v*) Lack of infrastructural facilities like proper storage and transport facilities etc.

In recent years, producing countries have identified new market niches for these distinctive fruit, transforming them from mainly domestic commodities to significant export markets (Tongdee, 1997). Demand for longans remains strong within these communities and there is preference for fresh Australian products over imported fruit from Thailand. There is low recognition and hence consumption of longan among Australians of Caucasian decent may change with time. The domestic demand for longan is limited as are export opportunities.

Storage

In particular, research on the storage of Longan was very weak although the growing areas and consumption trend increased dramatically. Among the a few references, the first, by Ye and Wei (1981), demonstrated effects of controlled atmosphere on storage of Longan, while the others, by Chen and Hong (1982) and by Cai (1988), showed some preliminary work on the decay control during storage of Longan fruits. Low temperature storage at 1–5 °C is used to reduce pathological decay, but has only a limited role in reducing pericarp browning. Longan fruit, treated with preservatives, *i.e.* sulfur dipping, sulfur fumigation or thiabendazide plus phosthyl-aluminium, were packed with polychloroethylene (PE) film (0.03 mm thick) bags and stored at 3°C for 46 days and studied in the way that the fruits were then removed from storage in three ways *i.e.* 1.Under gradual temperature rising; 2. Addition of oxidation retardant; 3. Direct removal. During cold storage, and it was found that the respiration rate of fruit decreased rapidly, but increased quickly after removal. The permeability of the skin cell membrane rose during and after cold storage. The cold-stored fruit removed from storage under gradual temperature increase or with addition of oxidation retardant maintained a shelf life for 32 hours at ambient temperature. Sulfur fumigation was favorable in retaining the original skin color and flesh eating quality. Fruit fumigated with sulfur and with oxidation retardant maintained 100 per cent marketability after 78 hours at room temperatures. In one Controlled Atmosphere (CA) considerations for longan fruits storage, it was reported, that MAP in 0.03 mm (1/1000 in) polyethylene bags at 4 °C with 3 per cent O_2 delayed browning and maintained SSC and vitamin C content for 35 days (Zhang and Quantick, 1997). Another recommendation for Optimum Storage Conditions is 4 to 7 °C at 90 to 95 per cent RH (Paull and Chen, 1987). Fruit can be held for 2 to 3 weeks, although the skin loses it yellowish coloration and becomes brown. At lower temperatures, there is a rapid loss of eating quality, and above 10 °C postharvest diseases are a concern. Longan fruits were stored in controlled atmospheres (CA) with the high O_2 concentration significantly decreased ethanol production in the flesh, maintained a lower pH value in the peel and maintained peel color. Higher CO_2 atmosphere more effectively reduced decay and extended storage life of longan of cv Shixia (Shipping *et al.*, 2002). The effects of high O_2 atmosphere (HOA) on physiological changes in longan fruit cv. Daw harvested at the commercial stage showed changes in respiration, ethylene production weight loss and ethanol content and HOA-stored fruits had lower rates of respiration, minimum weight loss and rise in ethylene production (Kaewsuksaeng *et al.*, 2007).

Commercial Processing of Longan Fruits

The lack of supporting services and infrastructural facilities is only posing hurdles in increasing the economic profits of longan industry. The preserving technology is not widely used in longan even in China, fruits must be sold in short time, which lowers the profit. Because of the high sugar content, longan fruit has a short shelf-life. Under ambient temperature (25°-31°C) harvested longan fruit rind turns brown within 3-4 days and the aril turns rotten within a week and loses its commercial value. An extended a few days of keeping is possible when temperature is lowered to 18°C. Prior to storage, longan fruit can be subjected to hydro-cooling or forced air cooling which can lead to longer storage duration when combined with low storage temperature. To prevent fungal growth on the fruit surface during storage, the fruit can be fumigated with sulphur dioxide. Short term sulphur dioxide fumigation combined with low storage temperature is currently the ideal and practical method of storage of longan fruit. Fumigation of sulphur dioxide for 20 minutes and storing the Shixia longan fruits at 4°C produced 100 per cent of good fruits after 28 days of storage without chilling injury (Han *et al.*, 2000). However, improper handling can cause high sulphur dioxide residue and reduce the flavour of the aril. Sulphur dioxide fumigation can affect both the external and the internal qualities of the longan fruit.

In a study on the shelf-life of cold-stored longan fruits, Lin *et al.* (2000) reported that longan fruits fumigated with sulphur dioxide and then packed in polychloroethylene (PE) film (0.03 mm thick) bags and stored at 3°C for 46 days are able to retain the original skin colour and aril eating quality when moved out of cold storage to room temperature for as long as 78 hours. When the fruits were moved out of cold storage, an oxidation retardant was added to them. Sulphur dioxide injury to the rind depends on the application concentration and the fumigation period (Tongdee, 1992). Injury to the rind occurs at intermediate concentrations. There is an important balance between the amount of sulphur dioxide applied, the sorption of the fruit and the residual sulphur dioxide detected on the fruits.

Recent studies showed that modified atmosphere storage (50 per cent air + 50 per cent nitrogen) of longan fruits treated with antifungal agent resulted in a 92.3 per cent marketable fruit ratio and high (per cent) of nutrient composition in fruits stored at 24°–26°C for seven days (Liu and Ma, 2000).

Longan fruit products are gaining importance as per the demands of the modern health conscious consumer. The main reasons for this growing demand are the awareness with regard to the health promoting phyto-chemical constituents like antioxidants, vitamins, minerals and dietary fibre. However, their contribution towards certain non-nutritive components like phenolics and flavonoides, with antioxidative properties are no less significant. Very little information is available on the processing of longan fruits except for canning pulp and juice preparation. Wine is prepared from longan fruits in China and Thailand. In view of many medicinal and aromatic properties of longan fruits and because of its short availability period and difficulty in storage conditions, the untapped potentials are still to be exploited.

Very negligible quantities of longan fruits are presently used for processing. Various processed products like canned fruit, squash, cordial, juice, RTS, syrup, wine, vinegar, jam, jellies and dehydrated products can be manufactured from longan. Dehydrated longan nuts are very common in Chinese diet. Fresh longan can be quick frozen and stored for more than one year. Longan industry may have far prospective with more innovative and shelf–stable items like osmo-dried longan, fermented longan juice in honey etc. Longans are mainly processed as limited tinned/canned fruits or dried fruits. Only very limited quantity is processed as wine or ointment. In Fujian region of China about 15.5 per cent longan produce is canned, 2.6 per cent is dried, 0.8 per cent is brewed (Shi and Li, 2010).

Due to the short shelf-life of fresh longan fruits, their uses can be extended by processing them into various products. In Thailand, the longan fruits are processed into canned longan, dried 'longan nuts', longan nectar and frozen longan (Subhadrabandhu, 1990). There are substantial canning factories for longan in Thailand, China and Taiwan Province of China. Large fruits are generally used for canning. The cultivars 'Biew Khiew' and 'Daw' are preferred for this purpose in Thailand. Due to the high brix content, little sugar additive is required as the fruits are canned in their own juice. Canned longan retained their individual flavour better than other commercial sapindaceous fruits.

Subhadrabandhu (1990) described in detail the processing of longan into dried fruits in Thailand. According to him longan fruits are dried either intact or with pericarp removed. Small and medium fruits are used for drying. The process of drying longan fruits can be done by boiling the fruit for five minutes, followed by sun or oven drying at 55°C. When the fruits start to dry, the temperature is raised to 70°C until they are completely dried. The total drying period is 19-20 hours. Dried intact fruits retain better their flavour and aroma when compared to fruits which are dried without the pericarp. The dried longan fruits contain moisture of about 18-19 per cent and have very high sugar content of 60–65° brix.

Conclusion

The subtropical group of fruits play a vital role in human health and nutrition. The longan fruit is though the least exploited for its potential outcome but is becoming very popular among Asians, particularly the Chinese. There are claims among some people in China and Thailand that the taste of longan equals or is superior to that of litchi. The edible portion of export quality fruit ranges from 67 to 78 per cent of the whole fruit. The energy value averages 458 kJ/100 g. The sugar content is very high. The Longan fruit itself is round and about 3 cm in diameter. Longans grow into a well developed and structured tree usually found around Southeast Asia. According to taste and physicochemical properties longan fruit is very sweet and juicy, and it can be eaten either raw, dry, or cooked. Asian people usually use it in soups, desserts, snacks, or in canned foods. Longan fruit is used through many years as a herbal medicine or therapy. About 100 g. of raw longan contains many nutrients like Vitamin B_1 (Thiamin), Vitamin B_2 (Riboflavin), Vitamin B_3 (Niacin), Vitamin C (Extreme amount), Protein, Iron, Magnesium, Phosphorus, Potassium, Copper, Manganese, Natural Fats, Dietary Fiber, Calories and Carbohydrates. To extend the uses of the fruit, the longan can be frozen, canned or dried. Longan fruit can be frozen in its skin in airtight containers. Upon thawing, the fruit can be used in similar manner as freshly picked fruit without any loss of quality. The fruit can be canned in its own juice with little or no added sugar. This is possible because the longan flesh contains a high level of soluble solids. For canning, cultivars with large fruit and small seed are preferably used. Canned longan retains their individual flavour better than litchi. Drying the fruit, either intact or after removal of the pericarp, is a practical way of preserving the longan fruit. The dried aril is black, leathery and smoky in flavour and is used mainly to prepare a refreshing drink. This drink is very popular among the Chinese. A liqueur is made by macerating the longan flesh in alcohol. Dried longan flesh is also an ingredient in herbal medicine used for stomach ache, insomnia and as an antidote for poison. Longans can be successfully grown in sub-tropical regions specially the litchi growing areas of this country and its production will be an aid to crop diversification. The benefits of high yielding ability and late harvesting season *i.e.*, August to September (after litchi) can also be enjoyed by the consumers consuming fresh longan. Processing/packaging industries will be engaged for longer period even after end of litchi harvesting season for processing and value addition. Proper R and D programmes, certainly will create new areas of its feasibility and fast area expansion. All these beneficial factors render longan fruit a highly nutritive potential crop for forthcoming period in the Indian subcontinent too.

References

Blumenfeld, A., Shaya, F. and Goren, M., 2000. Developing longan culture in Israel. Poster presented at the First International Symposium on Litchi and Longan, Guangzhou, China, 19–23 June, 2000.

Campbell, R.J. and Campbell, C.W., 2001b. Longan evaluation and selection in Florida., USA. *Acta Horticulturae* 558: 119–123.

Campbell, R.J. and Campbell, C.W., 2000. Evaluation and introduction of Litchi and Longan in South Florida. Paper presented at the First International Symposium on Litchi and Longan, Guangzhou, China, 19–23 June, 2000.

Chacko, E.K. and Downton, W.J.S., 1995. Current status and future potential for tropical fruit and nut crops in Australia. *In:* Commercial Production of Fruits, Vegetables and Flowers, Universiti Putra Malaysia, Serdang, pp. 163–182.

Chattopadhyay, P.K. and Ghosh, A., 1991. Studies on the physiology of fruit grown of longan. *Horticultural Journal*. 4: 27–31.

Chen, J.Z., 2003 Special demonstration of pomology (South section). China Agriculture Press Beijing.

Chen, L. Q., 2003. A experiment on application of $KClO_3$ to control shoot growth and promote flowers in longan. Guangxi Forestry Science 32: 209–210.

Chen, J.Y., Chen. J.Y., and Xu, X.D., 2001. Advances in research of longan witches' broom disease. pp. 413–416. In: Huang, H.B. and Menzel, C. (eds.) *Proceedings of the First International Symposium on Litchi and Longan*. Guangzhou, China, June 2000, ISHS Acta Horticulturae 558, 446 pp.

Chen, J.Y., Li, K.B., Chen, J.Y. and Fan, G.C., 1996. A preliminary study on litchi witches' broom and its relations to longan witches' broom. *Acta Phytopathologica Sinica* 26(4): 331–335. (In Chinese).

Chen, Jingying, Chen Jingyao and Xu, Xiudan 2000. Advances in the research of longan witches' broom disease. Paper presented at the First International Symposium on Litchi and Longan, Guangzhou, China, 19–23 June, 2000.

Coates, L.M., Sangchote, S., Johnson, G.I., and Sittigul, C., 2003. Diseases of lychee, longan and rambutan. pp. 307–325 In: Ploetz, R.C. (ed.) Diseases of Tropical Fruit Crops. CABI Publishing, Wallingford, UK. 527pp.

Crous, P.W. and Wingfield, M.J., 1993. A re–evaluation of *Cylindocladiella*, and a comparison with morphologically similar genera. *Mycological Research* 97 (4): 433–448.

Drinnan, J., 2004. Longans– Postharvest handling and storage. A report on the Rural Industries Research and Development Corporation (RIRDC). *RIRDC Publication No.* 30/125, RIRDC Canberra.

Farr, D.F., Bills, G.F., Chamuris, G.P. and Rossman, A.Y., 1989. *Fungi on Plants and Plant Products in the United States*. St. Paul, Minnesota, USA: American Phytopathological Society (APS) Press, 1252 pp.

Gu, X.L., Li, Y.P. and Liang, W.H., 2008. The survey of longan industry development in China. *Chinese Agricultural Science Bulletin*. 24: 470–474.

Groff, G. W., 1921. The Lychee and Longan. Orange Judd Company, New York, 188pp.

Han, D.M., Cai, C.H. and Wu, Z.X., 2001. Postharvest physiology and storage characteristics of longan fruit. *Guangdong Agricultural Science*.6: 20–21.

Han, D.M., Wu, Z.X, Chen, W.X. and Dai, H.F., 2002. Studies on the physiological and biochemical changes of longan (cv. Shixia and cv. Chuliang) pulp after harvest. *Journal of South China Agricultural University*. 23: 20–23.

Han, D.M., Wu, Z.X, Ji, Z.L., Cai, C.H,, Yi, G.J., Huo, H.Q. and Wang, X.R., 2000. Effects of SO_2 treatment on the overall quality of longan fruits (cv. 'Shixia') during cold storage. Paper presented at the First International Symposium on Litchi and Longan, Guangzhou, China, 19–23 June, 2000.

He, D.P., Zeng, M.L., Zhou B.P., Lin, S.X., Li, J.Y, and Peng, J.X., 2000. The cause of witches' broom in longan–discovery of a new gall–mite species. Paper presented at the First International Symposium on Litchi and Longan, Guangzhou, China, 19–23 June, 2000.

He, D.P., 2001. An overview of integrated management of insect pests in litchi orchards of Guangdong. pp. 401–405. In: Huang, H. and Menzel, C. (eds). *Proceedings of the First International Symposium on Litchi and Logan*. Guangzhou, China, June 2000. ISHS Acta Horticulturae 558. 446 pp.

http: / /www.google.co.in/hort.purdue.edu/longan+fruit+culinary+uses&aq/html#Food Uses

Hwang, J.S. and Hung, C.C., 1996. Gracillariid insect pests attacking litchi and longan in Taiwan. *Plant Protection Bulletin Taipei* 38(1): 75–78.

Huang, Huibai 1999. Personal communication.

Huang, H.B., 1995. Advances in fruit physiology of the arillate fruit litchi and longan. *Annual Review of Horticultural Science*. 1: 107–120.

Huang, Jinsong, Xu Xiudan, Zhen, Shaoquan and Xu, Jiahui 2000. Selection for aborted–seeded longan cultivars. Paper presented at the First International Symposium on Litchi and Longan, Guangzhou, China, 19–23 June, 2000.

Jian, S., John, S., Yueming, J., Sophia, J.X. and Xiaoyi, W., 2007. Identification of Two Polyphenolic Compounds with Antioxidant Activities in Longan Pericarp Tissues, *Journal of Agricultural and Food Chemistry*, 55 (14): 5864–5868.

Kaewsuksaeng, S.; Uthairattanakij, A.; Srilaong, V. and Kanlayanarat, S., 2007. High O_2 effects on physiological changes in longan (*Dimorcarpus longan* Lour) fruits. Asia Symposium on Quality Management in Postharvest Systems–Eurasia 2007. ISHS *Acta Horticulturae* 804: Europe.

Ke, Guangwu, Wang, Changchun and Tang, Zifa 2000. Palynological studies on the origin of longan cultivation. Paper presented at the First International Symposium on Litchi and Longan, Guangzhou, China, 19–23 June, 2000.

Ke, G.W., Wang, C.C. and Huang, J.H., 1992. The aril initiation and ontogenesis of longan fruit. *Journal of Fujian Academy of Agricultural Science*.7(1): 22–26

Lai, Z.X., 1997. Plant regeneration system of protoplast culture in longan. Ph.D. dissertation. Fujian Agricultural University. Fouzhou, China.

Lai, Z.X. and Chen, Z.G, 1998. Plant regeneration at high frequency from somatic embryos of longan. Journal of Fujian Agricultural University, Fouzhou, China. 27: 31–36.

Lai, Z.X. and Chen, Z.G, 1997. A study on protoplast fusion between longan and litchi and the primary culture. In Proceedings of the 5th Congress of the Fujian Society of Genetics, Fouzhou, China. P 38 (Abstract)

Lai, Z.X., Chen, C.L. and Chen, Z.G, 2000. Progress in biotechnology research in longan. Paper presented at the First International Symposium on Litchi and Longan, Guangzhou, China, 19–23 June, 2000.

Lai, Z.X., Chen, C.L. and Chen, Z.G, 2000. Somatic embryogenesis in longan (*Dimocarpus longan* Lour). *Forestry Sciences*. 67: 415–432.

Lai, Z.X., Chen, C.L. and Chen, Z.G, 2001. Progress in biotechnology research in longan (*Dimocarpus longan* Lour). ISHS *Acta Horticulturae*. 558: 137–141.

Leenhouts, P.W., 1971. A revision of *Dimocarpus* (Sapindaceae). Blumea 19: 113–131

Leenhouts, P.W., 1973. A new species of *Dimocarpus* (Sapindaceae) from Australia. Blumea 21: 377–380.

Li, Y.H., Chen, X.S. and Zheng, Z., 2003. Recent progress in the studies of distant hybridization breeding in fruit crops. *Journal of Shandong Agricultural University*. 34(1): 139–143.

Liao, J.Q., Cai, S.M. and Zhu, D.B., 2006. The selective breeding of longan with novel late ripening cultivar and large fruit size. *China Fruits*. 48: 7–8.

Lin, H.T., Chen, S.J. and Hong, Q.Z., 2000. A study of the shelf–life of cold–stored longan fruits. Paper presented at the First International Symposium on Litchi and Longan, Guangzhou, China, 19–23 June, 2000.

Lin, H.T., Xi, Y.F., Chen, S.J., Chen, J. Q. and Hong, Q.Z., 2000d. Research advances of postharvest physiology, postharvest pathology and storage and transport technologies for longan fruit. *Transactions of the Chinese Societies of Agricultural Engineering*. 18: 185–190.

Lin, H.T., Xi, Y.F., Chen, S.J., Chou, R. S., Wu, Y.J. and Li, W.L., 2002e. Effect of storage temperature on postharvest physiology and quality of longan fruit. *Transactions of the Chinese Societies of Agricultural Engineering*. 18(supplement): 115–120.

Lin, S.M. and Chi, P.K., 1992. Some new species and new records of genus *Phomopsis* in China. Journal of South China Agricultural University 13: 93–97.

Litz, R.E., 1988. Somatic embryogenesis from cultured leaf explants of the tropical tree *Euphoria longan* Stend. Journal of Plant Physiology 132: 190–193.

Liu, Weiwen, Huang, Fengxin and Luo, Junsheng 2000. A superior quality longan cultivar–'Chuliang'. Paper presented at the First International Symposium on Litchi and Longan, Guangzhou, China, 19–23 June, 2000.

Liu, X. H. and Ma, C.L., 2001. Production and research of longan in China. *Acta Horticulturae* 558: 73–82.

Liu, A.Q., Huang, G.D., Wu, C.K., Zhang, C.L. and Tang, L.H., 2001. Dynamics of *Adoxophyes cyrtosema* Meyrick in fields and its control. *Plant Protection* 27(1): 25–26.

McConchie, C.A., Vithanage, V. and Batten, D.J., 1994. Intergeneric hybridization between litchi (*Litchi chinensis* Sonn.) and longan (*Dimocarpus longan* Lour). *Annals of Botany* 74: 111–118.

Menzel, C., 2002. *The Lychee Crop in Asia and the Pacific*. Food and Agriculture Organization of the United Nations. Regional Office for Asia and the Pacific. Bangkok, Thailand. RAP Publication: 2000/16. http: / /www.fao.org/DOCREP/005/ AC681E/ac681e00.htm#Contents

Menzel, C.M., Watson, B.J. and Simson, D.R 1989. Longan: a place in Queensland's horticulture? *Queensland Agricultural Journal* 115: 251–265.

Menzel, C.M., Watson, B.J. and Simpson, D.R., 1990. Longan. *In:* T.K. Bose and S.K. Mitra (eds.) Fruits: Tropical and Subtropical. Naya Prokash, Calcutta, India, pp.522–546.

Morton, J., 1987. Longan. p. 259–262. In: Fruits of warm climates. Julia F. Morton, Miami, FL.

Nguyen, Minh Chau 2000. Personal communication.

Noller, J., 2001. Australlian Longans– Market opportunity identification and analysis of the domestic market. A report on the Rural Industries Research and Development Corporation (RIRDC). *RIRDC Publication No.* 01/25, RIRDC Canberra.

Pan, D.M.; Zhong, F.L.; Guo, Z.X.; Lin, L. and Zheng, Z.C. 2010. Analysis of Dongbi Longan with isoenzyme and RAPD technologies Proc. 3rd International Symposium on Longan, Litchi and Other Fruit trees in Sapindaceae Family. ISHS *Acta Horticulturae* 863:

Paull, R.E. and Chen, N.J., 1987b. Changes in longan and rambutan during postharvest storage. *Hort. Science*. 22: 1303–1304

Poerwanto, R., 2005. Rambutan and longan production in Indonesia. Proc. 2nd Inernational Symposium on Litchi, Longan, Rambutan and Other Sapindaceae plants. ISHS *Acta Horticulturae* 665: 81–86

Qiu, J.D., Luo, X.Z. and Wu, D.Y., 2000. Flower bud differentiation of longan (*Dimocarpus longan* Lour) and its regulation. Paper presented at the First International Symposium on Litchi and Longan, Guangzhou, China, 19–23 June, 2000.

Rangkadilok, N., Worasuttayangkurn, L., Bennett, R.N. and Satayavivad, J., 2005. Identification and Quantification of Polyphenolic Compounds in Longan (*Euphoria longana* Lam.) Fruit. Journal of Agricultural and Food Chemistry, 53 (5): 1387–1392.

Sardsud, V., U. Sardsud, C. Sittigul and T. Chaiwangsri., 1992. The effects of post–fumigation washing treatments and storage temperature on disease development in fresh longan.(In): ACIAR Proj. 8844, Wkshp Postharv. Hand. Trop. Fruit, Bangkok, Thailand, pp. 21.

Schuetz, P., Sauerborn, J., Martin, K. and Hengsawad, V., 2002. Consequences of pesticide use and weed management to arthropod communities in litchi orchards in northern Thailand. *International Symposium: Sustaining Food Security and Managing Natural Resources in Southeast Asia–Challenges for the 21st Century*. Chiang Mai, Thailand, 8–11 January 2002.

Shiping Tian, Yong Xu, Aili Jiang and Qinqin Gong 2002. Physiological and quality responses of longan fruit to high O_2 or high CO_2 atmospheres in storage. *Postharvest Biology and Technology*. 24(3): 335–340

Singh, N, Campbell, T.P. and Nicholls, B., 2000. The Australian longan industry. Paper presented at the First International Symposium on Litchi and Longan, Guangzhou, China, 19–23 June, 2000.

Subhadrabandhu, S., 1990. Lychee and Longan Cultivation in Thailand. Rumthai Publication, Bangkok, Thailand. 40 p.

Subhadrabandhu, S., 2000. Personal communication.

Subhadrabandhu, S. and Yapwattanaphun, C., 2000a. Lychee and longan production in Thailand. Paper presented at the First International Symposium on Litchi and Longan, Guangzhou, China, 19–23 June, 2000.

Subhadrabandhu, S. and Yapwattanaphun, C., 2001. Regulation of flowering time for off–season production of longan in Thailand. Paper presented at the First International Symposium on Litchi and Longan, Guangzhou, China, 19–23 June, 2000. ISHS *Acta Horticulturae* 558: 193–198

Tongdee, S. C., 1992. Postharvest handling and technology of tropical fruit. *Acta Horticulturae* 321: 713–717.

Van Welzen, P.C., Lamb, A. and Wong, W.W.W., 1988. Edible Sapindaceae in Sabah. Nature Malaysiana 13: 10–25.

Visitpanich, J., Sittigul, C. and Chanbang, Y., 2000. Longan and leaf blight and fruit drop. *House Agricultural Magazine* (Bangkok, Thailand). 24(1): 144–148.

Waite, G.K., 1999. New evidence further incriminates honey–bees as vectors of lychee erinose mite *Aceria litchii* (Acari: Eriophiidae). *Experimental and Applied Acarology* 23(2): 145–147.

Waite, G.K. and Hwang, J.S., 2002. Pests of Litchi and Longan. Chapter 11. In: Pena, J.E., Sharp, J.L. and Wysoki, M. (eds.) *Tropical Fruit Pests and Pollinators: Biology, Economic Importance, Natural Enemies and Control.* CABI Publishing, Wallingford, UK, 430 pp.

Wall, Marisa M., 2006. Ascorbic acid and mineral composition of longan (*Dimocarpus longan*), lychee (*Litchi chinensis*) and rambutan (*Nephelium lappaceum*) cultivars grown in Hawaii. *Journal of food composition and analysis.*, 19(6–7): 655–663.

Wang, Jiafu 2000. Shoot tip culture and production of vims–free plantlets of longan. Paper presented at the First International Symposium on Litchi and Longan, Guangzhou, China, 19–23 June, 2000.

Wei, W., 1990. Longan: anther and embryo culture (In) Chen, Z., Evans, D.A., Sharp, W.R., Ammiarto, P.V. and Sondahl, M.R. (eds). Handbook of Plant Cell Culture. Vol.–6. Perennial Crops. McGraw Hill, New York. Pp–385–396.

Wong, K.C., 2000. Variations of local longan in Malaysia. Paper presented at the First International Symposium on Litchi and Longan, Guangzhou, China, 19–23 June, 2000.

Wong, K.C. and Kesta, S., 1991. *Dimocarpus longan* Lour. In Verheji, E.W.M. and Corone, R.E. (eds.). Plant Resources of South East Asia No.2 Edible Fruit and Nuts. Pudoc Wageningen. The Netherlands. Pp–146–151.

Wong, K.C. and Gan, Y.Y., 1992. The diversity of *Dimocarpus longan* ssp. *malesianus* var. *malesianus* in Sarawak. Acta Horticulturae, 292: 29–39.

Wong, K.C., 2000. Longan production in Asia. FAO Regional Office for Asia and the Pacific. Bangkok, Thailand

Wong, K. C. & Saichol, K., 1991. *Dimocarpus longan* Lour. (*In*): E.W.M. Verheij and R.E. Coronel (eds.) Edible fruits and nuts. Plant Resources of Southeast Asia, Pudoc, Wageningen. Pp. 146–151.

Yang, Y.Q. and Wei, W.X., 1984. Induction of anther derived plantlet in longan. *Acta Genetica Sinica* 11: 288–293

Yen, Chung–Ruey 2000. Personal communication.

Yen, Chung–Ruey and Chang, Jer–Way 1991, Variation of fruit characters among longan (*Euphoria longana* Lam) varieties in Taiwan Province of China. J. Chinese Soc. Hort. Sci., 37 (I): 21–34. (in Chinese with English summary).

Yonemoto, Y., Choudhary, A.K. and Kato, H., 2006. Cultivar identification and their genetic relationships in *Dimocarpus longan* subspecies based on RAPD markers. *Sci. Hort*. 109: 147–152.

Zeng, L.H., Chen, Z.G., Lu, L.X., Zhou, P. and Zheng, X.Q., 2001. *In vitro* transformation mediated by Ri plasmid of *Agrobacterium* rhozogenes and transgenes plant regeneration of longan. ISHS *Acta Horticulturae*. 558: 149–154.

Zhang, D. and P. C. Quantick., 1997. Preliminary study on the effects of modified atmosphere packaging on postharvest storage of longan fruit. Proc. 7th Intl. Cont. Atmos. Res. Conf., Univ. Calif, Davis CA.

Zhang, Q., and Zhang, Q., 1999. Investigation of the occurrence of longan witch–broom and its control. *South China Fruits* 28: 24.

Zhong, F.L., 2006. POD isoenzyme and RAPD analysis of longan (*Dimocarpus longana* L.) genetic resource. D. Fuzhou: Fujian Agriculture and Forestry University 26–27.

Zhu, W.S., Huang, H.Y., Huang, T.L., Lei, H.D. and Jiang, Y.H. (eds)., 1994. *The Handbook of Diseases and Pests of Fruits in Southern China*. Agricultural Press, Beijing 258 pp.

Zeng, L.H., Chen, Z.G., Lu, L.X., Zhou, P. and Zheng X.Q., 2000. In vitro transformation mediated by Ri plasmid of *Agrobacterium rhizogenes* and transgenic plant regeneration of longan. Paper presented at the First International Symposium on Litchi and Longan, Guangzhou, China, 19–23 June, 2000.

Zhang, Z. (ed.) 1997. *Litchi Pictorial Narration of Cultivation*. Pomology Research Institute, Guangdong Academy of Agricultural Science, 189 pp. (In Chinese).

2011, Future Crops Vol. 1 ***Pages 143–171***
***Editor:* Professor K.V. Peter**
***Published by:* DAYA PUBLISHING HOUSE, NEW DELHI**

Chapter 9
NONI
(*Morinda citrifolia* L.)

D.R. Singh, Shrawan Singh and R.C. Srivastava
Central Agricultural Research Institute,
Port Blair – 744 101, Andaman and Nicobar Islands
E-mail: drsingh1966@yahoo.com

The genus *Morinda* to which Noni belongs comprises approximately 154 species distributed all over tropical regions of the world. These species are trees, shrubs or vines. *Morinda citrifolia* known vernacularly as noni or Indian mulberry is a small evergreen shrub or tree growing from 2 to 6 m. tall and belongs to family Rubiaceae. Fruits are green until maturity. The fruit is covered with reddish-brown eyes which contain seeds. Number of seeds/fruit varies from 150-200 (Singh *et al.*, 2007).

Noni fruit has made a remarkable transition from a traditional Polynesian herbal medicinal plant to a modern natural remedy over the last a few decades and has become incredibly popular even before sufficient evidence accumulated to establish its efficacy according to biomedical criteria. Noni juice has become increasingly popular in recent years as a health food drink in Western and Asian countries where there are substantial markets. Over 200 food and beverage companies came in the market with different products like beverages (ready to serve drinks), fruit powders (for manufacture of reconstituted juice or juice drink products made from dried ripe or unripe fruits), toiletries (lotions, soaps, etc.), oil (from seeds), leaf powders (for encapsulation or pills), biopsesticides, etc., although the fruit juice being the predominant form.

The various products specially 'Noni juice' is being promoted for its nutraceutical and therapeutic properties thereby emerging as a major product of health and wellness industry. It has recently been the object of many claims concerning its therapeutic properties. One of popular claims is by Heinicke (1985) who claimed that noni fruit has presence of an active "alkaloid" namely xeronine and said to derive from a precursor perxeronine, for which no structure was given. He described a wide range of potential uses for noni juice including against high blood pressure, menstrual crams, hypertension, gastric ulcers, sprains, injuries, mental depression, atherosclerosis, blood vessel problems, drug addiction, relief of pain and many others. Many publications have reported that noni juice can be used to relieve different diseases and disorders. The observed beneficial effects result from compounds like alkaloids, scopoletin, damnacanthal and other reported molecules. As a result, consumption of this fruit is currently high, not only in USA but also in Japan, Europe and India. Although, reliable sale figures are not available, it is claimed that the market has reached US $ 1.3 billion in annual sales (Olivier and Matthias, 2007). A few companies produce flavored Noni juices by addition of other fruit juices to render the products more embellish and acceptable in markets.

In response to this demand, countries including India have increased the commercial cultivation of noni. Recently, this plant has come up as a source of livelihood for farmers of Andaman and Nicobar Islands after NoniBiotech Pvt.Limited,a Chennai based company entered into an agreement of contract farming with them (Singh *et al.*, 2005). Noni juice has already been accepted in the European Union as a novel food (European Food Safety Authority, 2002). Despite the real market opportunities, there has been little scientific research to review the actual nutritional and functional properties of noni products. This prompted a concomitant increase in research on the phytochemical constituents and biological activity.

Nutritional Value of Noni

Proximate Composition

Juice content in Morinda fruits is the most economically useful parameter and influenced by genotypes, climatic and soil factors and cultural practices. It was the highest in Andaman collection HD-6a (65.5 per cent). The TSS content has negative correlation with juice content as the highest TSS value was recorded in JGH-5 (9.0°B) (Singh *et al.*, 2011). The ripe fruit contains 90 per cent water and the pH is 3.72. The edible part of the fruit has a high level of soluble dietary fiber (9.8µ0.4 per cent), soluble solids (8°Brix), proteins (2.5 per cent) and low level of lipids (0.15 per cent). Singh *et al.* (2011) estimated the highest protein in MEM-1(5.5 per cent), crude fibre in MEM-1 (10 per cent), fat in MAN-1 (0.24 per cent) and free sugar in ABH-1 (13.0mg/100g).

Glucose and fructose are more abundant monosaccharides in fruit (Chunhieng, 2003) while hexose and pentose are reported in root bark. Chemical composition of unfermented noni juice contains approximately 10 per cent dry matter consisting mainly of glucose and fructose (3-4 per cent each), protein (0.2-0.5 per cent) and lipids (0.1-0.2 per cent). An ethanol-insoluble, high molecular weight fraction of fruit juice is primarily composed of monosaccharides like arabinose, galactose, galacturonic acid, rhamnose and pectic polysaccharides *viz.* homogalacturonan, rhamnogalacturonan I, arabinan, type I and type II arabinogalactan (Bui *et al.*, 2006).

Micro-nutrient Analysis

The major minerals in Noni fruits are potassium (30-150mg/100g), calcium (20-25mg/100g), sodium (15-40mg/100g) and magnesium (3-12mg/100g) (Shovic and Whistler 2001). Chunhieng (2003) reported higher contents of potassium (3900 mg/100g), sodium (214 mg/100g), calcium (28 mg/100g) and magnesium (14 mg/100g). Noni germplasm of Andamans contains comparatively higher levels of micronutrients like Cu in HBAY-11 (95.44 ppm/100g), Zn in MANJ-9 (984.86 ppm/100g), Mn in JGH- 5 (739.88 ppm/100g), Fe in CAL-10 (8168.14 ppm/100g), K in MEM-3 (1182 ppm/100g), Ca in HD-6 (9352.98 ppm/100g) and Mg in PBAY-7 (3036.76 ppm/100g) (Singh *et al.*, 2011).

The fruit juice contains a large concentration of anthraquinone and a variety of other phytochemicals as summarized in Table 9.1. A number of anthraquinones and other class of molecules

Figure 9.1: Noni tree.

Figure 9.2: Noni fruits.

are reported for their biological activity (Table 9.2). The chemical structures of these compounds are given in Figure 9.3. Several studies showed that cell cultures are capable of accumulating substantial amounts of anthraquinones at a higher level than those in native plants through optimization of cultural conditions. This technique has immense potential for commercial exploitation of these valuable anthraquinones (Lucilla and Johannes, 1995; Lucilla *and* Johannes,1995 and Marc, *et al.*, 2003).

Table 9.1: Isolated compounds from *Morinda citrifolia*.

Name	*Class*	*Plant Part*	*References*
(–)-Pinoresinol	Lg	Fruits	Deng *et al.*, 2007
(+)-3,4,3',4'-Tetrahydroxy-9,7'α-epoxylignano-7 α,9'-lactone	Lg	Fruits	Deng *et al.*, 2007
(*E*)-6-Dodeceno-γ-lactone	L	Fruits	Farine *et al.*, 1996
(*Z*)-6-Dodeceno- γ–lactone	L	Fruits	Farine *et al.*
(*Z,Z*)-2,5-Undecadien-1-ol	Ao	Fruits	Farine *et al.*
(*Z,Z,Z*)-8,11,14-Eicosatrienoic acid	A	Fruits	Farine *et al.*
1,3α,4,7a-Tetrahydro-6-(hydroxymethyl)-3*H*-furo [3,4-*c*]pyran-4-carboxylic acid	I	Fruits	Su *et al.*, 2005
1,6-dihydroxy-5-methoxy-2-methoxymethylanthraquinone	Aq	Fruits	Lin *et al.*, 2005
1,5,15-Trimethylmorindol	Aq	Leaves	Takashima et *al.*, 2007
1,5,15-tri-*O*-methylmorindol	Aq	Fruits	Akihisa *et al.*, 2007
1, 5, 7-trihydroxy-6-methoxy-2-methoxymethylanthraquinone	Aq	Fruits	Lin *et al.*, 2005
6,10,14-Trimethyl-2-pentadecanone	K	Leaves	Brett *et al.*, 2008
13^2(*R*)-Hydroxypheophorbide a methyl ester	Cd	Leaves	Takashima *et al.*, 2007
13^2(*S*)-Hydroxypheophorbide a methyl ester	Cd	Leaves	Takashima *et al.*, 2007
13-*epi*-Phaeophorbide a methyl ester	Cd	Leaves	Takashima *et al.*, 2007
13-Hydroxy-9,11,15-octadecatrienoic acid	A	Leaves	Takashima *et al.*, 2007
15^1(*R*)-Hydropurpurin-7 lactone dimethyl ester	Cd	Leaves	Takashima *et al.*, 2007
15^1(*S*)-Hydropurpurin-7 lactone dimethyl ester	Cd	Leaves	Takashima *et al.*, 2007
1-Butanol	Ao	Fruits	Farine *et al.*, 1996
1-Hexanol	Ao	Fruits	Farine *et al.*, 1996
1-Hydroxy-2-methylanthraquinone	Aq	Roots	Sang and Ho, 2006
1-Methoxy-3-hydroxyanthraquinone	Aq	Roots	Sang and Ho, 2006
1-*n*-Butyl-4-(5'-formyl-2'-furanyl) methyl succinate	E	Fruits	Samoylenko *et al.*, 2006
1-*n*-Butyl-4-methyl-2-hydroxysuccinate	E	Fruits	Samoylenko *et al.*, 2006
1-*n*-Butyl-4-methyl-3-hydroxysuccinate	E	Fruits	Samoylenko *et al.*, 2006
1-*O*-(3'-Methylbut-3'-enyl)-β-D-glucopyranose	S	Fruits	Samoylenko *et al.*, 2006
1-Palmitin	Mc	Fruits	Pawlus *et al.*,2005
1,2-Dihydro-1,1,6-trimethyl-naphthalene	K	Leaves	Brett *et al.*, 2008
α-Ionone	K	Leaves	Brett *et al.*, 2008
19*a*-Methylursolic acid	Ts	Fruits	Sang *et al.*, 2002
24-Methylenecholesterol	Ts	Cell culture	Dyas *et al.*, 1994

Contd...

Table 9.1–Contd...

Name	*Class*	*Plant Part*	*References*
24-Methylenecycloartanol	Ts	Cell culture	Dyas *et al.*, 1994
24-Methylenecycloartanyl linoleate	Ts	Cell culture	Dyas *et al.*, 1994
2-*O*-(β-D-glucopyranosyl)-1-*O*-hexanoyl-β-D-gluropyranose	E	Fruit	Akihisa *et al.*, 2007
2-*O*-(β-D-glucopyranosyl)-1-*O*-octanoyl-β-D-gluropyranose	E	Fruit	Akihisa *et al.*, 2007
2-Ethoxymethyl-3-methoxy-1,5,6-trihydroxyanthraquinone	Aq	Cell culture	Leistner, 1975
2-Formylanthraquinone	Aq	Roots	Sang and Ho, 2006
2-Heptanone	K	Fruits	Farine *et al.*, 1996
2-Hydroxy-1-methoxy-7-methylanthraquinone	Aq	Roots	Rusia and Srivastava,1989
2-Hydroxyanthraquinone	Aq	Stems	Siddiqui *et al.*, 2006
2-Methoxy-1,3,6-trihydroxyanthraquinone	Aq	Fruits	Pawlus *et al.*, 2005
2-Methoxyanthraquinone	Aq	Stems	Siddiqui *et al.*, 2006
2-methyl-3,5,6-trihydroxyanthraquinone	Aq	Cell culture	Kenichiro *et al.*, 1981
2-Methyl-3,5,6-trihydroxyanthraquinone-6-β-primeveroside	Ag	Cell culture	Kenichiro *et al.*, 1981
2-Methylbutanoic acid	A	Fruits	Farine *et al.*, 1996
2-Methylpropanoic acid	A	Fruits	Farine *et al.*, 1996
2,5-dihydroxy-4-methoxybenzaldehyde	Ad	Fruits	Lin *et al.*, 2007
2,6,10,14,18,22-Tetracosahexaene	Ae	Leaves	Deng *et al.*, 2007
3,19 Dihydroxyursolic acid	Ts	Fruits	Sang *et al.*, 2002
3,3'-Bisdemethylpinoresinol	Lg	Fruits	Deng *et al.*,2007,29,39,
3,3'-Bisdemethyltanegool	Lg	Fruits	Deng *et al.*,2007
3-Hydroxy-2-butanone	K	Fruits	Farine *et al.*, 1996
3-Hydroxymorindone	Aq	Cell culture	Kenichiro *et al.*, 1981
3-Hydroxymorindone-6-β- primeveroside	Ag	Cell culture	Kenichiro *et al.*, 1981
3-Methyl-2-buten-1-ol	Ao	Fruits	Farine *et al.*,,1996
3-Methyl-3-buten-1-ol	Ao	Fruits	Farine *et al.*, 1996
3-Methylthiopropanoic acid	A	Fruits	Farine *et al.*, 1996
3-*O*-Acetylpomolic acid	T & S	Leaves	Takashima *et al.*, 2007
4-*epi*-Borreriagenin	I	Fruits	Samoylenko *et al.*, 2006
4-hydroxy-3-methoxybenzaldehyde	Ad	Fruits	Lin *et al.*, 2007
4-Hydroxy-3-methoxycinnamaldehyde	Ad	Fruits	Lin *et al.*, 2007; Pawlus *et al.*, 2005
5-Methylfurfural	Ad	Leaves	Brett and Bing-nan, 2008
5,15-Dimethylmorindol	Aq	Fruits, leaves	Takashima *et al.*, 2007; Kamiya *et al.*, 2005
5,6-Dihydroxylucidin	Aq	Cell culture	Kenichiro *et al.*, 1981
5,6-Dihydroxylucidin-3-β- primeveroside	Ag	Cell culture	Kenichiro *et al.*, 1981
5,7-Dimethyl apigenin 4'-*O*-β-D-galactopyranoside	F	Flowers	Singh and Tiwari, 1976

Contd...

Table 9.1–Contd...

Name	*Class*	*Plant Part*	*References*
6,8-Dimethoxy-3-methylanthraquinone-α-L-*O*-β-rhamnosyl glucopyranoside	Ag	Flowers	Tiwari and Singh, 1977
6α-Hydroxyadoxoside	I	Fruits	Su *et al.*, 2005
6β,7β-Epoxy-8-*epi*-splendoside	I	Fruits	Su *et al.*, 2005
6-Hydroxyanthragallol-1,3-di-*O*-methyl ether	Aq	Fruits	Kamiya *et al.*, 2005
6-Methyl-anthrapurpurin	Aq	Cell culture	Kenichiro *et al.*, 1981
α- and β-Glucose	S	Fruits	Samoylenko *et al.*, 2006; Levand and Larson, 1979
β-Ionone	K	Leaves	Brett and Bing-nan,2008
Acacetin 7-*O*-β-D-glucopyranoside	F	Flowers	Singh and Tiwari, 1976
Acetic acid	A	Fruits	Farine *et al.*, 1996
Alizarin	Aq	Cell culture, heartwood	Leistner, 1975; Thomson, 1971
Alizarin 1-*O*-methyl ether	Aq	Roots	Pawlus *et al.*, 2005; Simonsen, 1920
Americanin A	Lg	Fruits	Lin *et al.*, 2007; Su *et al.*, 2005; Kamiya *et al.*, 2004
Americanin D	Lg	Fruits	Lin *et al.*, 2007
Americanoic acid	Lg	Fruits	Kamiya *et al.*, 2004
Americanol A	Lg	Fruits	Kamiya *et al.*, 2004
Anthragallol 1,2-di-*O*-methyl ether	Aq	Roots	Thomson, 1971
Anthragallol 1,3-di-*O*-methyl ether	Aq	Fruits	Kamiya *et al.*, 2005
Anthragallol 2,3-di-*O*-methyl ether	Aq	Heartwood	Thomson, 1971
Anthragallol 2-*O*-methyl ether	Aq	Fruits	Pawlus *et al.*, 2005; Kamiya *et al.*, 2005
Ascorbic acid	A	Fruits	Presada *et al.*, 1990
Asperuloside	I	Fruits, leaves	Su *et al.*, 2005; Sang[b] *et al.*, 2001
Asperulosidic acid	I	Fruits, leaves	Su *et al.*, 2005; Samoylenko *et al.*, 2006; Kamiya *et al.*, 2005; Sang[b] *et al.*, 2001; Wang *et al.*, 1999
Asperulosidic acid methyl ester	I	Fruits	Sang *et al.*, 2002
Austrocortinin	Aq	Fruits	Kim *et al.*, 2005
Balanophonin	Lg	Fruits	Pawlus *et al.*, 2005
Barbinervic acid	Ts	Leaves	Takashima *et al.*, 2007
β-Carotene	C	Leaves	Aalbersberg *et al.*, 1993
Benzaldehyde	Ald	Leaves	Brett and Bing-nan,2008

Contd...

Table 9.1–Contd...

Name	*Class*	*Plant Part*	*References*
Benzoic acid	A	Fruits	Farine *et al.*, 1996
Benzyl alcohol	Ao	Fruits	Farine *et al.*, 1996
Benzeneacetaldehyde	Ad	Leaves	Brett and Bing-nan,2008
Borreriagenin	I	Fruits	Su *et al.*, 2005; Kamiya *et* al., 2005; Schripsema *et al.*, 2006
β-Sitosterol	Ts	Leaves, cell culture	Pawlus *et al.*, 2005; Dyas *et al.*, 1994; Ahmad and Bano 1980; Saludes *et al.*, 2002
β–Hydroxypropiovanillone	Mc	Fruits	Pawlus *et al.*, 2005
Butanoic acid	A	Fruits	Farine *et al.*, 1996
Campesta-5,7,22-trien-3β-ol	Ts	Leaves	Saludes *et al.*, 2002
Campesterol	Ts	Cell culture	Dyas *et al.*, 2004
Campesteryl linoleate	Ts	Cell culture	Dyas *et al.*, 2004
Campesteryl palmitate	Ts	Cell culture	Dyas *et al.*, 2004
Citrifolinin A-1	I	Leaves	Sang *et al.*, 2003
Citrifolinin Ba	I	Leaves	Sang *et al.*, 2003
Citrifolinin Bb	I	Leaves	Sang *et al.*, 2003
Citrifolinoside A	I	Leaves	Sang *et al.*, 2003
Citrifoside	I	Leaves	Takashima *et al.*, 2007
Clethric acid	Ts	Leaves	Takashima *et al.*, 2007
Cycloartenol	Ts	Cell culture, leaves	Dyas *et al.*, 1994; Saludes *et al.*, 2002
Cycloartenyl linoleate	Ts	Cell culture	Dyas *et al.*, 1994
Cycloartenyl palmitate	Ts	Cell culture	Dyas *et al.*, 1994
Cytidine	N	Fruits	Sang *et al.*, 2002; Su *et al.*, 2005
Damnacanthal	Aq	Heartwood, roots	Thomson, 1971; Hiramatsu *et al.*, 1993
Damnacanthol	Aq	Roots	Sang and Ho, 2006; Thomson, 1971
Damnacanthol-3-*O*-β-D-primeveroside	Aq	Fruits	Kamiya *et al.*, 2008
Deacetylasperuloside	I	Fruits	Su *et al.*, 2005; Takashima *et al.*, 2007
Deacetylasperulosidic acid	I	Fruits	Samoylenko *et al.*, 2006; Kamiya *et al.*, 2005
Deacetylasperulosidic acid methyl ester	I	Fruits	Sang *et al.*, 2002
Decanoic acid	A	Fruits	Farine *et al.*, 1996

Contd...

Table 9.1–Contd...

Name	*Class*	*Plant Part*	*References*
Dehydroepoxymethoxygaertneroside	I	Leaves	Schripsema *et al.*, 2006; Sang *et al.*, 2001
Dehydromethoxygaertneroside	I	Fruits	Su *et al.*, 2005
Elaidic acid	A	Fruits	Farine *et al.*, 1996
Ethyl decanoate	E	Fruits	Farine *et al.*, 1996
Ethyl hexanoate	E	Fruits	Farine *et al.*, 1996
Ethyl octanoate	E	Fruits	Farine *et al.*, 1996
Ethyl palmitate	E	Fruits	Farine *et al.*, 1996
Ethylthiomethyl benzene	Mc	Fruits	Farine *et al.*, 1996
Eugenol	Ao	Fruits	Farine *et al.*, 1996
E-Phytol	Mc	Leaves	Brett and Bing nan, 2008
Geranyl acetone	Ac	Leaves	Brett and Bing nan, 2008
Hederagenin	*Ts*	*Leaves*	*Takashima et al., 2007*
Heptanoic acid	A	Fruits	Farine *et al.*, 1996
Hexanamide	Mc	Fruits	Farine *et al.*, 1996
Hexanedioic acid	A	Fruits	Farine *et al.*, 1996
Hexanoic acid	A	Leaves	Legal *et al.*, 1999
Ibericin	Aq	Cell culture, Roots	Sang and Ho, 2006
Isoamericanoic acid A	Aq	Fruits	Lin *et al.*, 2007
Isofucosterol	Ts	Cell culture	Dyas *et al.*, 1994
Isofucosteryl linoleate	Ts	Cell culture	Dyas *et al.*, 1994
Isoprincepin	Lg	Fruits	Kamiya *et al.*, 2004
Isoscopoletin	Mc	Fruits	Deng *et al.*,2007
Kaempferol	F	Fruits	Deng *et al.*,2007
Kaempferol 3-*O*-β-D-glucopyranosyl-(1→2)-[α-l-rhamnopyranosyl-(1→6)]-β-D-galactopyranoside	F	Leaves	Sang[a] *et al.*, 2001; Shixin *et al.*, 2008
Kaempferoi-3-*O*-α-l-rhamnopyranosyl-(1 → 6)-β-d-glucopyranoside	F	Leaves	Sang[a] *et al.*, 2001; Shixin *et al.*, 2008
Lauric acid	A	Fruits	Farine *et al.*, 1996
Limonene	Mc	Fruits	Farine *et al.*, 1996
Linoleic acid	A	Fruits	Farine *et al.*, 1996
Linoleic acid	A	Leaves	Brett and Bing nan,2008
Lucidin	*Aq*	*Cell culture*	*Kenichiro et al., 1981*
Lucidin ϖ-methyl ether	Aq	Cell culture	Kenichiro *et al.*, 1981
Lucidin 3-*O*-β-D-primeveroside	Ag	Fruits	Kamiya *et al.*, 2008
Lucidin-3-β-primeveroside	Ag	Cell culture	Kenichiro *et al.*, 1981
Luteolin	Aq	Fruits	Lin *et al.*, 2007

Contd...

Table 9.1–Contd...

Name	*Class*	*Plant Part*	*References*
Methyl 3-methylthio-propanoate	E	Fruits	Farine *et al.*, 1996
Methyl α-D-fructofuranoside	S	Fruits	Su *et al.*, 2005
Methyl β-D-fructofuranoside	S	Fruits	Su *et al.*, 2005
Methyl decanoate	E	Fruits	Farine *et al.*, 1996
Methyl elaidate	E	Fruits	Farine *et al.*, 1996
Methyl hexanoate	E	Fruits	Farine *et al.*, 1996
Methyl octanoate	E	Fruits	Farine *et al.*, 1996
Methyl oleate	E	Fruits	Farine *et al.*, 1996
Methyl oleate	E	Leaves	Brett and Bing-nan, 2008
Methyl palmitate	E	Fruits	Farine *et al.*, 1996
Methyl plamitate	E	Leaves	Brett and Bing nan,2008
Methyl pheophorbide a	Cd	Leaves	Takashima *et al.*, 2007
Methyl pheophorbide b	Cd	Leaves	Takashima *et al.*, 2007
Morenone-1	Aq	Roots	Jain and Srivastava, 1992
Morenone-2	Aq	Roots	Jain and Srivastava, 1992
Morindicininone	Aq	Stems	Siddiqui *et al.*, 2006
Morindicinone	Aq	Stems	Siddiqui *et al.*, 2006
Morindin			
	Ag	Cell culture	Kenichiro *et al.*, 1981; Simonsen, 1920
Morindolin	Lg	Fruits	Kamiya *et al.*, 2004
Morindone	Aq	Heartwood, root, cell culture	Leistner, 1975; Kenichiro *et al.*, 1981; Thomson, 1971; Simonsen, 1920; Srivastava and Singh, 1993
Morindone-5-*O*-methyl ether	Aq	Fruits	Pawlus *et al.*, 2005; Kamiya *et al.*, 2005
Morindone-6-*O*-β-D-primeveroside	Aq	Fruits	Kamiya *et al.*, 2008
Morindone-6-β-primeveroside	Aq	Cell culture	Kenichiro *et al.*, 1981
Myristic acid	A	Fruits	Farine *et al.*, 1996
Narcissoside	F	Fruits	Su *et al.*, 2005
Nicotifloroside	F	Fruits, leaves	Su *et al.*, 2005; Sanga *et al.*, 2001
Nonanoic acid	A	Fruits	Farine *et al.*, 1996
Nonioside A	S	Fruits	Wang *et al.*, 2000; Dalsgarrd *et al.*, 2006
Nonioside B	S	Fruits	Wang *et al.*, 1999; Dalsgarrd *et al.*, 2006

Contd...

Table 9.1–Contd...

Name	Class	Plant Part	References
Nonioside C	S	Fruits	Wang *et al.*, 2000; Dalsgarrd *et al.*, 2006
Nonioside D	S	Fruits	Wang *et al.*, 2000;
Nonioside E	S	Fruits	Dalsgarrd *et al.*, 2006
Nonioside F	S	Fruits	Dalsgarrd *et al.*, 2006
Nonioside G	S	Fruits	Dalsgarrd *et al.*, 2006
Nonioside H	S	Fruits	Dalsgarrd *et al.*, 2006
Nordamnacantha	Aq	Root, cell culture	Sang and Ho, 2006; Leistner, 1975; Thomson, 1971
Octanoic acid	A	Fruits	Farine *et al.*, 1996
Oleanolic acid	Ts	Leaves	Takashima *et al.*, 2007
Oleic acid	A	Fruits	Farine *et al.*, 1996
Palmitic acid	A	Fruits	Farine *et al.*, 1996
Palmitic acid	A	Leaves	Brett and Bing nan,2008
p-cresol	Ao	Fruits	Lin *et al.*, 2007
p-hydroxybenzoic acid	A	Fruits	Lin *et al.*, 2007
p-hydroxybenzaldehyde	Ad	Fruits	Lin *et al.*, 2007
Peucedanocoumarin III	Mc	Leaves	Takashima *et* al., 2007
Pheophorbide a	Cd	Leaves	Takashima *et al.*, 2007
Physcion	Aq	Heartwood, root	Srivastava and Singh, 1993
Physcion-8-*O*-[{α-L-arabinopyranosyl (1→3)}-β-D galactopyranoside]	Ag	Heartwood + –	Srivastava and Singh, 1993
Phytol	Mc	Leaves	Takashima *et al.*, 2007
Pteryxin	Mc	Leaves	Takashima *et al.*, 2007
Quercetin	F	Fruits, Leaves	Deng *et al.*, 2007; Shixin *et al.*, 2008
Quercetin 3-*O*-β-D-glucopyranoside	F	Leaves	Sang[a] *et al.*, 2001
Quercetin 3-*O*-β-D-glucopyranosyl-(1→2)-[α-Lrhamnopyranosyl-(1→6)]-β-D- galactopyranoside	F	Leaves	Sang[a] *et al.*, 2001
Ricinoleic acid	A	Seeds	Daulatabad *et al.*, 1989
Roseoside II	Mc	Leaves	Takashima *et al.*, 2007
Rotungenic acid	Ts	Leaves)	Takashima *et al.*, 2007
Rubiadin	Aq	Root, cell culture	Sang and Ho, 2006; Leistner, 1975; Kenichiro *et al.*, 1981; Thomson, 1971
Rubiadin-1-*O*-methyl ether	Aq	Root, heartwood	Thomson, 1971; Simonsen, 1920

Contd...

Table 9.1–Contd...

Name	*Class*	*Plant Part*	*References*
Rutin	F	Fruits, leaves	Sang[a] *et al.*, 2001; Wang *et al.*, 1999
Scopoletin	Mc	Fruits	Deng *et al.*, 2007; Lin *et al.*, 2007; Samoylenko *et al.*, 2006; Pawlus *et al.*, 2005; Saludesa *et al.*, 2002
Sitosteryl linoleate	Ts	Cell culture	Dyas *et al.*, 1994
Sitosteryl palmitate	Ts	Cell culture	Dyas *et al.*, 1994
Soranjidiol	Aq	Roots	Thomson, 1971
Stigmasta-4-22-dien-3-one	Ts	Leaves	Saludes *et al.*, 2002
Stigmasta-4-en-3-one	Ts	Leaves	Saludes *et al.*, 2002
Stigmasterol	Ts	Cell culture, leaves	Dyas *et al.*, 1994; Saludes[b] *et al.*, 2002
Stigmasteryl linoleate	Ts	Cell culture	Dyas *et al.*, 1994
Tectoquinone	Aq	Roots	Sang and Ho, 2006
Undecanoic acid	A	Fruits	Farine *et al.*, 1996
Ursolic acid	Ts	Leaves	Takashima *et al.*, 2007; Ahmad and Bano., 1980
Vanillin	Mc	Fruits	Deng *et al.*,2007; Pawlus *et al.*, 2005
Vomifoliol	Mc	Fruits	Farine *et al.*, 1996

A: Acids; Ao: Alcohols; Ad: Aldehydes; Lg: Lignans; Aq: Anthroquinone; Ag: Anthroquinone glycosides; C: Carotenoids; Cd: Chlorophyll derivatives; E: Esters; F: Flavonoids; I: Iridoids; K: Ketones; L: Lactones; N: Nucleosides; S: Sacharides; Ts: Triterpenoid and sterols; Mc: Miscellaneous compounds.

Amino Acids

The reported amino acids in leaves is alanine (Srivastava and Singh, 1993; Sang *et al.*, 2002) serine, threonine, tryptophan, tyrosine, valine, arginine, aspartic acid, cysteine, cystine, glutamic acid, glycine, histidine, isoleucine, leucine, methionine, phenylalanine, and proline (Dittmar, 1993; Elkins, 1998).

Vitamins

Noni is a rich source of vitamins and particularly ascorbic acid (24–158 mg/100 g) (Morton, 1992; Shovic and Whistler 2001; Chunhieng, 2003) and provitamin A (Dixon *et al.*, 1999). Genotype and climatic factors greatly influence its content in fruit portion. Maximum ascorbic acid content was observed in Andaman accessions GAH-1 (98.20 mg/100g) and TRA-2(94.9mg/100g) (Singh *et al.*, 2011). Morinda fruits produced under Arecanut + Noni system contains more vitamin C (126.09mg/100g) (Singh *et al.*, 2011).

Table 9.2: *In vitro* biological activity of isolated compounds.

Bioactive Compound	*Part*	*Class of Compound*	*Bioactivity*	*Reference(s)*
E-phytol, cycloartenol and stigmasterol	Fruit	Sterol	Anti-tuberculosis	Saludes[b] *et al.*, 2002
Damnacanthal	Fruit	Anthraquinone	Anti-HIV	Masakazu *et al.*, 2006
Isoscopoletin, aesculetin and quercetin	Fruit	Phenolics	Antioxidant	Chang *et al.*, 2007
Americanin A	Fruit	Lignans	Antioxidant	Su *et al.*, 2005
Narcissoside	Fruit	Flavonoids	Antioxidant	Su *et al.*, 2005
2-methoxy-1,3,6-trihydroxyanthraquinone	Fruit	Anthraquinone	Phase II enzyme inducer	Pawlus *et al.*, 2005
Lucidin 3-O-β-D-primeveroside (xi) and morindone-6-O-β-D-primeveroside	Roots	Anthraquinone	Anti-diabetic	Kamiya *et al.*, 2008
2-O-(β-D-glucopyranosyl)-1-O-octanoyl-β-D-gluropyranose	Fruit	fatty acid ester	Anti-inflammatory	Akihisa *et al.*, 2007
1, 4-dihydroxy-2-methoxy-7-methylanthraquinone	Fruit	Anthraquinone	Wound healing activity	Kim *et al.*, 2005
(+)-3,4,3',4'-tetrahydroxy-9,7'alpha-epoxylignano-7 alpha,9'-lactone and (+)-3,3'-bisdemethyltanegool, quercetin and Kaempferol	Fruit	Lignans	Cardiovascular activity	Deng *et al.*, 2007
Americanol A, 3,3'-bisdemethylpinoresinol, isoprincepin and morindolin	Fruit	Lignans	Cardiovascular activity	Kamiya *et al.*, 2004
Damnacanthal	Root	Anthraquinone	Anti-cancer, Anti-HIV	Hiramatsu *et al.*, 1993; Faltynek *et* al., 1995; Hiwasa *et al.*, 1999; Tosa *et al.*, 1998; Masakazu *et al.*, 2006
Morindone	Root	Anthraquinone	Anti-cancer	Tosa *et al.*, 1998
6-*O*-(β -D-glucopyranosyl)–1-*O*-octanoyl-β-D-glucopyranose and asperulosidic acid	Fruit juice	Glycosides	Anti-cancer	Guangming *et al.*, 2001

Phenolics

Phenolics content in a plant species is influenced by genotypes, climatic and edaphic factors and biotic stresses. Germplasm from Andaman showed great range of polyphenols (165.40- 370 mg GAE/ 100g fresh weight basis). TRA-1 contains THE highest polyphenols (370mg/100g) followed by CHTAP-13 (365.89 mg GAE/100g FW) and HD-6 (365.23 mg GAE/100g FW). Anthocyanin and phenol contents (146.11mg/100g and 340.0 mg/100g) were maximum in Noni fruits produced from Arecanut+Noni system. Fruits from sea water ingressed areas contain less anthocyanin and phenol (100.19 mg/100g and 230.0mg/100g respectively) as compared to fruits from cultivated land (116.89mg/100g and240.0mg/100g) (Singh *et al.*, 2011a). In grafted plants, anthocyanin content was less (104.36mg/ 100g) as compared to seeded plants (116.89mg/100g) while total phenol content was more in grafted plants (270.0mg/100g) as compared to seeded plants (240.0mg/100g) (Singh *et al.*, 2011a). Another study examined the carotenoid contents in leaves, bark and fruits and found that the leaves were a substantial source of carotenoids and had the potential to treat vitamin A deficiency (Aalbersberg *et al.*, 1993).

Figure 9.3: Structure of Isolated Bioactive compounds in Nonia.

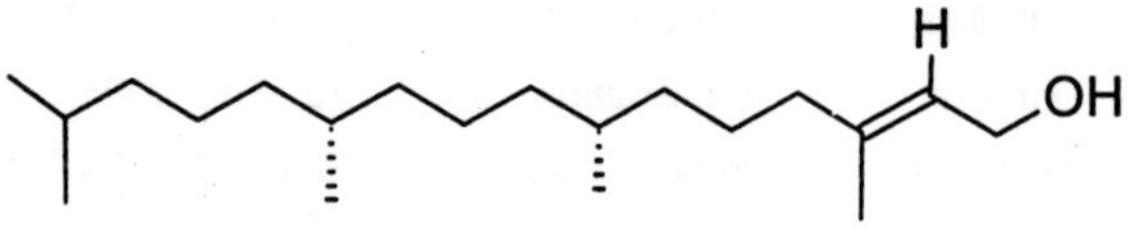

E-Phytol

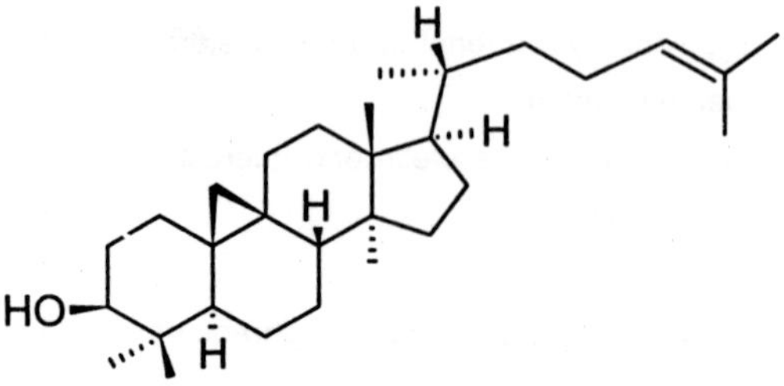

Cycloartenol

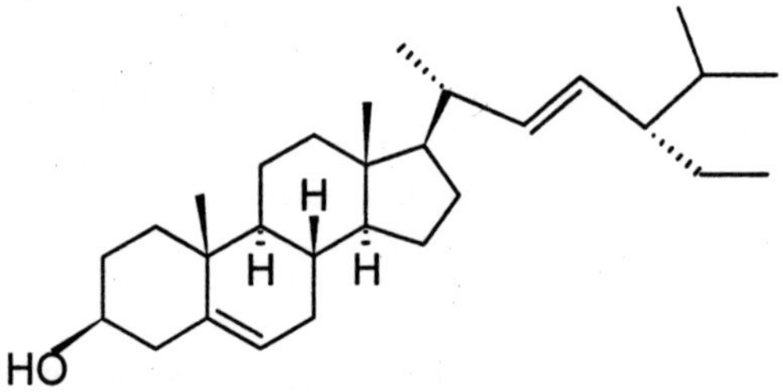

Stigmasterol

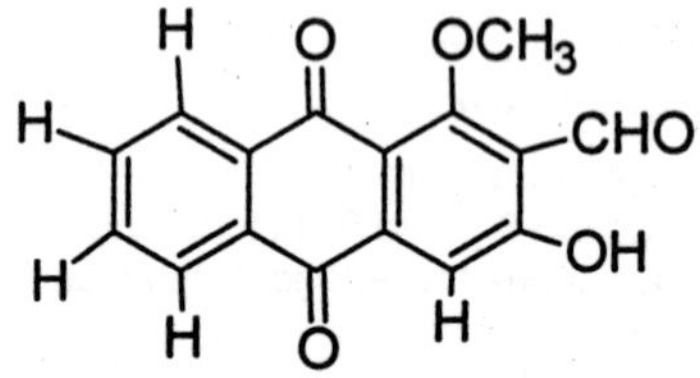

Damnacanthal

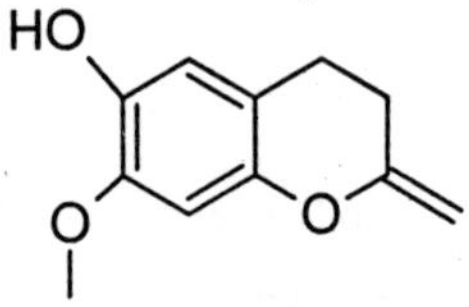

Isoscopoletin

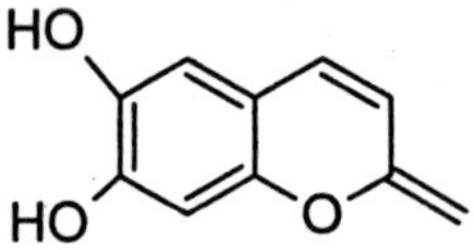

Aesculetin

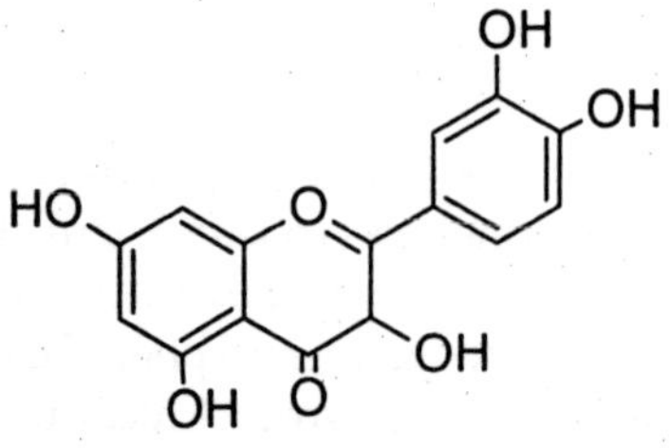

Quercetin

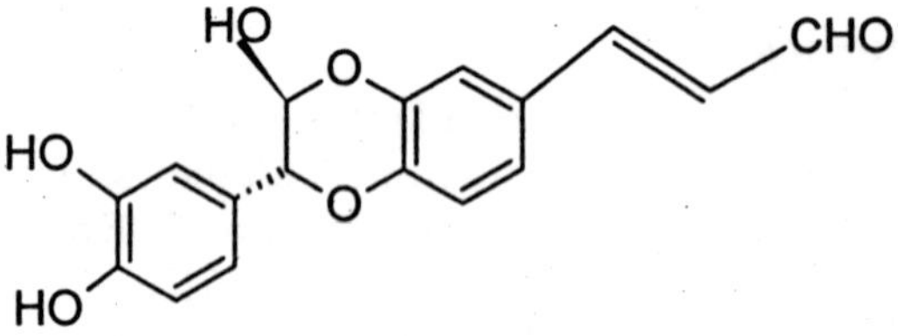

Americanin A

Contd...

Figure 9.3–Contd...

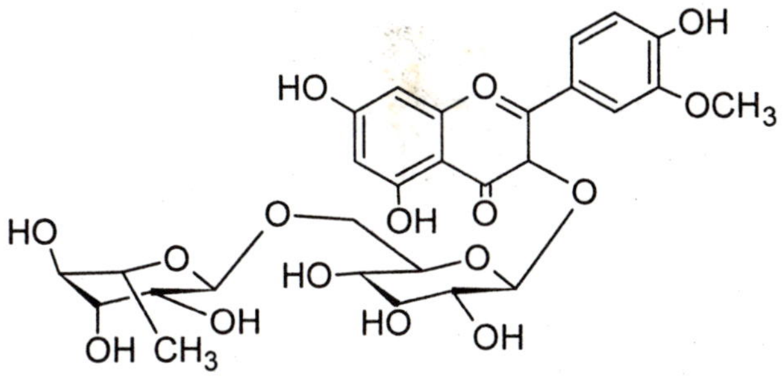

Narcissoside

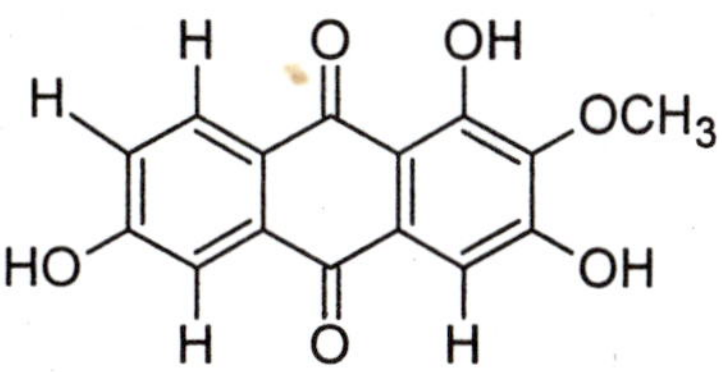

2-methoxy- 1, 3, 6-trihydroxyanthraquinone

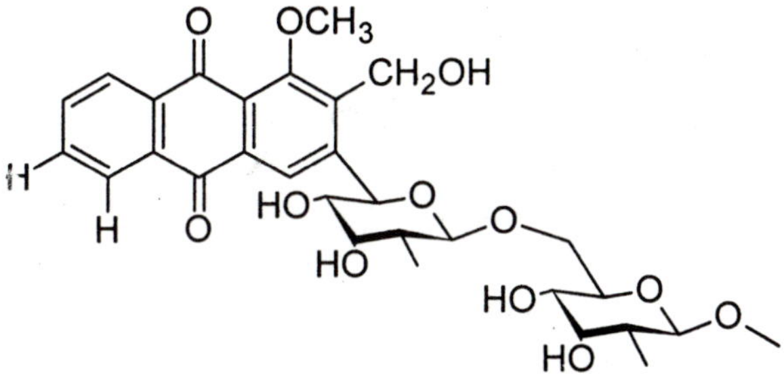

Lucidin 3-O-β-D-primeveroside

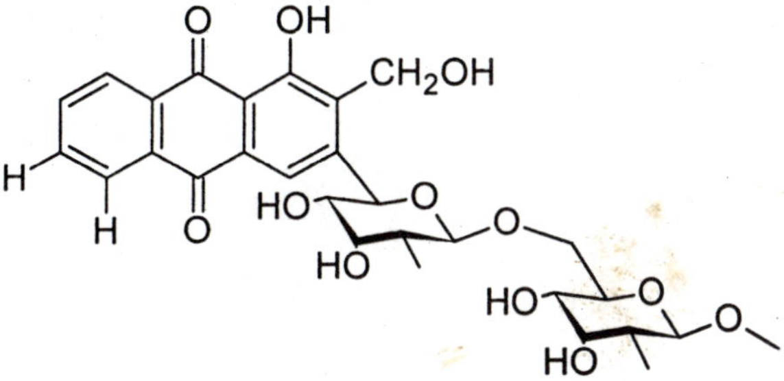

Morindone-6-*O*-β-D-primeveroside

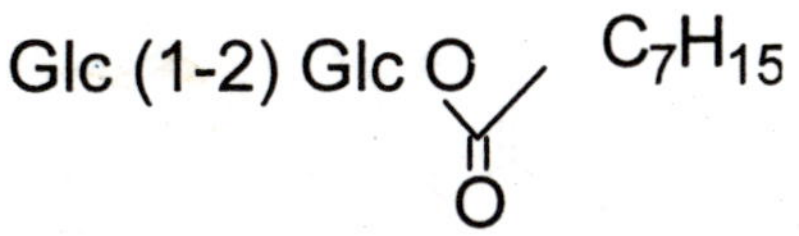

2-O-(β-D-glucopyranosyl)-
1-O-octanoyl-β-D-gluropyranose

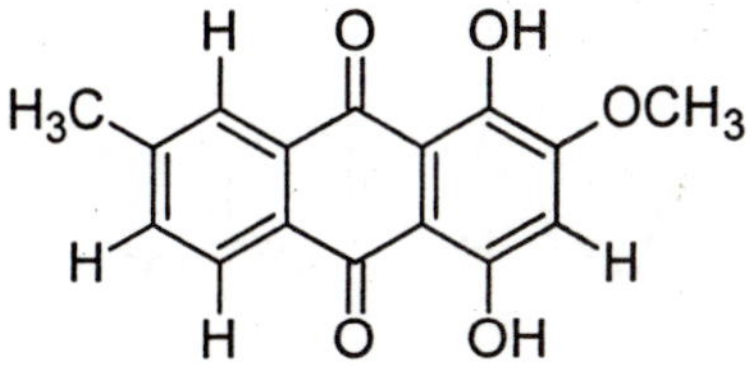

1, 4-dihydroxy-2-methoxy-7-methylanthraquinone

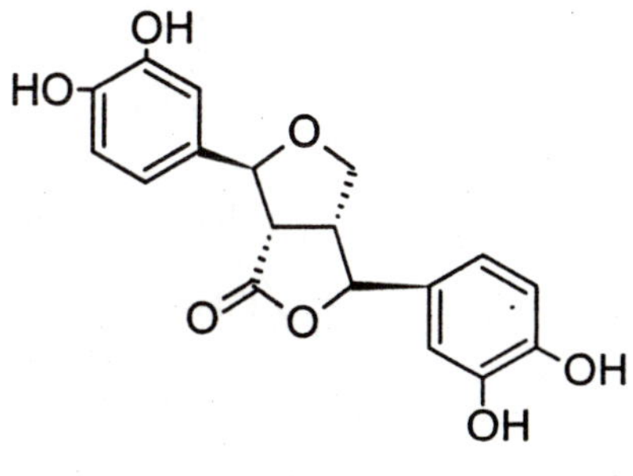

(+)-3, 4, 3', 4'-tetrahydroxy-9, 7'alpha-epoxylignano-7 alpha, 9'-lactone

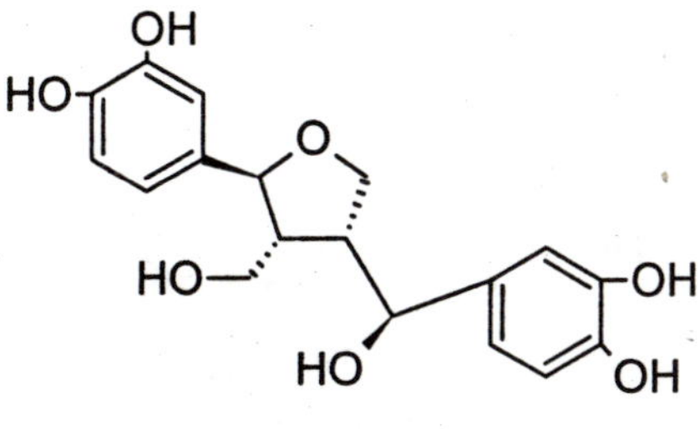

(+)-3, 3'-bisdemethyltanegool

Contd...

Figure 9.3–Contd...

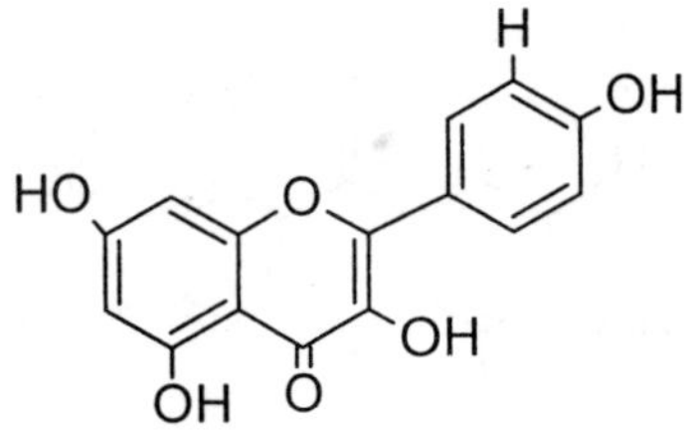

Kaempferol

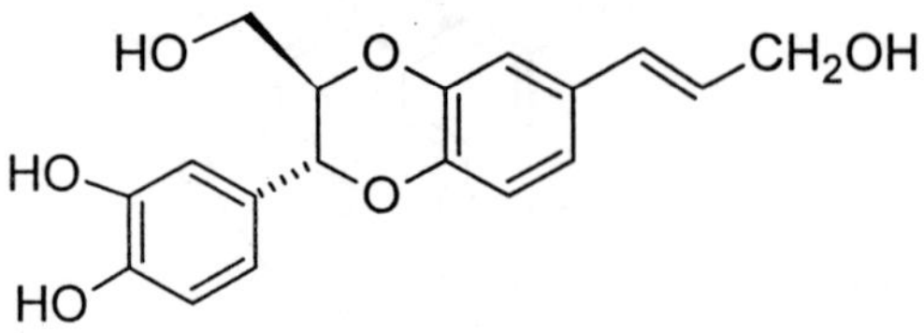

Americanol A

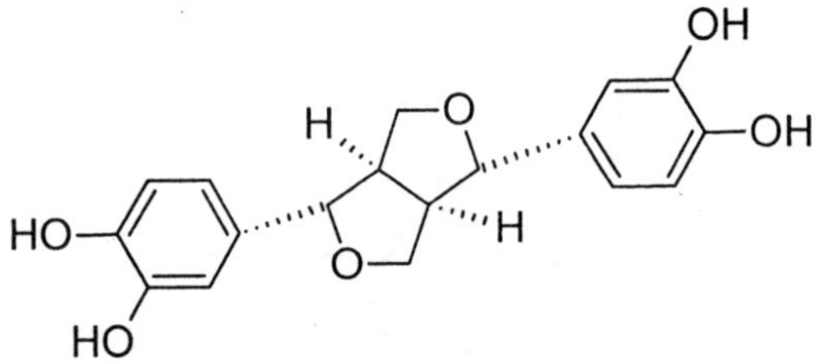

3, 3'-bisdemethylpinoresinol (xix)

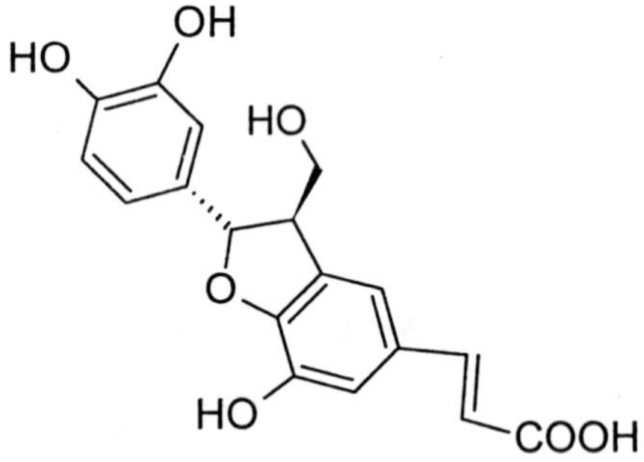

Morindolin

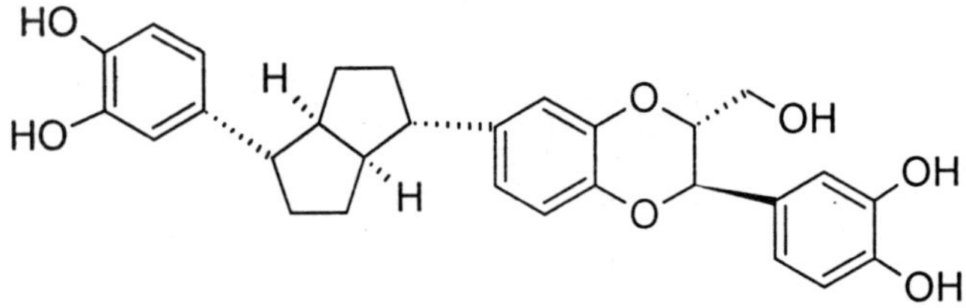

Isoprincepin

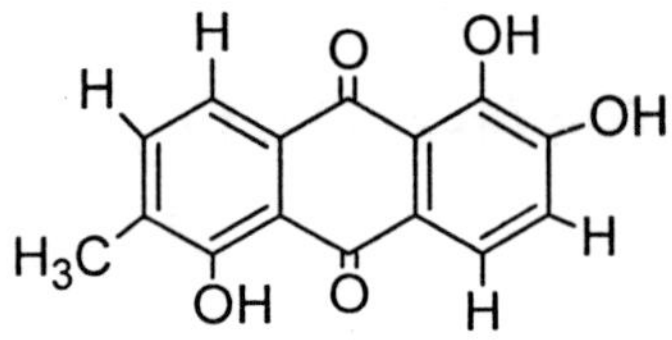

Morindone

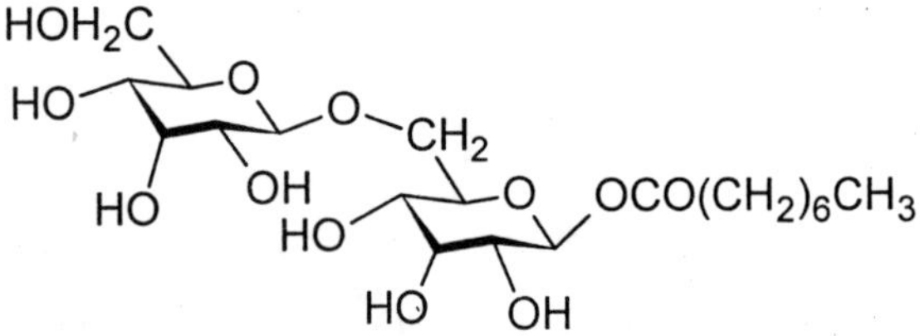

6-*O*-(β -D-glucopyranosyl)-1-*O*-octanoyl-β-D-glucopyranose

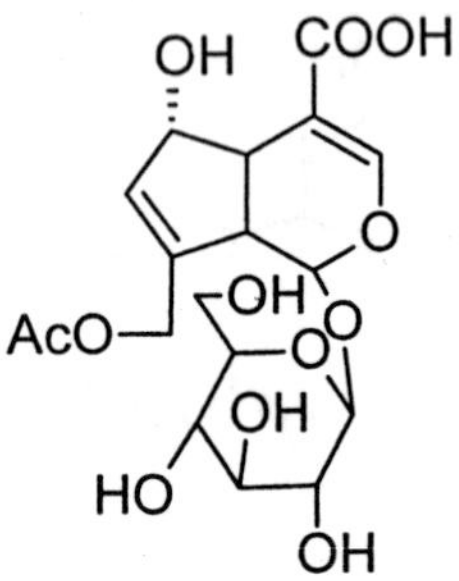

Asperulosidic acid

Figure 9.4: Noni product.

Seed Composition

The average oil content of noni seeds is 124.9mg/kg.and reported useful as a new source of vegetable oil (West *et al.*, 2008). Fatty acids were extracted from the whole seeds and analysed using GC-MS after methyl esterification and it was found that major acid was Linolic acid followed by Oleic acid and Palmatic acid.It has more of unsaturated fatty acids when compared to saturated fatty acids. In the three accessions CARI-1, TRA-1 and TRA-2, Lauric acid and Erucic acid were not detected (Singh *et al.*, 2011b).

Fatty Acids

Fatty acid glycosides and alcohols are another class of phyto-chemicals reported in the fruits. Due to their structure, they possess more or less pronounced amphiphilic properties and may be responsible for the soapy taste of ripe fruits (Wang *et al.*, 1999; Wang *et al.*, 2000; Samoylenko *et al.*, 2006; Dalsgarrd *et al.*, 2006).

The ripe fruits contain 51 volatile compounds which include 20 acids, 7 alcohols, 11 esters, 2 ketones, 2 lactones and 9 other miscellaneous compounds (Sang *et al.*, 2001)

The major component is octanoic acid and the minor component is hexanoic acid.The biological effects of ripe fruits and their main acids were investigated with behavioral studies.It was reported that Octanoic acid is very toxic to most *Drosophila* species, (except *Drosophila sechellia*) (Farine *et al.*, 1996). Lately, Brett and Bing-nan (2008) identified 14 major aromatic compounds in leaf. With the exception of E-phytol, all of the known volatile compounds identified in leaf samples were done so for the first time.

Biological Activities of *Morinda citrifolia*

Widespread claims of the therapeutical effectiveness encouraged the researchers worldwide to investigate further some of these possibilities. By virtue of this, one can easily find several research articles on the biological activities of *Morinda citrifolia* based on modern scientific investigations.

Antibacterial Activity

In the late 1950's Atkinson(1956) reported its antibacterial activity against certain infectious bacterial like *Pseudomonas aeruginosa, Proteus morgaii, Staphylococcus aureus, Bacillus subtilis, Escherichia coli, Salmonella,* and *Shigela*. He accentuated that the anti-microbial effect may be due to presence of phenolic compounds like Acubin, *L*-asperuloside, and alizarin in the fruit, and other anthraquinone compounds in roots. A related showed antimicrobial effects on different strains of *Salmonella, Shigella, and E. coli* (Bushnell *et al.*, 1950; Dittmar, 1993). Another study by Locher *et al.* (1995) showed that an acetonitrite extract of the dried fruit inhibited growth of *Pseudomonas aeruginosa, Bacillus subtilis, Escherichia coli* and *Streptococcus pyrogene*. Furthermore, it also helps in stomach ulcer through inhibition of the bacteria *H. pylori* (Duncan *et al.*, 1998). It was also found that ethanol and hexane extract of fruit provide protection against *Mycobacterium tuberculosis*. The major components identified in the hexane extract were E-phytol, cycloartenol and stigmasterol (Saludes *et al.*, 2002).

The methanol extract of fruit demonstrated zones of inhibition in a range of 7.7-26 mm against *Vibrio cholerae, Klebsiella, Bacillus subtitis, Lactobacillus lactis,Pseudomonas aeruginosa,, Salmonella typhi, E. coli, Staphylococcus aureus,Streptococcus thermophilus,Shigella flexneri, Vibrio harveyi, Chromobactrium violaceum,Aeromonas hydrophila* and *Salmonella paratyphi A* (Lee *et al.*, 2008). Ethyl acetate extract demonstrated zones of inhibition in range of 5.7.7-15.7 mm against *Lactobacillus lactis, Salmonella typhi, Bacillus subtitis, E. coli, Vibrio harveyi, Staphylococcus aureus, Shigella flexneri, Vibrio cholerae, Aeromonas hydrophila, Chromobactrium violaceum* and *Salmonella paratyphi A* (Jayaraman *et al.*, 2008). Lately, antibacterial activities of fruit juice were compared with that of sodium hypochlorite (NaOCl) and chlorhexidine gluconate (CHX) to remove the smear layer of *Enterococcus faecalis* from the canal walls of endodontically instrumented teeth. The data from this experiment indicated that fruit juice and sodium hypochlorite (NaOCl) treatments have similar effects (Murray *et al.*,2008).

Antifungal Activity

Candida albicans lives in 80 per cent of the human population with no harmful effects, although overgrowth from cellular yeast to a filamentous form results in candidiasis, which often observed in immunocompromised individuals like AIDS, cancer chemotherapy and organ or bone marrow transplantation. Recent research demonstrated that it contained a water-soluble component or components which interfere with the morphological conversion of *Candida albicans* and may have potential therapeutic value with regard to candidiasis (Banerjee *et al.*, 2006). Another study showed that methanol extract of the dried fruit exhibited maximum percentage of inhibition against *Tricophyton mentagrophytes* (79.3 per cent), while approximately 50 per cent activity was recorded against *Penicillium* sp., *Fusarium* sp. and *Rhizopus* spp. (Jayaraman *et al.*, 2008).

Antiviral Activity

Viral protein R (Vpr) is one of the human immunodeficiency virus type 1 (HIV-1) accessory proteins and contributes to multiple cytopathic effects, G2 cell cycle arrest and apoptosis. A team from National Institute of Health (USA), identified damnacanthal a component of noni fruit, as an inhibitor of Vpr induced cell death. The mechanism(s) of damnacanthal, which inhibits Vpr induced apoptosis is currently unknown and need to be elucidated (Masakazu *et al.*, 2006).

Antioxidant Activity

Many degenerative human diseases are recognized as consequences of free radical damage. All the chromatographic fractions obtained from the root, fruit and leaf demonstrated high antioxidative activity compared to either BHT or α-tocopherol (Zin *et al.*, 2006).Zin *et al.* (2002) previously reported that methanol extract of root and ethyl acetate partitions of all parts of the plant exhibited higher antioxidant activity in ferric thiocyanate (FTC) and thiobarbituric acid test (TBA)· Another study showed that, ethyl acetate extract of fruit exhibited higher antioxidative activity than mannitol or vitamin C, while the petroleum ether and n-butanol extracts showed lower activities compared to mannitol. Three antioxidant phenolic compounds, isoscopoletin, aesculetin and 3, 3', 4', 5, 7-pentahydroxyflavone were isolated from the ethyl acetate extract of fruit by several chromatography techniques for the first time (Chang *et al.*, 2007). Americanin A and Narcissoside isolated from fruits were potent antioxidants in terms of both DPPH and ONOO (-) bioassays (Su *et al.*, 2005).

Free radicals species influence normal body mechanism and need to be neutralized by supplementing endogenous homeostasis mechanisms through exogenous supply of antioxidants. The health refreshing capacity of Noni juice is due to its high antioxidant providing capacity which scavenges large number of free radicals. Singh *et al.* (2011) observed variations in antioxidant activity with genotypes as juice extract SPG-2 has high capacity (81.86 per cent) while HD-6 only 71.3 per cent. It also got influenced by light intensity and quality as fruits produced under coconut plantation was high (85.5 per cent) as compared to open cultivation (82.4 per cent).

Antihelmintic Activity

Alcoholic extracts of tender leaves showed good *in vitro* anthelmintic activity against human *Ascaris lumbricoides* (Raj *et al.*, 1975).

Hepatoprotective Activity

Fruit juice from Tahiti was examined on carbon tetrachloride (CCl_4) induced liver injury in female Sprague-Dawley (SD) rats. This study indicated that noni juice may be effective in protecting the liver against acute CCl_4 toxicity and it was observed that necrosis of hepatocytes was minimized if rats were pretreated with juice. It was also reported that fruit juice is able to normalize liver function after acute exposure to CCl_4 (Mian *et al.*, 2008).

A few reports suggested involvement of noni juice in the development of chemically induced hepatitis in a limited number of cases in Europe. An official European investigation of these cases determined that no relationship between noni juice and hepatitis was evident. Therefore consumption of noni juice is unlikely to induce adverse human liver effects (European Food Safety Authority, 2006). It may be an efficacious natural hepatoprotective nutritional supplement even at very high doses (West *et al.*, 2006).

Anti-obesity and Hypoglycemic Effects

While most of the research focus on cancer, inflammation and antimicrobial functions, there are s other aspects also. One of them is the rate of gastric emptying in rats (Pu *et al.*, 2004). In effect, noni juice reduced the rate at which food exited the stomach when taken as a juice for seven consecutive days. Although the mechanism is not clear, it acts like an appetite suppressant, useful for weight loss. Another study showed that Noni juice reduced body weight by 40 per cent, when a mice fed with a control diet and whereas 25 per cent in high-fat-diet (HFD). It also reduced adipose tissue weights, plasma triglyceride levels and improved glucose tolerance in these animals (Adrienne and Pratibha,

2007). A Japanese research team studied more specifically the hypoglycemic effects of the anthraquinone lucidin 3-*O*-β-*D*-primeveroside and morindone-6-*O*-β-*D*-primeveroside isolated by roots on streptozotocin (STZ)-induced diabetic mice and results showed that these molecules are responsible for the hypoglycemic effects (Kamiya *et al.*,2008). A related report revealed that leaf and fruit extracts are inhibiting lipoprotein lipase (LPL) in an *in vitro* system and may be used as antiobesity agent (Pak *et al.*, 2008).

Analgesic Activity

The lyophilised aqueous extract of roots was screened for analgesic activities in mice through writhing and hot plate tests. The data showed that analgesic efficacy of the extract was 75 per cent as strong as morphine, yet non addictive and also proved non-toxic (Chafique *et al.*, 1990). Similar findings were also reported by Punjanon and Nandhasri (1999) in the alcoholic extract of fruits.

Anxiolytic Activity

Methanol crude extract of fruit showed significant affinity to the gamma-amino butyric acid A (GABAa), the commonest inhibitory neurotransmitter in the central nervous system and displayed 75 per cent binding inhibition as an agonist and thus induces its anxiolytic and sedative effects (Deng *et al.*, 2007).

Anti-inflammatory Activity

The aqueous extract of fruit was investigated for anti-inflammatory activity against bradykinin and carrageenan induced oedema in the rat paw. The results showed that the bradykinin-induced inflammatory response was inhibited and subsided rapidly in rats pretreated either orally or intraperitoneally with fruit juice extract, whereas a higher dose of extract was required to completely inhibit the inflammatory response to carrageenan (Marsha-lyn *et al.*, 2002).

Another study showed that ethanol extract of fruit powder has a selective inhibition effect on cyclooxygenase-1 (COX-1). Its IC50 value (163 μg/ml) was lower than those produced by Aspirin (241 μg/ml), and higher than indomethacin (1.2 μg/ml) used as the reference COX-1 inhibitor in this study (Rachel *et al.*, 2003), By contrast, it should be noted that it did not exhibit (*in vtro and in vivo*) nitric oxide (NO*) scavenging activity, a key mediator in the phenomenon of inflammation (Basu and Hazra, 2006). Akihisa *et al.* (2007) reported that new saccharide fatty acid ester 2-*O*-(β-D-glucopyranosyl)-1-O-octanoyl-beta-D-gluropyranose extracted from the fruits exhibited potent anti-inflammatory activity.

Wound Healing Activity

The ethanol extract of leaves (150mg $kg^{-1}day^{-1}$) was used to evaluate the wound-healing activity in rats, using excision and dead space wound models. The result showed that the extract-treated animals exhibited 71 per cent reduction in the wound area when compared with controls which exhibited 57 per cent. The mass and hydroxyproline content in the dead space wounds of granulation tissue also increased significantly in noni-treated animals compared with controls (Shivananda *et al.*, 2007). In a similar study, wound-healing activity of fruit juice was evaluated in streptozotocin-induced diabetic rats. The data indicated that fruit juice significantly reduced blood sugar level and hastened wound healing activity (73 per cent) in diabetic rats as compared to diabetic controls (63 per cent) (Nayak *et al.*,2007).

An anthraquinone was isolated from fruit and identified as 1, 4-dihydroxy-2-methoxy-7-methylanthraquinone. It significantly increased elaboration of procollagen type I C-terminal peptide and glycosaminoglycans and reduced or inhibited expression of the collagenase matrix metalloproteinase-I in primary cultures of human fibroblasts. Furthermore, a clinical trial found that this molecule increased the dermal type I procollagen in nude mouse skin. These results suggest that identified anthraquinone is a good indispensable candidate for use as a new anti-wrinkle agent due to its strong induction of biosynthetic activity of extracellular matrix components (Schripsema *et al.*, 2006).

Hypotensive Activity

In early 60s, Youngken *et al.* (1960) found that total extract of roots has a hypotensive effect. Later on, Moorthy and Reddy (1970) found that ethanol extract of roots lowered the blood pressure in an anesthetized dog. It was reported that noni juice contains angiotensin-I-converting enzyme inhibitor (ACE-I), commonly prescribed to treat high blood pressure (Yamaguchi *et al.*, 2002).

Cardiovascular Activity

The 5-lipoxygenase and 15-lipooxygenase are responsible for production of leukotrienes which instigate asthmatic and allergic reactions and act to sustain inflammatory reactions. Recent research showed the role of 5-lipoxygenase in cardiovascular and neuropsychiatry illnesses (Radmila and Hari, 2004).

Deng *et al.*, 2007 isolated two new lignans, (+)-3, 4, 3′, 4′-tetrahydroxy-9, 7′alpha-epoxylignano-7 alpha, 9′-lactone and (+)-3, 3′-bisdemethyltanegool, as well as seven known compounds (-)-pinoresinol, (-)-3, 3′-bisdemethylpinoresinol, quercetin, kaempferol, scopoletin, isoscopoletin, and vanillin, which are responsible for inhibition of 5- or 15-lipoxygenase.Another study showed that methanol and ethyl extract of fruit showed 88 and 96 per cent inhibition, respectively, of copper–induced oxidation of low density lipoprotein particles *in vitro*. The main classes of compounds responsible for activity isolated from ethyl extract of fruit were lignans namely, americanol A, 3, 3'-bisdemethylpinoresinol, isoprincepin and morindolin (Kamiya *et al.*, 2004).

Estrogenic Activity

A German research team studied the estrogenic properties of the hexane extracts from the fruit in two *in vitro* assays, the estrogen receptor binding assay with both estrogen receptors, ER-α and ER-β, and the estrogen-receptor dependent induction of alkaline phosphatase in Ishikawa cells. The result suggested that it exhibited high activity in both systems (Basar *et al.*, 2006). Another study showed that, alcohol and water extracts of fruit exhibited very low potency in comparison to estradiol (Chearskul *et al.*, 2004). Besides above investigations; the chemical nature of phytoestrogen in noni is still mysterious and needs further investigations.

Immunological Activity

Recent research examined the mechanisms involved in immunological properties of Tahitian Noni Juice (TNJ) and Noni fruit juice concentrates (NFJC) in mice. The results showed that both modulate the immune system via activation of the CB2 receptors, and suppression of the IL-4, but increasing the production of IFN-gamma cytokines (Palu *et al.*,2008).

A related report found that ethanol precipitation of fruit contained a polysaccharide-rich fraction which showed antitumour activity in the Lewis lung carcinoma model in mice. It also stimulated the release of several potential mediators, including TNF-α, IL-1β, IL-10, IL-12 p70, IFN-γ and nitric oxide but had no effect on IL-2 and suppressed IL-4 (Hirazumi and Furusawa, 1999). *Morinda citrifolia* fruit juice influenced growth and production performance of poultry as observed in Nicobari Fowl with supplement of normal diet with 1.5ml/bird/day (Singh *et al.*, 2011b).

Anti-cancer Activity

It was reported that methanol extract of fruit at a concentration of 0.1 mg/l exhibited cytotoxic activity against breast cancer (MCF7) and neuroblastoma (LAN5) cell lines at 29 per cent and 36 per cent, respectively in MTT assay (Arpornsuwan and Punjanon, 2006).Another study showed that the ethanol precipitated Polysaccharide-rich substance of noni juice possesses immunomodulatory and antitumour activity against Sarcoma 180 ascites tumour in mice (Furusawa *et al.*, 2003).

Wang and Su (2001) studied cancer preventive effect of Tahitian Noni Juice (TNJ), in Sprague-Dawley (SD) rats. In this study, a chemical carcinogen DMBA (7, 12-dimethylbenz [a] anthracene) was used to induce cancer in specific organs of rats. Results showed that 10 per cent Tahitian Noni Juice (TNJ) reduced DMBA-DNA adduct formation by 30 per cent in the heart, 41 per cent in the lung, 42 per cent in the liver, and 80 per cent in the kidney in one week.

One of the current strategies for the treatment of human cancer is to activate the cellular apoptotic death program. Along this line, Hornick *et al.* (2003) found that Noni juice in concentrations of 5 per cent or greater was highly effective for inhibiting the initiation of new vessel sprouts from model of placental vein explants. At higher concentration of 10 per cent it induced vessel degeneration and apoptosis in established capillary networks within a few days of its application.

An anthraquinone Damnacanthal isolated from the chloroform extract of the root was found as new inhibitor of *ras* function and thus suppressed *ras*-expressing tumors (Hiramatsu *et al.*, 1993). It also showed potent inhibitory activity towards tyrosine kinases like Lck, Src, Lyn and EGF receptors and might be related to the stimulatory effect on ultraviolet-induced apoptosis (Hiwasa *et al.*, 1999). Another anthraquinone 2-methoxy-1, 3, 6-trihydroxyanthraquinone was isolated from methyl extract of noni fruit and was reported as an Inducer of phase II detoxification enzyme Quinone reductase (QR). This new anthraquinone was nearly 40 times more potent than a positive control, L-sulforaphane and may help maintain the antioxidant function of the cell as well as Quinone reductase induction which protects breast cells against estrogen-induced oxidative DNA damage (Pawlus *et al.*,2005). Other studies reported that the two novel glycosides, 6-*O*-(β-D-glucopyranosyl)–1-*O*-octanoyl-β-D-glucopyranose and asperulosidic acid, extracted from the juice of fruits, were effective in inhibiting cell transformation induced by TPA or EGF in the mouse epidermal JB6 cell line (Guangming *et al.*, 2001)

New compounds with high levels of pharmacological activity are urgently required for a wide range of human disorders and diseases. A number of scientific publications showed that this plant contains several nutritional and functional compounds, but the current state of knowledge is still far from satisfactory. The most important compounds identified are damnacanthal and a few other phenolic compounds. Its activity as chemopreventive and anti-HIV needs further investigation. A drug development programme may be undertaken to develop modern drugs with the compounds isolated, but before that, a comprehensive phytochemical profiling, extensive investigation of its bioactivity, mechanism of action, pharmacotherapeutics, toxicity and clinical trials are needed to provide sufficient data. In addition, research should be initiated for identification of elite chemotypes, quality control of produces through marker development like TLC, HPTLC, standardization of cell culture techniques for production of bioactive compounds and identification of pathway related to the production of potent bioactive compounds. To realize these goals a number of projects are initiated by Academic Institutions and Pharmaceutical companies.

In a nut shell, combination of natural products chemistry, molecular and cellular biology; synthetic and analytical chemistry, biochemistry, and pharmacology are needed to exploit the vast diversity of chemical structures and their biological activities for the development of new pharmacological agents from Noni-a future crop.

References

Aalbersberg, W.G.L., Hussein, S., Sotheeswaran, S. and Parkinson, S., 1993. Carotenoids in the leaves of *Morinda citrifolia*. *J. Herbs Spices Med. Plants*, 2: 51–54.

Adrienne Nishioka and Pratibha Nerurkar, 2007. Effects of *Morinda citrifolia* (Noni) on Obesity and Glucose Tolerance in C57BL/6 Mice. *The FASEB Journal*.

Ahmad, V.U. and Bano, S., 1980. Isolation of β-sitosterol and ursolic acid from *Morinda citrifolia* Linn. *J. Chem. Soc. Pak.*, 2: 71.

Akihisa, T., Matsumoto, K., Tokuda, H., Yasukawa, K., Seino, K., Nakamoto, K., Kuninaga, H., Suzuki, T. and Kimura, Y., 2007. Anti-inflammatory and potential cancer chemopreventive constituents of the fruits of *Morinda citrifolia* (Noni). *J Nat. Prod.*, 70(5): 754–757.

Arpornsuwan, T. and Punjanon, T., 2006.Tumor cell-selective antiproliferative effect of the extract from *Morinda citrifolia* fruits. *Phytother. Res.*, 20(6): 515–517.

Atkinson, N., 1956.Antibacterial substances from flowering plants. 3. Antibacterial activity of dried Australian plants by rapid direct plate test. *Australian Journal of Experimental Biology*, 34: 17–26.

Banerjee, S., Johnson, A.D., Csiszar, K., Wansley, D.L. and McGeady, P., 2006.An extract of *Morinda citrifolia* interferes with the serum-induced formation of filamentous structures in *Candida albicans* and inhibits germination of *Aspergillus nidulans*. *Am. J. Chin Med.*, 34(3): 503–509.

Basar, S., Iznaguen, H., Zeglin, A. and Westendorf, J., 2006. Phytoestrogenic activity of *Morinda citrifolia* L. fruits. *Planta Med.*, p. 72–78.

Basu, S. and Hazra, B. 2006. Evaluation of nitric oxide scavenging activity, *in vitro* and *ex vivo*, of selected medicinal plants traditionally used in inflammatory diseases. *Phytother Res.*, 20(10): 896–900.

Brett, J. West and Bing-nan Zho, 2008. Identification of major aroma compounds in the leaf of *Morinda citrifolia* Linn. *J. Nat. Med*.

Bui, A.K.T., Bacic, A. and Pettolino, F., 2006.Polysaccharide composition of the fruit juice of *Morinda citrifolia* (noni). *Phytochemistry*, 67: 1271–1275.

Bushnell, O.A., Fukuda, M. and Makinodian, T., 1950. The antibacterial properties of some plants found in Hawaii. *Pacific Science*, 4: 167–183.

Chafique Younos, Alain Rolland, Jacques Fleurentin, Marie-Claire Lanhers, René Misslin and François Mortier, 1990. Analgesic and Behavioural Effects of *Morinda citrifolia*. *Planta Med*, 56: 430–434.

Chang hong, LIU, Ya rong, XUE, Yong Hang, YE, Feng-feng, YUAN, Jun-yan, LIU and Jing-lei, SHUANG, 2007. Extraction and characterization of antioxidant compositions from fermented fruit juice of *Morinda citrifolia* (Noni). *Agricultural Science in China*, 6(12): 1494–1501.

Chearskul Supornpim, Kooptiwut Suwattanee, Chatchawalvanit Santipong, Onreabroi, Churintrapun Smarn Malika, Saralamp Promchit, and Soonthornchareonnon Noppamas, 2004. *Morinda citrifolia* has very weak estrogenic activity *in vivo*. *Thai Journal of Physiological Sciences*, 17: 22–29.

Chunhieng, M.T., 2003. De´veloppement de nouveaux aliments sante´ tropicale: application a' la noix du Bre´ sil Bertholettia excelsa et au fruit de Cambodge *Morinda citrifolia*. *Ph.D. Thesis*, INPL, France.

Dalsgarrd, P.W., Potterat, O., Dieterle, F., Paululat, T., Kuhn, T. and Hamburger, M., 2006. Noniosides E–H, New trisaccharide fatty acid esters from the fruit of *Morinda citrifolia* (Noni). *Planta Med*, 72: 1322–1327.

Daulatabad, C.D., Mulla, G.M. and Mirajkar, A.M., 1989. Ricinoleic acid in *Morinda citrifolia* seed oil. *J. Oil Technol. Assoc. (India)*, 21: 26–27.

Deng[a], S., Palu, K., West, B.J., Su, C.X., Zhou, B.N. and Jensen, J.C., 2007. Lipoxygenase inhibitory constituents of the fruits of noni (*Morinda citrifolia*) collected in Tahiti. *J. Nat. Prod.*, 70(5): 859–862.

Deng[b], S., West, B.J., Palu, A.K., Zhou, B.N. and Jensen, C.J., 2007. Noni as an anxiolytic and sedative: a mechanism involving its gamma–aminobutyric acidergic effects. *Phytomedicine*, 14(7–8): 517–522.

Dittmar, A., 1993. *Morinda citrifolia* L.Use in indigenous Samoan medicine. *Journal of Herbs Spices and Medicine Plants*, 1: 77–92.

Dixon, A.R., McMillan, H. and Etkin, N.L., 1999. Ferment this: the transformation of Noni traditional Polynesian medicine (*Morinda citrifolia*, Rubiaceae). *Econ Bot.*, 53: 51–68.

Duncan, S.H., Flint, H.J. and Stewart, C.S., 1998.Inhibitory activity of gut bacteria against *Escherichia coli* O157 mediated by dietary plant metabolites. *FEMS Microbiology Letters*, 164: 258–283.

Dyas, L., Threlfall, D.R. and Goad, L.J., 1994.The sterol composition of five plant species grown as cell suspension cultures. *Phytochemistry*, 35: 655–660.

Elkins, R., 1998. Hawaiian Noni (*Morinda citrifolia*) Prize Herb of Hawaii and the south pacific. Woodland Publishing, Utah.

European Food Safety Authority, 2006. Opinion on a request from the Commission related to the safety of Noni juice (juice of the fruits of *Morinda citrifolia*). *EFSA J.*, 376: 1–12.

European Food Safety Authority, 2002. Opinion of the Scientific Committee on Food of Tahitian Nonis Juice. SCF/CS/DOS/18 ADD 2, Belgium.

Faltynek, C.R., Schroeder, J., Mauvais, P., Miller, D., Wang, S., Murphy, D., Lehr, R., Kelley, M., Maycock, A., Michne, W., Miski, M. and Thunberg, A.L. 1995.Damnacanthal is a highly potent, selective inhibitor of p56lck tyrosine kinase activity. *Biochemistry*, 34: 12404–12410.

Farine, J.P., Legal, L., Moreteau, B. and Quere, J.L.L., 1996. Volatile compounds of ripe fruits of *Morinda citrifolia* and their effects on Drosophila. *Phytochemistry*, 41: 433–438.

Furusawa, E., Hirazumi, A., Story, S. and Jensen, J., 2003.Antitumour potential of a polysaccharide–rich substance from the fruit juice of *Morinda citrifolia* (Noni) on sarcoma 180 ascites tumour in mice. *Phytother. Res.*, 17(10): 1158–1164.

Guangming, Liu, Ann Bode, Wei-Ya Ma, Shengmin Sang, Chi-Tang Ho, and Zigang Dong, 2001. Two Novel Glycosides from the Fruits of *Morinda citrifolia* (Noni) Inhibit AP–1 Transactivation and Cell Transformation in the Mouse Epidermal JB6 Cell Line. *Cancer Research*, 61: 5749–5756.

Heinicke, R.M., 1985.The pharmacologically active ingredient of Noni. *Pacific Trop. Bot. Gard. Bull.*, 15: 10–4.

Hiramatsu, T., Imoto, M., Koyano, T. and Umezawa, K., 1993. Induction of normal phenotypes in ras-transformed cells by damnacanthal from *Morinda citrifolia*. *Cancer Lett.*, 30: 161–166.

Hirazumi, A. and Furusawa, E., 1999.An immunomodulatory polysaccharide-rich substance from the fruit juice of *Morinda citrifolia* (noni) with antitumour activity. *Phytother Res.*, 13(5): 380–387.

Hiwasa, Takaki, Arase, Yoshiko, Chen, Zheng, Kita, Kazuko, Umezawa, Kazuo, Ito, Hisao and Suzuki, Nobuo, 1999. Stimulation of ultraviolet-induced apoptosis of human fibroblast UVr-1 cells by tyrosine kinase inhibitors. *FEBS Letters*, 444(2–3): 173–176.

Hornick, C.A., Myers, A., Sadowska-Krowicka, H., Anthony, C.T. and Woltering, E.A., 2003. Inhibition of angiogenic initiation and disruption of newly established human vascular networks by juice from *Morinda citrifolia* (noni). *Angiogenesis*, 6(2): 143–149.

Jain, R.K. and Srivastava, S.D., 1992. Two new anthraquinones in the roots of *Morinda citrifolia*. *Proc. Natl Acad. Sci. India*, 62: 11–13.

Jayaraman Kumar Sathish, Manoharan Saravanan Muthu and Illanchezian Seethalakshmi, 2008. Antibacterial, Antifungal and Tumor cell suppression potential of *Morinda citrifolia* fruit extracts. *International Journal of Integrative Biology*, 3(11): 44–49.

Kamiya, K., Hamabe, W., Harada, S., Murakami, R., Tokuyama, S. and Satake, T., 2008. Chemical constituents of *Morinda citrifolia* roots exhibit hypoglycemic effects in streptozotocin-induced diabetic mice. *Biol. Pharm. Bull.*, 31(5): 935–980.

Kamiya, K., Tanaka, Y., Endang, H., Umar, M. and Satake, T., 2005.New anthraquinone and iridoid from the fruits of *Morinda citrifolia*. Chem. *Pharm. Bull.*, 53: 1597–1599.

Kamiya, K., Tanaka, Y., Endang, H., Umar, M. and Satake, T., 2004.Chemical constituents of *Morinda citrifolia* fruits inhibit copper induced low–density lipoprotein oxidation. *J. Agric. Food Chem.*, 52: 5843–5848.

Kenichiro Inoue., Hidekazu Nayeshiro, Hiroyuki Inouyet and Meinhart Zenk, 1981. Anthraquinones in cell suspension cultures of *Morinda citrifolia*. *Phytochemistry*, 20(7): 1693–1700.

Kim, S.W., Jo, B.K., Jeong, J.H., Choi, S.U. and Hwang, Y.I., 2005. Induction of extracellular matrix synthesis in normal human fibroblasts by anthraquinone isolated from *Morinda citrifolia* (noni) fruit. *J. Med. Food*, 8: 552–555.

Lee Seong Wei, Najiah Musa, Chuah Tse Sengm, Wendy Wee and Noor Azhar Mohd Shazili., 2008. Antimicrobial properties of tropical plants against 12 pathogenic bacteria isolated from aquatic organisms. *African Journal of Biotechnology*, 7(13): 2275–2278.

Legal, L., Moulin, B. and Jallon, J.M., 1999.The relationship between structures and toxicity of oxygenated aliphatic compounds homologous to the insecticide octanoic acid and the chemotaxis of two species of *Drosophila*. *Pest. Biochem. Physiol.*, 65: 90–101.

Leistner, V.E., 1975.Isolation, identification and biosynthesis of anthraquinones in cell suspension cultures of *Morinda citrifolia*. *Planta Med.*, 32 (Suppl.): 214–224.

Levand, O., and Larson, H.O. 1979. some chemical constituents of *Morinda citrifolia*. *Planta Med.*, 36: 186–187.

Lin, Chwan Fwu, Ni, Ching Li, Huang, Yu Ling, Sheu, Shuenn Jyi and Chen, Chien Chih, 2007. Lignans and anthraquinones from the fruits of *Morinda citrifolia*. *Natural Product Research*, 21(13): 1199–204.

Locher, C.P., Burch, M.T., Mower, H.F., Berestecky, H., Davis, H., Van Polel, B., Lasure, A., Vander Berghe, D.A. and Vlieti–Nick, A.J., 1995.Anti–microbial activity and anti-complement activity of extracts obtained from selected Hawaiian medicinal plants. *Journal of Ethnopharmacology*, 49: 23–32.

Lucilla Bassetti and Johannes Tramper, 1995. Increased anthraquinone production by *Morinda citrifolia* in a two-phase system with Pluronic F-68. *Enzyme and Microbial Technology*, 17(4): 353–358.

Lucilla Bassetti, Marc Hagendoorn and Johannes Tramper, 1995. Surfactant-induced non-lethal release of anthraquinones from suspension cultures of *Morinda citrifolia*. *Journal of Biotechnology*, 39(2): 49–155.

Marc Stalman, Anne-Marie Koskamp, Rianne Luderer, Juanita H.J. Vernooy, Jobien C. Wind, George J. Wullems and Anton F. Croes, 2003. Regulation of anthraquinone biosynthesis in cell cultures of *Morinda citrifolia*. *Journal of Plant Physiology* 160(6): 607–614.

Mrimuthu, T. and Peter, P.I., 2010. Noni (*Morinda citrifolia* L.): A textbook. World Noni Research Foundation, Chennai.

Marsha-lyn G. Mckoy, Everton A. Thomas and Oswald R. Simon, 2002. Preliminary Investigation of the anti-inflammatory properties of an aqueous extract from *Morinda citrifolia* (Noni). *Proc. West. Pharmacol. Soc.*, 45: 76–78.

Masakazu Kamata, Raymond P. Wu, Dong Sung Anc, Jonathan P. Saxeb, Robert Damoiseauxd, Michael E., Phelps, Jing Huangb, and Irvin S.Y. Chena, 2006. Cell-based chemical genetic screen identifies damnacanthal as an inhibitor of HIV-1 Vpr induced cell death. *Biochem. Biophys. Res. Commun.*, 29: 1101–1106.

Mian-Ying Wang, Diane Nowicki, Gary Anderson, Jarakae Jensen and Brett West, 2008. Liver protective effects of *Morinda citrifolia* (Noni). *Plant Foods Hum. Nut.*, 63(2): 59–63.

Moorthy, N.K. and Reddy, G.S., 1970. Preliminary phyto-chemical and pharmacological study of *Morinda citrifolia* Linn. Antiseptic, 67: 167–171.

Morton, J.F., 1992. The ocean-going Noni, or Indian mulberry (*Morinda citrifolia*, Rubiaceae) and some of its "colourful" relatives. *Ecological Botony*, 46: 241–256.

Murray, P.E., Farber, R.M., Namerow, K.N., Kuttler, S and Garcia-Godoy, F., 2008. Evaluation of *Morinda citrifolia* as an endodontic irrigant. *J. Endod.*, 34(1): 66–70.

Nayak, B.S., Isitor, G.N., Maxwell, A., Bhogadi, V. and Ramdath, D.D., 2007. Wound-healing activity of *Morinda citrifolia* fruit juice on diabetes-induced rats. *J. Wound Care*, 16(2): 83–86.

Olivier Potterat and Matthias Hamburger, 2007. *Morinda citrifolia* (Noni) fruit: Phytochemistry, pharmacolgy, safety. *Planta Med*, 7.172862 40 (P.S.E.B.D)

Pak-Dek, M.S., Abdul-Hamid, A., Osman, A. and Soh, C.S., 2008. Inhibitory effect of *Morinda citrifolia* L. on lipoprotein lipase activity. *Journal of Food Science*, 73(8): 595–598(1).

Palu, A.K., Kim, A.H., West, B.J., Deng, S., Jensen, J. and White, L., 2008. The effects of *Morinda citrifolia* L. (noni) on the immune system: its molecular mechanisms of action. *J. Ethnopharmacol.*, 12, 115(3): 502–506.

Pawlus, A.D., Su, B.N., Keller, W.J. and Kinghorn, A.D., 2005. An anthraquinone with potent quinone reductase–inducing activity and other constituents of the fruits of *Morinda citrifolia* (noni). *J. Nat. Prod.*, 68(12): 1720–1722.

Peerzada, N., Renaud, S. and Ryan, P., 1990. Vitamin C and elemental composition of some bush fruits. *J. Plant Nutr.*, 13: 787–793.

Peter, P.I., 2007. *Monograph on Noni (Morinda citrifolia L.),* (Ed.) K.V. Peter. World Noni Research Foundation, Chennai, 852 p.

Peter, P.I., 2008. *All about Noni.* Noni Publications, B.G. Road, Bangalore, 52 p.

Peter, P.I. (Compiled) 2009. *Compendium on Noni Research,* (Ed.) K.V. Peter. World Noni Research Foundation, Chennai, 884 p.

Pu, H.F., Huang, W.J. Tseng, W.M., Wang, S.W., Liu, Y.W., Doong, M.L. and Wang, P.S., 2004. Effects of juice from *Morinda citrifolia* (Noni) on gastric emptying in male rats. *Chin. J. Physiol.*, 31, 47(4): 169–174.

Punjanon, T. and Nandhasri, P., 1999. Analgesic effect of the alcoholic extract from the fruits of *Morinda citrifolia. ISHS Acta Horticulturae,* 678 (4).

Rachel, W.Li., Stephen, P., Myers, David, N., Leach, G., David Lin and Greg Leach, 2003. A cross-cultural study: Anti-inflammatory activity of Australian and Chinese plants. *Journal of Ethnopharmacology,* 85(1): 25–32.

Radmila Maney and Hari Maney, 2004. 5-lipoxygenase as a putative link between cardiovascular and psychiatric disorders. *Crit. Rev. Neurobiol.*, 16: 181–186.

Raj, R.K., 1975. Screening of indigenous plants for anthelmintic action against human *Ascaris lumbricoides.* Part II. *Indian J. Physiol. Pharmacol.*, 19: 47–49.

Rusia, K. and Srivastava, S.K., 1989. A new anthraquinone from the roots of *Morinda citrifolia* Linn. *Current Science,* 58: 249.

Saludes[a], J.P.., Garson, M.J. and Franzblau, S.G., 2002. Chemical and antitubercular studies of *Morinda citrifolia* Linn. (noni) fruits. *Sci. Augustinian (Manila),* 2: 2–64.

Saludes[b], J.P., Garson, M.J., Franzblau, S.G. and Aguinaldo, A.M., 2002. Antitubercular constituents from the hexane fraction of *Morinda citrifolia* Linn. (Rubiaceae). *Phytother. Res.*, 16(7): 683–685.

Samoylenko, V., Zhao, J., Dunbar, D.C., Khan, I.A., Rushing, J.W. and Muhammad, I., 2006. New constituents from noni (*Morinda citrifolia*) fruit juice. *J. Agric. Food Chem.*, 54: 6398–6402.

Sang, S., He, K., Liu, G., Zhu, N., Cheng, X., Wang, M., Zheng, Q., Dong, Z., Ghai, G., Rosen, R. T. and Ho, C.T., 2001. A new unusual iridoid with inhibition of activator protein–1 (AP–1) from the leaves of *Morinda citrifolia* L. *Org. Lett.*, 3: 1307–1309.

Sang, S. and Ho, C.T., 2006. Chemical components of noni (*Morinda citrifolia* L.) root. In: Wang, M., Sang, S., Hwang, L. S. and Ho, C.T. (Eds) *Herbs: Challenges in Chemistry and Biology.* American Chemical Society, Washington, DC, p. 384.

Sang, S., Liu, G., He, K., Zhu, N., Dong, Z., Zheng, Q., Rosen, R. T. and Ho, C.T., 2003. New unusual iridoids from the leaves of noni (*Morinda citrifolia* L.) show inhibitory effect on ultraviolet Binduced transcriptional activator protein–1 (AP–1) activity. *Bioorg. Med. Chem.*, 11: 2499–2502.

Sang, S., Wang, M., He, K., Liu, G., Dong, Z., Badmaev, V., Zheng, Q. Y., Ghai, G., Rosen, R. T. and Ho, C.T. 2002. Chemical components in noni fruits and leaves (*Morinda citrifolia* L.). In: Ho, C.T., Zheng, Q. Y. (Eds) *Quality Management of Nutraceuticals,* ACS symposium series 803. American Chemical Society, Washington, DC, pp. 134–150.

Sang[a], S., Cheng, X., Zhu, N., Stark, R. E., Badmaev, V., Ghai, G., Rosen, R. T. and Ho, C.T., 2001.Flavonol glycosides and novel iridoid glycosides from the leaves of *Morinda citrifolia*. *J. Agric. Food Chem.*, 49: 4478–4481.

Sang[b], S., Cheng, X., Zhu, N., Wang, M., Jhoo, J.W., Stark, R. E., Badmaev, V., Ghai, G., Rosen, R. T. and Ho, C.T., 2001. Iridoid glycosides from the leaves of *Morinda citrifolia*. *J. Nat. Prod.*, 64: 799–800.

Schripsema, J., Caprini, G.P. and Dagnino, D. 2006. Revision of the structures of citrifolinin a, citrifolinoside, yopaaoside a, yopaaoside b, and morindacin, iridoids from *Morinda citrifolia* L. and *Morinda coreia* Ham. *Org. Lett.*, 8: 5337–5340.

Shivananda Nayak B., Sandiford Steve and Maxwell Anderson, 2007. Evaluation of the wound-healing activity of ethanolic extract of *Morinda citrifolia* L. leaf. *e CAM Advance Access Published*, 1–6.

Shixin Deng, Brett J.West and C.Jarakae Jensen, 2008. Simultaneous characterization and quantitation of flavonol glycosides and aglycones in noni leaves using a validated HPLC–UV/MS method. *Food Chemistry*, 111(2): 526–529.

Shovic, A.C. and Whistler, W.A., 2001. Food sources of provitamin A and vitamin C in the American Pacific. *Tropical Science*, 41: 199–202.

Siddiqui, B.S., Sattar, F.A., Begum, S., Gulzar, T. and Ahmad, F., 2006. New anthraquinones from the stem of *Morinda citrifolia* Linn. *Nat. Prod. Res.*, 20: 1136–1144.

Simonsen, J.L., 1920. Note on the constituents of Morinda citrifolia. *J. Chem. Soc.*, 117: 561–564.

Singh, D.R., Rai, R.B. and Singh, B., 2005. The Great Morinda: A potential under utilized fruit for tsunami affected areas in Bay Islands. *TheUTS'Voice*. Port Blair.pp–21.

Singh, D.R., Srivastava, R.C., Chand, Shubhash and Kumar, Abhay, 2007. *Morinda citrifolia* L. an evergreen plant for diversification in commercial horticulture. In: *Monograph on Noni*, (Ed.) P.I. Peter. World Noni Research Foundation, Chennai, p. 18–33.

Singh, J. and Tiwari, R.D., 1976. Flavone glycosides from the flowers of *Morinda citrifolia*. *J. Indian Chem. Soc.*, 53: 424.

Singh, D.R., Singh, S., Minj, D., Antia, Jincy, M. and Srivastava, R.C., 2011. Evaluation of *Morinda citrifolia* L. in challenge environment condition. (In) Abstract Book: *National Seminar on Potential of Underutilized Crops*, Feb., (In press).

Singh, D.R., Singh, S. and Srivastava, R.C., 2011. *Morinda citrifolia* L.: A phytochemical rich pioneer crop for sustainable livelihood security in tropical regions". In: *International Conference on Tropical Island Ecosystems* to be held at CARI, Port Blair, A & N Islands from 23–25 March.

Singh, D.R., Rai, R.B. and Singh, B., 2005. The Great Morinda: A potential under utilized fruit for tsunami affected areas in Bay Islands. *TheUTS'Voice*, Port Blair, pp. 21.

Singh, D.R., Srivastava, R.C., Chand, Shubhash and Kumar, Abhay, 2007. *Morinda citrifolia* L.: An evergreen plant for diversification in commercial horticulture. In: *Monograph on Noni*, (Ed.) P.I. Peter. World Noni Research Foundation, Chennai, p. 18–33.

Singh, J. and Tiwari, R. D., 1976. Flavone glycosides from the flowers of *Morinda citrifolia*. *J. Indian Chem. Soc.*, 53: 424.

Srivastava, M. and Singh, J., 1993. A new anthraquinone glycoside from *Morinda citrifolia. Int. J. Pharmacog*, 31: 182–184.

Su, B.N., Pawlus, A. D., Jung, H.A., Keller, W. J., McLaughlin, J. L. and Kinghorn, A. D., 2005. Chemical constituents of the fruits of *Morinda citrifolia* (noni) and their antioxidant activity. *J. Nat. Prod.*, 68: 592–595.

Takashima, J., Ikeda, Y., Komiyama, K., Hayashi, M. and Kishida, A., Ohsaki, A., 2007. New constituents from the leaves of *Morinda citrifolia. Chem. Pharm. Bull.*, 55: 343–345.

Thomson, R.H., 1971. *The Naturally Occurring Quinones.* Academic Press, London.

Tiwari, R.D. and Singh, J., 1977. Structural study of the anthraquinone glycoside from the flowers of *Morinda citrifolia. J. Indian Chem. Soc.*, 54: 429–430.

Tosa, H., Ilinuma, M., Asai, F., Tanaka, T., Nozaki, H., Ikeda, S., Tsutsui, K., Tsutsui, K., Yamada, M. and Fujimori, S., 1998 Anthraquinones from Neonauclea calycina and their inhibitory activity against DNA topoisomerase II. *Biol. Pharm. Bull.*, 21:641–642.

Wang, M. and Su, C., 2001. Cancer preventive effect of *Morinda citrifolia* (Noni). *Ann. N.Y. Acad. Sci.*, 952: 161–168.

Wang, M., Kikuzaki, H., Csiszar, K., Boyd, C.D., Maunakea, A., Fong, S.F.T., Ghai, G., Rosen, R.T., Nakatani, N. and Ho, C.T., 1999.Novel trisacccid ester identified from the fruits of *Morinda citrifolia* (noni). *J. Agric. Food Chem.*, 47: 4880–4882.

Wang, M., Kikuzaki, H., Jin, Y., Nakatani, N., Zhu, N., Csiszar, K., Boyd, C., Rosen, R. T., Ghai, G., Ho, C.T., 2000.Novel glycosides from noni (*Morinda citrifolia*). *J. Nat. Prod.*, 63: 1182–1183.

West, Brett J., Jarakae Jensen, Claude and Westendorf, Johannes, 2008 A new vegetable oil from noni (*Morinda citrifolia*) seeds. *International Journal of Food Science & Technology*, 5: 1988–1992.

West, B.J., Jensen, C.J. and Westendorf, J., 2006.Noni juice is not hepatotoxic. *World J. Gastroenterol.*, 12: 3616–3619.

West, B.J., Jensen, C.J., Westendorf, J. and White, L.D., 2006. A safety review of noni fruit juice. *J. Food Sci.*, 71: R100–R106.

Yamaguchi, S., Ohnishi, J., Sogawa, M., Maru, I., Ohta, Y. and Tsukada, Y. (2002): Inhibition of angiotensin I converting enzyme by noni (*Morinda citrifolia*) juice. *Nippon Shokuhin Kagaku Kogaku Kaishi*, 49: 624–627.

Youngken, H.W., Jenkins H J. and Butler CL., 1960. Studies on *Morinda citrifolia* L. II. *J. Am. Pharm. Assoc.*, 49: 271–273.

Zin Mohd, Z., Hamid Abdul, A., Osman, A. and Saari, N., 2002. Antioxidative activity of extracts from Mengkudu (*Morinda citrifolia* L.) root, fruit and leaf. *Food Chemistry*, 78: 227–231.

Zin Mohd, Z., Hamid Abdul, A., Osman, A. and Saari, N., 2006. Antioxidative activities of chromatographic fractions obtained from root, fruit and leaf of Mengkudu (*Morinda citrifolia* L.). *Food Chemistry*, 94 (2): 169–178.

2011, Future Crops Vol. 1 ***Pages 172–202***
***Editor:* Professor K.V. Peter**
***Published by:* DAYA PUBLISHING HOUSE, NEW DELHI**

Chapter 10
RAMBUTAN
(*Nephelium lappaceum* L.)

Rajesh Kumar
National Research Centre for Litchi,
Muzaffarpur – 842 002, Bihar
E-mail: rajeshkr_5@yahoo.com

The rambutan (*Nephelium lappaceum* L. syn. *Euphoria nephalium* DC, *Dimocarpus crinita* Lour) an evergreen tree with its abundant coloured fruits makes beautiful landscape specimens. In fact,English common name–Rambutan is derived from Malaya word "Rambut" which means hair. As colourful fruit is completely covered with thick hairy outgrowths and a closer look reveals that it's long red, yellow and green protuberances are really soft and harmless hairs (spinterns). Rambutans grow in clusters on evergreen trees, and are hairy-looking exotic wonders. Though a close relative of litchi and equally desirable fruit, this member of the same family Sapindaceae is not nearly well known to rest of the world, because it is adopted to somewhat tropical climate and found only in South East Asia. Most rambutans are red, but in Malaysia a smaller, yellow rambutan can also be found.

Over a considerable period of time, rambutan acquired immense popularity as a seasonally available fruit in many countries of tropical Asia and is widely grown in both home and commercial orchards. This has a significant future as domestic and export fruit crop. Rambutan is an economically important crop of many South East Asian countries (Thailand, Malaysia and Indonesia) and demands from Asian immigrant fans have increased the import of this rather rare fruit into Europe, the Middle East and North America. Traditional suppliers include Malaysia, Thailand and Indonesia but today it is also grown in Africa, Cambodia, the Caribbean island, India, Srilanka etc (Lim, 1992; Zee, 1993; Zee *et al.*, 1998). It can also be found in Ecuador, where it is known as achotillo. An anticipated worldwide demand is the result of area expansion and improved handling of postharvest fruits, increased production and faster long-distance transporting of this delicate fruit. Thailand exports the most canned Rambutan while Malaysia is the largest exporter of fresh Rambutan.

The rambutan fruit is round and oval in shape with dark red shell and red with yellow/green hairs. It could be consumed when ripe, the consuming part is white flesh inside which has a sweet and tasty scent. Rambutan is a juicy, very popular and tasty fruit in many countries due to its high sugar and vitamin C content, which stop bleeding gum. Calcium in rambutan also strengthens bones and tooth while vitamin B_2 helps to digest fat and build growing body. Niacine helps to activate digestive system, relieve dysentery, floated stomach, fever and nurture the body. Rambutan roots, bark, and leaves have various uses in medicine and in the production of dyes. Seeds are used in manufacture of soaps and medicines. Due to its excellent qualitative attributes (nutritional, medicinal and ornamental), it is currently a focal crop of a rapidly expanding tropical specialty fruit industry.

Origin and Spread

The origin of rambutan is uncertain but is a native to West Malaysia and the islands of Western Indonesia (van Welzen and Verheij, 1992; Watson, 1984). The origin ranges from Southern China (Yunnan and Hainan) through Indo-China, Malaysia, Indonesia, to the Philippines. Being native of Malaysia, it is mostly found cultivated in Southeast Asia. A few rambutan trees are also grown in the coastal lowlands of Colombia, Ecuador, Honduras, Costa Rica, Trinidad and Cuba. Some trees are also cultivated in Australia (Laksmi, *et al.*, 1987). It is in the Borneo island particularly Sabah, Sarawak and Brunei Darussalam that maximum diversity occurs. Most species are edible but not all have the desired quality of *N. lappaceum.* Even within *N. lappaceum,* the fruit quality is highly variable. The wild types possess less pulp and are more sour. *N. lappaceum* and *N. ramboutan-ake* are cultivated for their fruits.

It is now cultivated in most Southeast Asian countries, Sri Lanka and in small areas in the humid tropics of America, Africa and Australia. In Vietnam, it is grown in the Mekong Delta where the climate is hot and humid. Rambutan is widely cultivated in the lowland and humid tropics of south-east Asia. Later, Arab, Indian and Malaysian traders carried rambutan far from its native range, and it was cultivated since early times from Zanzibar to Philippines. Rambutan was taken to the Philippines from Indonesia in 1912. Further introductions were made in 1920 (from Indonesia) and 1930 (from Malaya), but until the 1950's its distribution in the country was rather limited. Then popular demand brought about systematic efforts to improve the crop and resulted in the establishment of many commercial plantations in the new provinces of South East Asian Countries of Warm Climates. Today, rambutan is one of the best known fruits of Southeast Asia and is also widely cultivated elsewhere in the tropics. Rambutan is relatively a new crop in China. The crop was introduced to Taiwan as early as 1915. Many years ago, Arab traders introduced it into Zanzibar and Pemba. Then popular demand brought about systematic efforts to improve the crop and resulted in the establishment of many commercial plantations in the provinces of the United States.

Botany and Taxonomy

Rambutan belongs to family *Sapindaceae.* Its common botanical name is *Nephelium lappaceum,* but there are two other synonyms: *Euphoria nephelium* and *Dimocarpus crinita* (Morton, 1987a). The family *Sapindaceae* includes many fruit trees of about 37 genera and 72 species but only a few species are important like rambutan, litchi (*Litchi chinensis*) and longan (*Euphoria longana*) which have commercial importance (Rehm and Espig, 1984). The species like *N. glabrum* Cambess, *N. chryseum* Blume, *N. sufferrugineum* Radlk. are synonyms to *Nephelium lappaceum* var. lappaceum. Two botanical varieties of *N. lappaceum* are identified, they are pallens (Hiern) Leenh, and xanthioides (Radlk.) Leenh. There are over 22 species of *Nephelium* with edible fruits recorded, including 5 in Myanmar, Thailand and Indo-China; 13 in Peninsular Malaysia; 16 in Borneo; 4 in the Philippines and 3 in Western Java

(Tindal, 1994). Other species include pulasan (*N. ramboutan-ake*), korlan (*N. hypoleucum*), giant rambutan (*N. cuspidatum*), sungkit (*N. maingayi*) and arut (*N. xerospermoides*). It is botanically classified as follows:

- ✰ Division–Magnoliophyta
- ✰ Class–Magnoliopsida
- ✰ Order–Sapindales
- ✰ Family–Sapindaceae
- ✰ Genus–Nephelium
- ✰ Species–*N. lappaceum*

Cytogenetic studies on the chromosome number of rambutan gave varying results as Ramirez (1961) considered the haploid (n) number of chromosome to be 11, where as Jong *et al.* (1973) found haploid (n) number to be 16.

Morphological Attributes of Rambutan

Tree

The medium sized trees have an erect, dense habit with a straight trunk and grow to a height of 8 to 10 m. The height of the rambutan tree varies among different varieties from 8 to 25 m (Baertels, 1990, Franke, 1994 and Morton, 1987).Grafted cultivars are usually more compact reaching a height of only 3 to 5 m. Principal trunk is erect with an open crown of large branches; bark is slightly rugose, greyish or red. Rambutan trees are either dioecious or monoecious with male or hermaphrodite flowers and need cross pollination for fruit set.

Leaves

The leaves are alternate, 10–30 cm long, pinnate, with 3-11 leaflets, each leaflet 5–15 cm wide and 3-10 cm broad, with an entire margin. The leaflets are elliptic to oblong, pale shining with a dark-green colour. On the lower surface of each leaflet are the domatia, small crater-like hills located in the axils between the mid and secondary veins.

Flowers

The flowers are small, 2.5–5 mm, apetalous, discoidal and borne in erect terminal panicles, 15–30 cm wide. The petalless flowers are very inconspicuous because of their tiny size and their greenish colour. They are growing on long shafted and much branched, 10 to 30 cm long panicles and have sweet smelling fragrance.

Rambutan trees are either male (producing only staminate flowers and, hence, produce no fruit), female (producing flowers which are only functionally female), or hermaphrodite (producing flowers that are female with a small percentage of male flowers). Inflorescence pseudo-terminal to usually terminal; flowers either male (only stamens well developed; trees dioecious) or hermaphrodite (trees monoecious), the latter either effectively female (stamens small, anther not dehiscing) or male (stigma not opening), actinomorphic, whitish, yellowish or greenish; sepals 4-5(7), nearly free to more than halfway connate, 0.7-2.1 mm long; petals usually absent, sometimes up to 4 reduced ones, not exceeding 1.6 mm; disk complete, hairy or glabrous. The perfect flowers are functionally pistillate or staminate. Most commercial cultivars behave hermaphroditically and are self fertile, with 0.05-0.9 per cent of the functional females possessing functional stamens. Male flowers have yellow nectaries and 5-7 stamens and a non functional ovary in male panicles and greenish-yellow flowers with functional ovary in

Figure 10.1: Rambutan flower panicle.

Figure 10.2: Fruit set in rambutan.

Figure 10.1: Rambutan flower panicle.

Figure 10.2: Fruit set in rambutan.

each hermaphroditic panicle. During the different studies, it was reported that in rambutan, 3 types of flowers are produced, and trees are classified into 3 groups based upon flower morphology (Tindall *et al.*, 1994; Valmayor *et al.*, 1970). One group produces only staminate flowers, which lack a functional ovary and possesses 5 to 7 well-developed stamens with anthers which dehisce and shed pollen at anthesis. A second group produces only hermaphroditic functionally female (HFF) flowers which set fruit but do not shed pollen. Insect pollination is necessary. Depending on the cultivar, flowering may spread over a period of 23-38 days, with an average of 3.4 per cent setting fruit.

Pollination and Fruit Set

Honey Bees foraging for nectar routinely contact the stigmata of female flowers and gather significant quantities of sticky pollen from male blossoms. Little pollen is seen on bees foraging female flowers. Although male flowers open at 6 am, foraging by *A. cerana* is the most intense between 8 am and 11 am, tapering off rather abruptly thereafter. In Thailand, *A. cerana* is the preferred species for small scale pollination of rambutan. Fruits may be produced in large bunches, with 40-60 fruits/ panicle, but most often only 12-13/panicle are retained to maturity. Final fruit set is usually between 0.7-1.45 per cent. Time required from fruit set to harvest is about 105-115 days. Each flower has six anthers, usually a bi-lobed stigma, and one ovule in each of its two sections (locules). The flowers are receptive for about one day but may persist if pollinators are excluded.

Fruits

The fruit is a round to oval drupe 3–6 cm (rarely to 8 cm) tall and 3-4 cm broad, borne in a loose pendant cluster of 10-20 together. The leathery skin is reddish (rarely orange or yellow), and covered with fleshy pliable spines, hence the name *rambutan*, derived from the Malay word *rambut* which means hairs (Ito and Hamilton, 1990; Tindall, 1994). The fruit weighs 20-95 g, usually consisting of only 1 nutlet, yellowish to purplish-red, hardly stalked, often finally dehiscing (at least the apical part) and glabrous, Fruit has numerous, characterizing, long, soft, red or red and green colored spine-like protuberances (spinterns) on surface of the peel. The pericarp of this attractive oval-shaped fruit can be red, orange, pink, or yellow in coloir and is removable by a twist of hands. The edible, pearlish white, juicy, crispy, sweet and subacid flavoured flesh (sarcotesta) conceals a single seed with a thin, fibrous seed coat (testa).

The fruit matures in 15 to 18 weeks after flowering and gets an ovoid, round or ellipsoid shape with a pinkish-red, bright- or deep-red, orange-red, maroon or dark-purple, yellowish red or yellowish colour. The size of the rambutan fruit varies 3.4 to 8 cm (Baertels, 1990, Franke, 1994 and Morton, 1987a). The white or rose-tinted and translucent 0.4 to 0.8 cm thick flesh is very juicy and covers one ovoid to oblong almond-shaped dark seed which is 2.5 to 3.5 cm long and 1 to 1.5 cm wide. Rambutan fruits have an exotic aroma which results out of the interaction of fruity-sweet and fatty-green odours, with the possible contribution of civet-like sweetish, spicy and woody notes (Ong *et al.*, 1998).

Seeds

The single seed is glossy brown, 2–3 cm, with a white basal scar. The seed is soft and crunchy. Seeds are mildly poisonous when raw, but may be cooked and eaten. The single light brown seed rich in certain fats and oils (primarily oleic and arachidic acids) is valuable to industry, and used in cooking and manufacture of soap.

Climate and Soil

In most countries, distribution of rambutan cultivation seems to be more influenced by climatic than by other factors. The rambutan tree prefers a tropical climate (within 15º of the equator) on land

Figure 9.3: Rambutan plant in full bloom.

below about 500 m elevation. The climate should be warm and wet with a few dry periods (van Welzen and Verheij, 1992; Yaacob and Subhadrabanhu, 1995). A moist, hot climate with a fairly evenly distributed rainfall pattern is the main requirement and regions with precipitation ranging from 2,000-5,000 mm/annum are generally suitable for rambutan cultivation. Rambutan, grown commercially within 15° of the equator (Nichols and Christie 1993), is adapted to warm tropical climates and sensitive to low temperatures (below 10°C). The major rambutan cultivation areas in the Pennisular region of Malaysia coincide closely with agro-climatically dry zones located in areas with dominant and distinct dry period lasting 2-3 months a year (Nieuwolt *et al.*, 1982). The commercial rambutan is adapted to warm tropical climates and is sensitive to temperatures below 10°C, and is grown commercially within 15° of the equator. In Thailand, the rambutan is a tropical tree the best grown in temperatures ranging 22°C to 35°C, with 2000 to 3000 mm well distributed rainfall. The annual precipitation and rainfall distribution patterns have a direct effect on the vegetative growth, flower initiation and fruit development of rambutan. It is intolerant to frost, especially during the juvenile stage. Mature tree survives a brief period of temperature as low as 4°C, but with severe defoliation. *N. lappaceum* thrives in humid tropical lowlands within about 17 degrees from the equator. The trees occur in the lower or middle storey in different types of primary and secondary forest ranging from dryland to swamp. Exposure to dry winds leads to browning of the leaf margins; sheltered locations or wind screens are recommended. Flower induction is independent from photo period, but can be induced by water stress. Low relative humidity and wind during fruiting could cause excessive moisture loss from fruit spinterns and result in poor fruit appearance. It is intolerant to frost, especially during the juvenile stage. Rambutan production in Hawaii occurs in environments, very different from traditional growing areas of the humid tropics. As a consequence, production is very erratic due to inconsistent flowering and the occurrence of deformed or aborted fruit that lacks a fleshy aril. The dry season should not last more than three months (Rehm and Espig, 1984). According to Morton (1987) the Oriental Mindora region of the Philippines with an average temperature of about 27.3 °C, a relative humidity of 82 per cent and about 165 rainy days is the ideal environment. In Indonesia, rambutan production is successful when cultivated in locations having a temperature of 20-30°C and needed altitude is <500msl. The plantations in locations of >600msl have low productivity due to low temperature affecting pollination.

The rambutan can be grown successfully in a wide range of soils but optimum growth and fruiting occur on rich, well drained sandy loam, rich in organic matter. The soil should generally be well-drained and slightly to moderately acidic (van Welzen and Verheij, 1992; Yaacob and Subhadrabanhu, 1995). It also prefers clay loam soil, pH 5 to 6.5, but can be grown in a wide range of soil types, even ones with poor drainage, but not water-logged. The trees do the best on deep soils rich in organic matter and thrive on hilly terrain as they require good drainage.

Production and Productivity

Rambutan is an emerging tropical fruit crop and is profitable to grow (Ngo and Baker, 1990). Majority of this development has occurred recently after realizing the potential of this crop. There is not much information available about economics of rambutan in many of its growing countries. Rambutan is grown on medium scale commercial cultivation and the fruit has a high local demand (Doasl, 2007 and Barry, 2007). Statistics show significant growth of cultivation of rambutan (Doasl, 2007).

Production varies from year to year due to biennial nature of the crop with a gradual upward trend resulting from new plantings coming into production. At the same time, smaller older orchards are being abandoned due to high management requirements of the crop. Most orchards are less than 500 trees in size. In North Queensland there are several orchards with tree numbers in the 2,000 to 4,000 range. Average tree yields range from 30 to 60 kg, with yields up to 200 kg/tree being recorded on older trees at widr spacings. Australian production varies from 500 to 1,000 tons/annum with approximately 50 to 150 tons being exported.

Composition and Uses

The principle biochemical constituents in fruit pulp are carbohydrates, sugars, organic acids, vitamins and proteins. The nutritive value and pleasant aroma of rambutan fruits make its juice a delicacy as its pulp contains about 80 per cent moisture. Apart from vitamin C and calcium, rambutan fruit provides niacin, iron and fiber. There are 59 calories of energy from an average fruit. It is observed from the various sources that a 100 g sample of rambutan flesh is composed of 79-80 per cent water, 0.9 to 1.0 per cent protein, 0.3 per cent fat, 0.3 per cent ash, 13-16 g carbohydrates, 5-6 g sucrose, 1.0-2.5 g dietary fiber, 0.05 per cent malic acid, 0.31 per cent citric acid, 0.5 mg of niacin, 7-15 mg of calcium, 0.1 to 2.5 mg of iron, 40-70 mg of vitamin C, 140 mg of potassium, 2–6 mg sodium and 10–16 mg of magnesium (Morton, 1987; Nakasone and Paul, 1998; Anon, 2003; and Sivakumar *et al.*, 1997).

Table 10.1: Food value/100g edible portion of rambutan fruit.

Sl.No.	*Edible Portion*	*I*	*II*	*III*	*IV*
1	Moisture (g)	82.3	79.5	84.7	–
2	Protein (g)	0.46	1.05	0.7	–
3	Lipid (g)	–	0.45	0.1	–
4	Total carbohydrates (g)	16.02	15.7	13.9	–
5	Reducing sugars (g)	2.9	–	–	–
6	Sucrose (g)	5.8	–	–	–
7	Fibre(g)	0.24	2.8	0.3	–
8	Calcium(mg)	10.6	8.22	7.9	–
9	Magnesium(mg)	10	10	–	16
10	Potassium (mg)	–	140	–	179.7
12	Iron (mg)	–	0.1	2.5	0.5
13	Phosphorus (mg)	12.9	12.9	30	16.5
14	Sodium (mg)	–	–	–	6.3
15	Ascorbic acid (mg)	30	70	38.6	36.4
16	Energy	–	297kJ	59cal	–
17	Ash (g)	–	0.3	0.3	–

Cf. Pohlan *et al.* (2008).

I. Morton, 1987; II. Nakasone and Paul, 1998; III. Anon, 2003; IV. Sivakumar *et al.*, 1997.

Food Value

The values above show that rambutan is a valuable contributor to our daily nutritional needs. The calcium, magnesium and potassium are all important essential elements in a balanced diet. The acidity or pH of 4.66 combines with a relatively high sugar levels to give the fruit great intensity of flavour. There is a natural sweet and sour effect as the acids play against the sugars to great effect. The trees are cultivated for their very popular fruit. The usually juicy sarcotesta around the seed is eaten. The sweet-tasting fruits are consumed fresh, the more sour ones are eaten stewed. The sarcotesta can be canned or used in jam, but loses much of its flavour. Seeds are edible when roasted, they are otherwise bitter and said to be narcotic.

Medicinal Uses

The fruit (perhaps unripe) is astringent, stomachic and is taken to relieve diarrhea and dysentery. In Java, the toxic saponin extracted from the seeds is dried and used as medicine. A source of vitamin C and calcium, rambutan provides niacin, iron, protein and fiber. In Malaysia, roots are used in a decoction for treating fever; and the bark as an astringent for tongue diseases. The leaves are poulticed on temples to alleviate headache. In Malaya the dried fruit rind is sold in drugstores and employed in local medicine. The astringent bark decoction is a remedy for thrush. A decoction of the roots is taken as a febrifuge. Eating five daily servings of fruits and vegetables lowers the chances of cancer. A recent study found eating nine or ten daily servings of fruits and vegetables, combined with three servings of low-fat dairy products, were effective in lowering blood pressure.

Other Uses

There is usually a single light brown seed which is high in certain fats and oils (primarily oleic and eicosanoic acids) valuable to industry, and used in cooking and the manufacture of soap (Almeyda, *et al.*, 1979; Kalayasiri, 1996; Tindall, 1994). Rambutan roots, bark and leaves have various uses in medicine and in the production of dyes (Lim, 1984). The wood of *N. lappaceum* is liable to splitting during seasoning. It is moderately hard to very hard, strong and tough. The wood is easy to work and can be finished well. It is durable under cover and generally resistant to insect attacks, but susceptible to fungal attacks. The reddish coloured rambutan wood is usually too small to be valued as timber and generally used as firewood. Young shoots are used to produce a green colour on silk that is first dyed yellow with turmeric. The fruit walls are used, together with tannin-rich parts of other plants, to dye silk black after a preliminary red staining. Leaves are used, together with mud, as an impermanent black dye. Seeds contain an oil formerly used for illumination and a fat used formerly for soap. The seed kernel can be used for production of rambutan tallow, a solid fat similar to cacao butter, which is used for soap and candles. The fruit wall contains a toxic saponin, which is poisonous to be consumed. The rambutan has ornamental values too, as it is beautiful, especially when in full fruit, making it a popular ornamental tree for backyards and public streets. The colourful fruits are frequently used in displays with flower and fruit arrangements.

Breeding and Genetic Improvement

Breeding and crop improvement attempts in rambutan appear to be limited as compared to other members of its family and as a result only insignificant genetic study in rambutan is reported. Present cultivars are mainly obtained either from chance seedlings or somatic mutation of wild species. Each country has selected materials suited to its growing climate and local palette.

Breeding Objectives

The major breeding objective of rambutan remains high and stable yield, but now it focuses on the cultivation of rambutan cultivars producing more number of flowers as hermaphrodite functioning as female (hff), early or late ripening periods, special qualities, storability (increasing shelf-life), processability, resolving problem of alternate/irregular bearing and stress (biotic as well as abiotic) resistance etc.

Rambutan Genetic Resources

The genetic resources of rambutan have always been taken care off alongwith the genetic variability of tropical fruit crops of Southeast Asian region. There are records of the genus *Nephelium* (Sapindaceae), which contains 22 species, with 13 species occurring in Malaysia and three of these are endemic species. This genus is distributed in Indo-Malaya, the Philippines and Borneo (Verheij and Coronel 1991). Detailed information on genetic resources of rambutan are very limited till recently, but realizing the significance of genetic variability as natural resource and the increased threats of crop gene pools due to urbanization and deforestation in Southeast Asian region, IBPGRI and E&NRRC, MARDI, Malaysia have taken lead to carry out systematic collection and conservation of *Nephalium* species. Salma (1993) identified and developed a key to 31 cultivars of rambutan grown in Malaysia out of a collection list that exceeds 65. Apart from Malaysia, a few institutes in China, Thailand, Indonesia, Australia, Mexico and Srilanka are also maintaining germplasm of *Nephalium* species for future crop improvement. Till date most of the cultivars originated from wild variety *Nephelium lappaceum* var. *lappaceum*.

Breeding Methods

Introduction

To increase the genetic resources variability in rambutan, importation of superior cultivars from Asia and elsewhere to the non-traditional areas of similar growing condition should be encouraged and the process of exchange of germplasm should continue among the important growing region/ provinces in different countries. It takes years for widespread distribution of superior clonal material. Introductions of selected clones are now available in tropical America including R-132, R-162 and R-167 from Malaysia; JT-Li from Singapore; Si Chomphu and Rong Rian from Thailand and Smooth from Hawaii.

Selection

Since rambutan is a cross pollinated crop, large genetic variation has occurred in nature over generations and numerous varieties are identified through selection. But their nomenclature is confused. A cultivar selection programme was initially started in the late 1970s in Malaysia with the objective of testing promising clones under different agroclimatic conditions. Later in Malaysia, Thailand and Indonesia there are many superior selections made by both public and private entities. Cultivars are clonally propagated and selected to serve specific target markets. Selection criteria are based on fruit characteristics mainly spine length, fruit wall colour, aril thickness, and aroma, adherence of aril to the seed, vitamin C content and fruit set. The Australian industry has the opportunity to select material from over 50 imported clones. Most of the selected clones are marketed as Classic Red which includes six varieties (Binjai, Jitlee, R134, R156 (red), R162 and R167). These varieties are similar in colour and shape, and the flesh is easily removed from the seed. Further two varieties R 9 and Rongrien are marketed separately due to shape and colour differences.

Progeny Selection

In 1960, 10 outstanding rambutans were selected in an evaluation of 100 seedling trees of the unsurpassed Indonesian 'Seematjan', also 'Seenjonja', 'Maharlika', 'Divata', 'Marikit', 'Dalisay', 'Marilag', 'Bituin', 'Alindog', and 'Paraluman'. Seedling orchards have low production due to existence of 50 per cent unproductive trees, irregular growth, production and fruit quality, but showing good amount of variability.

Hybridization

Both outcrossing and selfing occur in rambutan even though breeding and crop improvement through hybridization are limited. Report of Sarip *et al.* (1996) reported outcome of a large-scale evaluation of F_1 hybrids based on two maternal parents (R99 and R134) and 14 popular Malaysian cultivars. Six years after the establishment of 7,000 open-pollinated seedlings, 50 per cent of the population had flowered and about 40 per cent of them were males. Seven(per cent) of the population flowered two weeks earlier than both maternal parents and less than 1 per cent produced high-quality fruits with the combined attributes of good appearance, high recovery and cling free. In Thailand, Seetong or Seetongchoke is a relatively new cultivar which has arisen from hybridization between existing cultivars. Seedless varieties of pulasan (*Nephelium mutabile*) are known in cultivation and could be used to hybridize with rambutan to produce new/better type of fruits.

Biotechnology

Sixty four accessions involving 6 taxa (4 species and 2 taxonomic varieties) of *Nephelium* were fingerprinted, classified and evaluated for interspecific genetic diversity using isozyme polymorphism (Aradhya *et al.*, 1996). Multivariate analysis of the isozyme data suggests that *N. hypoleucum* is the closest relative of *N. lappaceum var. lappaceum* among 5 distinct clusters and the isozyme results dispute among the taxonomic relationships in *Nephelium* species. Randomly amplified polymorphic DNA (RAPD) markers were successfully employed to reveal genetic diversity and genetic relatedness among 211 *Nephelium lappaceum* accessions planted at the MARDI-IPGRI rambutan field genebank, Kemaman, Terengganu. A total of 270 clear, consistent and reproducible fragments were generated for Nephelium accessions by using eleven single random oligonucleotide primers. Genetic variation is high among the *N. lappaceum* accessions, with an average degree of polymorphism of 98.45 per cent and mean Shannon index, H0 = 6.00. The RAPD technique was able to differentiate all the *N. lappaceum* accessions studied as well as two TV. rambutan-ake accessions (*i.e.* PR155 and PR189) found in the same field genebank. The Jaccard index of genetic similarity ranged 0.232 to 0.900 among the accessions, with a mean value of 0.444 (Chew *et al.*, 2005). Buerki *et al.* (2009) carried out phylogenetic analysis among genera in Sapindaceae family using plastid and nuclear markers and showed the relationships of the genera. Nevertheless, the molecular phylogenetic relationships of *Nephelium* species have not been investigated, and this study is to examine the species relationships within the genus *Nephelium* using sequence data from the chloroplast (*trn*L-*trn*F, *trn*T-*trn*L and *trn*D-*trn*T) and nuclear (ITS and UFGT) genomes.

Varieties

There are more than 200 cultivars of rambutan in Southeast Asia and in the Asian region (Indonesia, Malaysia, Singapore, the Philippines, Thailand and Brunei Darussalam), the principle commercial cultivars are showing in Table 10.2.

Rambutan is grown for its attractive red colour hairy fruits, rich in quality attributes. Some attributes of a good variety/cultivar are:

(*a*) Fruit weight over 40 g with flesh recovery ratio over 50 per cent (*i.e.* fruit is more than 50 per cent flesh by weight). (*b*) Red/orange colour which is resistant to insect attack. Yellow fruit is more susceptible. (*c*) Resistance to stem canker. (*d*) Fruit with a fewer than 5.5 spinterns/cm^2 fruit surface and a spintern length under 12 mm to reduce moisture loss. (*e*) Fruit with a freestone and testa-free flesh which is firm and crisp. (*f*) Early and contracted flowering to reduce fruit loss from bird attack. (*g*) Yields of 12-15 tons/ha based on 8 year old trees at a population of 100 trees/ha, and (*h*) Tolerance of

Table 10.2: List of rambutan growing countries alongwith important cultivars/varieties.

Sl.No.	*Country*	*Cultivars/Varieties*	*Reference*
1.	Malaysia	R134, Gula Batu of Peng Thing Cheng (R3), Muar Gading (R156), Khaw Tow Bak (R160), Lee Long (R161), and Daun Hijau or Ong Heok (R162), Deli Cheng (R170), Azimal, Kelip and Singapura, Chooi Ang, Peng Thing Bee, Ya Tow, Azimat, Atjeh koonig (yellow) and Ayer Mas.	Salma (1986)
2.	Thailand	Lebakbooloos, Seematjan, Seenjonja, Sectangkooweh, Seelengkeng, Seekonto, Rongrien, Seechompoo, 'Bangyeekhan, Seetong or Seetongchoke, Namtangruad and Jemong.	Muchjajib (1990)
3.	Indonesia	Atjeh Kering, Manis, Silengkeng, Binjai, Lebak Bulus, Aceh Rapiah, Pao Pao, Sinyonya (Seenjonja), Aceh Gading, Aceh 6B, S.K.W.L., Aceh Pandang, Biji, Garuda, Aceh Gundul, Aceh Kuning, Sitangkue, Simacan (Seematjaan), Aceh Gendut, Aceh Gedong, Pandang Bulan and Gula Batu.	Watson (1984); Nazaruddin and Muchlisah (1994).
4.	Philippines	Seematjan, Seenjonja, Mahalika, Diwata, Marikit, Dalisay, Marilag, Bituin, Alindog and Paraluman	Salazar (1956) and Valmayor *et al.* (1961)
5.	Australia (Hawaii and Puerto Rico)	R3 (Gula Batu), R9, R134, R156 (Muar Gadang-red), R156 (Muar Gadang- yellow), R162 (Daun Hijau), R167 (Chai Tow Cheng), Silengkeng, Binjai, Jitlee, Rongrien,	Watson (1988)
6.	Srilanka	Malaysian Red, Malaysian Yellow and Malwana	Doasl (2007)
7.	China	BR-1, BR-2, BR-3, BR-4, BR-5, BR-6, BR-7, BR-8, BR-9.	Ruan (1989)
8.	Singapore	Deli Cheng, Jitlee, R37, Ah Kow Cheng and Kim San Cheng	
9.	USA	Introductions from Malaysia, Thailand and Indonesia.	
10.	India	Lebakbulus, Seelengkeng, Seedling selections	

temperatures below 15°C. (*i*) Clonal trees are shorter in stature and tend to bear earlier than seedling trees. The description of some important cultivars/varieties grown in different rambutan growing countries:

Lebakbooloos

The cultivar has a broad-topped tree with dark-red, dense fruits having uncrowded spines 3/5 in (1.5 cm) long, and greyish-white, tough, subacid flesh 1/5 in (0.5 cm) thick, frequently difficult to separate from the seed and often takes pieces of the testa with it. Ships well over long distances.(Cultivated in Thailand, Indonesia and India).

Rapiah

This cultivar does not produce dense fruit but it has good quality. The hair of the fruit skin is gradation of yellow and red, rugged and rather long. The flesh is sweet and slightly dry, rubbery easily and thick separable from seed and thick. The yield is 18-30kg/tree. The fruit weighs as much as 18.9g. The fruit can last six days after being harvested (Cultivated in Thailand and Indonesia).

Garuda

The fruit of this cultivar has red hairs with yellowish tips and is rather oval in shape. The flesh is sweet and the cultivar is high yielder (200-270kg/tree) (Cultivated in Indonesia).

Seematjan

Tree has an open crown and long, flexible branches. Fruits are dark-red with spines to 2 cm long. In Java, the tree is especially prone to attack by various insects. It is cultivated also in India and in the Philippines where it has averaged 16 kg/ha. There are 2 forms: 1) 'Seematjan besar' with small fruit, thin rind, spines fairly far apart; very sweet, somewhat coarse, fairly juicy flesh to which the coarse, fibrous testa tightly adheres; 2) 'Seematjan ketjil' (or 'Koombang')–the fruit has soft, tough and less sweet flesh to which the seed coat does not tightly adhere.

'Seenjonja'

Tree low-growing; has a drooping crown. Fruit nearly ovoid, about 4 cm long and 3 cm wide; dark wine-red with slender, flexible spines about 1 cm long. Flesh clings firmly to the seed. In Philippines, it has yielded on an average 41 kg/ha.

'Sectangkooweh'

Tree broad-topped. Fruit flattened ellipsoid, about 5 cm long, 4 cm wide with slim spines 1 cm long. Rind is thin, pliable, tough, flesh yellowish-white, sweet, clings tightly to the thick testa which separates from the seed. Fruits stand long-distance shipment.

'Seelengkeng'

Tree low-growing with drooping crown. Fruit ovoid, 3 cm long, 2 cm wide, with very fine, soft spines. Flesh slightly glossy, tough, moderately sweet and separates from the seed with a few particles of testa clinging to it. Air-layers are unsatisfactory, so it is rare in cultivation and expensive on the market. Much favoured by Chinese because of its resemblance to the lychee. (Cultivated also in India.)

'SiChomphu'

This is one of the old cultivars that stood the test of time (from Thailand). Fruit ellipsoid, red to reddish orange, with medium fruit flesh density and good taste. The hairs are of medium length, medium density, with high hair hardness. It has good aroma with sweetness of 18-20 per cent.

'Rong Rian'

A chance seedling derived from SiChomphu and named Rong Rian (from Thailand). It has exactly the same general characters as that of See Chomphu, except for colour of fruit, which is reddish, and the easier loosening of the flesh from the seed.

'SiThong'

A new cultivar was recently released by the Department of Agriculture, Thailand. It was derived through a breeding programme. It is a late cultivar, fruit ellipsoid, reddish, with medium fruit flesh density and good taste. The hairs are of medium length, medium density, with medium hair hardness. It has medium aroma with the sweetness of 18-20°Brix.

'BR-4'

The tree is of medium vigour with an open umbrella–shaped crown (from China). It grows to 4.1m in height and 5.6x5.6m in spread. The cultivar matures from early July to early August and is considered a mid season one. The fruit weighs 30g, oval shaped and red. The skin is thick (4.4mm) with long (1.1mm) and dense hairs. The flesh which separates easily from seed is sweet and slightly sour and soft to crispy, with a recovery of 42 per cent and TSS 18 per cent. The seed weighs 9.5g, compressed and long oval shape.

'BR-7'

The cultivar from China. The tree is vigorous with a spherical crown and compact branches. The fruit is red and nearly round. The fruit weighs 44g. The skin is thick with sparse, short and thick hairs. The flesh is crispy, juicy and sweet with strong fragrance. The nearly round seed separates easily from the flesh. The cultivar is high yielding.

Propagation

Rambutans can be propagated by means of both vegetative as well as by seeds. Normally vegetative propagation is adopted.

Seed/Sexual Propagation

Rambutan seeds are recalcitrant (short-lived and should be sown directly after they are extracted and washed) and rapidly lose viability after removal from fruit. Seed storage behaviour is recalcitrant, whole seed moisture content (mc) at shedding is 36 per cent, viability is reduced by 12 per cent on desiccation from 36 per cent to 25 per cent mc, and no seeds remain viable when the mean mc is reduced to 13 per cent (Teng, 1977). Excised embryos tolerate desiccation to 14-15 per cent mc, 67 per cent survive desiccation to 8-9 per cent mc; and 40 per cent survive overnight in liquid nitrogen when excised embryos are treated with appropriate chemicals plus sucrose followed by partial desiccation (1-2 hours). Progenies from seeds are highly variable, but seeds are used for the purpose of raising seedlings as rootstocks for grafting. Various forms of rambutan, including some cultivars as well as seedlings, have used as rootstocks, having been selected for their adaptability to various ecological conditions. The most common trees grown from seed often produce sour fruit and that too bear after 7-9 years. Freshly harvested seeds germinate readily within 14-15 days but seed position for sowing also affects germination period and percentage. It was the fastest if seeds were sown horizontally on the flattened side.

Vegetative Propagation

Rambutan is propagated by various vegetative method of propagation (Walter, 1976) mainly air layering, grafting and budding.

Marcotting or Air-Layering

This method produces well-rooted propagules, but survival rate is always inconsistent/poor. The success (per cent) in this traditional method can be enhanced by mixing IBA @ 1000-2000ppm in the rooting media, seemed to have better root formation and success. In Srilanka, air-layering in November showed maximum rooting. However, layering in July coinciding with high humid condition caused the earliest rooting in 30 days (Heenkenda, 1988). In Thailand, SeeChompoo cultivar was very responsive to air-layering with or without use of growth promoting chemicals. Relatively immature branches upto pencil (10mm) thickness and 30-40 cm long, have given good results with this technique producing rooting (per cent) more that 80. Young shoots with a fast growing potential in a position on the parent tree where there is adequate light, respond the best to marcotting/air-layering. Successful root initiation largely depends on the downward flow of essential substances from developing leaves and shoots which are growing above the marcot. Use of moist sphagnum moss grass as rooting medium is better as compared to use of soil as rooting medium in most of the cases.

Rootstocks for Budding and Grafting

Studies were undertaken (Indonesia) on root-scion relationships. Attempts were also made to identify dwarfing rootstocks for rambutan (Perron, 1987) and it was suggested that the use of such rootstock would reduce the costs of production and harvesting. Rootstocks may be propagated vegetatively, using the etiolation method of layering, or from seed.

Grafting and Budding

Approach grafting or inarching is highly successful, but it is more labour intensive. The method of T-budding is more popular in Thailand, Philippines and Indonesia. Patch budding and Shield budding were also tried with success in some rambutan growing countries (mainly Thailand and China). A popular method of budding of Rambutan is the modified Forket system, which is used in Malaysia. Bud grafting succeeds the best in dull, cloudy weather but in countries without marked dry season. It is suggested that it can be done any time except during or immediately after rain. Bud wood separated should be handled with proper care before completion of the budding operation. The time of budding operation and enriched soil medium of the rootstock play very significant role with enhanced success and growth rate in this operation. The best sources of budwood are trees which are normally good bearers, but which have borne little fruit in current season. Budded trees fruit after 2–3 years with optimum production occurring after 8–10 years. Well-grown rootstocks are bud grafted at 8 to 12 months. Dormant buds with well-healed petiole scars from one to two year old branches, averaged 80 per cent success between May and October. Rootstock should be cut back 25 cm above the bud union and all foliage removed at two weeks after budding. This cutting back and defoliation promote bud break of the new graft 14 to 17 days later. In Philippines, the rambutan is successfully cleft grafted on Babua (*Nephalium intermedium* Radlk.) using non-petioled green to light brown budsticks.

Cultivation

Planting

Rambutan is usually planted at 6 to 10 m distance and more commonly 8x8m spacing, with sufficient wind protection. Other spacings are also observed, depending on the vigour of the stock-scion combination and on the growing conditions like soil depth, fertility and irrigation. The most common planting distance for rambutan is 8-10 m within rows and 10 m between rows. A square or triangular planting system is used, giving densities of 100-150 trees/ha. A few orchards have wider tree spacing of 12 m x 10 m giving a density of 80 trees/ha. Inter-cropping is quite common in commercial orchards. Planting of different cultivars in one orchard, to enhance pollination, is advocated. Economic life of a tree is about 15-20 years and may be up to 30 years. Depending on the location, rambutan can produce up to 2 crops a year.

Nutrition and Irrigation

Application of manures and fertilizers is an integral part for ensuring sustained quality production from perennial fruit trees. For growing trees, a fertilizer rate of 200 g nitrogen, 25 g of phosphate and 100 g potassium/tree/year of age is recommended. For the first 4 years, the fertilizers should be applied in 4 equal dressings, every 3 months. For fruiting trees, 200 g N, 25 g P and 130 g K/tree/year of age is recommended. Maximum fertilizer rate is reached at 12 years, and should remain constant thereafter during the commercial bearing period. Detailed nutritional investigations have not been carried out on rambutan and the most fertilizer recommendations are empirical (Tindal, 1994). Rambutan trees are shallow rooted and very sensitive to drought. Irrigation is a necessary part of production in the wet-dry tropics. In the growing areas, the dry season is the main period during which irrigation is required. Irrigation rates and frequency vary from orchard to orchard. Occurring wet season although reliable in terms of total rainfall, is still a period in which irrigation is required due to in-season variability (Mollah, 1986). Still there is little scientific exercises on the knowledge of

Figure 10.4: Rambutan plantation in commercial bearing stage.

water requirements at various growth stages and climatic seasons (Tatt, 1976) except a few attempts by Diczbalis (1997) who developed an irrigation management strategy for rambutan production in the Darwin region. Rambutan irrigation requirements during fruit filling are also important. When low rainfall or dry conditions occur, the observations suggest that fruit evaporation rates are higher and are prominent in the Northern Territory of Australia (Lim and Diczbalis, 1998).

Weed Control

Control of weeds is very important for orchard sanitation and clean cultivation and is a recurring cost in rambutan orchard budget. Weed control measures like mechanical methods by garden tractors and power tillers are useful. Two main types of chemical herbicides *viz.* (i) Paraquat (selective) contact herbicide can be used in rainy season to control annual grasses/weeds and (ii) Glyphosate a non selective systemic herbicide now widely used to control perennial weeds in many orchards. One should be careful about the use of weedicides. At any stage, do not use glyphosate herbicide near the drip line of rambutan, it can cause a severe yellowing and abscission of the lower leaves.

Mulching

The application of dried weeds, grass and other organic residues as a mulch around the base of both young and mature rambutan tree is a commonly used means of retaining soil moisture, preventing weed growth, reducing soil temperature through incorporation of the decayed residues and contributing soil fertility. Mulches are normally placed around the base of the tree, to a depth of 10-15cm, leaving the area around the base of the trunk free. Decomposed mulches should be replaced or added to at regular intervals. Mulching is essential during establishment and dry periods. No mulching should be applied prior to flowering.

Inter-cropping

The planting of cash crops or green manure crops among young trees can be beneficial, but crops should not be planted too near to the trees. Legumes with low growth habits like broad bean, sunhemp and cowpea can also be intercropped with rambutan for additional benefit.

Plant Growth Regulators and Horticultural Treatments

In Thailand, rambutan growers do not keep male trees in the orchards, since these male trees do not bear fruits, and is a waste to grow them. Such practice results in low fruit setting as a fewer amount of pollens are brought in by honeybees to pollinate the female flowers. To overcome this problem, the farmers are spraying a hormone NAA, to a few plants to alter the female to male flowers so that they can be used as source of pollen and this really helps in increased fruit setting of Rambutan.

Pruning and Training for Canopy Management

Little research was done on the need and effectiveness of pruning in rambutan. Rambutan trees exhibit strong apical dominance and have a tendency to produce long, upright growth. Early pruning and training aim to achieve a strong framework of branches to encourage secondary lateral branches. Annual pruning is designed to remove water sprouts, pests and diseases infected and dead or crossing branches (Buisson, 1986). In some areas, panicle is pruned after harvest to induce vigorous canopy growth. The effect of pruning on cropping of rambutan in the wet/dry tropics of Australia was studied (Menzel *et al.*, 2000). Pruning delays harvest and reduces fruit yield the following year if done late after harvesting and severe pruning for corrective structural pruning to form an open centre tree or semicircular tree. After harvesting, fruited twigs are pruned back to stimulate new growth of upto 4 new side shoots, of which 22 per cent of the shoots bear fruitssss in the following season. At many a places, in rambutan plants little or no pruning is practiced. If the plant become "leggy" due to poor light conditions then cut it back greatly to force it to bush out. If the top become too large for the root structure, again severe pruning is needed. Also when you start to get leaf shed and twigs die back, this is an indication that the root structure is not large enough for the top foliage and pruning is needed. Timely pruning treatments also improve synchronization of harvest. Presently modified leader system is in practice in Thailand (Tindal, 1994).

Physiological Disorders

Skin Splitting

Pre-harvest disorders include skin splitting (O'Hare, 1992). Skin splitting occurs in thin skinned cultivars often following heavy rains during the last phase of fruit growth. High temperature coupled with drought during fruit growth was found associated with abnormal skin development and increased cracking in thin skinned rambutan cultivars. In poor managed orchards receiving heavy irrigation at once or heavy rains abruptly after a long gap creating imbalance between temperature and moisture in the environment, cracking becomes more pronounced. It can be managed by proper irrigation and nutrition management practices.

Retarded Fruits (Poor Filling of Fruit)

Production of deformed rambutan fruits is attributed to poor pollination and an insufficient amount of male flowers in Hawai'i orchards (O'Hare, 1992). Poor filling is associated with poor nutrition and dry conditions just after flowering. Poor fruit set and development of fruits which lack an aril are recurring problems in many rambutan orchards in Hawai'i. Studies in affected orchards showed that exclusion of insect pollinators during anthesis resulted in development of a high (per cent) of these deformed fruits and indicated that insufficient pollination was the cause of the problem (Nagao *et al.,* 2002). Production of deformed fruits was similar to reports from Thailand where parthenocarpic type fruits failed to develop fully and contained little aril (Kosiyachinda and Salma, 1987). A shortage of male flowers results in poor fruit set and development of parthenocarpic type fruits which contain little aril (Kosiyachinda and Salma, 1987; Salakpetch, 2005). Establishing male trees into the orchard, suspending male panicles from branches during flowering and using napthaleneacetic acid (NAA) to increase male flower development are methods used to aid pollination in Thailand (Salakpetch, 2005).

Chilling Injury

Darkening of spines (spinterns) and skin are major postharvest disorders. Darkening is due to dehydration and mechanical injury (Landrigan *et al.,* 1996). Low temperature storage with high humidity reduce the extent of damage.

Insect Pests and Diseases

Insect Pests

Rambutan is host to 118 different species of insects, but only 17 identified as attacking rambutan fruits and out of those only the following pests are important in rambutan.

Rambutan Stink Bugs (*Tessaratoma javanica* Thnb.)

This is the major pest and affects almost all the plant parts like fruit, inflorescence, flowers, stem and leaves of rambutan. It is found widespread throughout Malaysia and Philippines. It is also found distributed in China, India, Indonesia, Sri Lanka, Taiwan and Thailand. The adult large bugs develop in cluster and feed on the sap of young leaves, shoots and young inflorescences producing toxic saliva. They eject an odorous liquid when they are disturbed and they may cause necrotic spots which develop on young leaves. It also sucks sap from flower panicles, young fruits and newly developed shoots, causing withering and later abscission of those infested organs. The infested fruit ceases to develop. In many countries of rambutan production, this stink bug can be of epidemic level during March to April, which coincides with flowering and fruit setting of rambutan. The population of the bug's natural enemies in the orchard should be maintained at a satisfactory level. The natural predators of this stink bug are identified as *Anastatus* sp., *Micropanurus* sp. and *Eupelmid* sp.

Control

Spraying solution of 20 per cent Sumicidin (fenvalerate) gave good chemical control (Li *et al.,* 2001). The application of oil soap and/or insecticide reduced number of scale insects and mealybugs in rambutan.

Litchi Fruit and Shoot Borer (*Acrocercops cramerella*)

The larvae of this insect penetrate and feed on rambutan fruit. Adults feed externally on the fruit (He, 2001; Waite and Hwang, 2002). In rambutan, damaged fruit often falls from the tree. It is predominantly found in China, India (Waite and Hwang, 2002); Taiwan (Hwang and Hung, 1996); Thailand (Schuetz *et al.*, 2002; Waite and Hwang, 2002). They prefer orchards with shady, humid and closed conditions (Zhang *et al.*, 1997). During the off-season, when fruit is not available, the larvae survive by feeding on young leaves and shoots (Waite and Hwang, 2002). Adults feed externally on longan fruit and their lifespan is from 5-8 days (Waite and Hwang, 2002). The moths are attracted to leaf flushes which emerge during the rainy season from June to October in Thailand (Menzel, 2000). Affected shoots often wilt.

Control

In Thailand, fruits are inspected weekly from fruit set to detect eggs of *C. sinensis*, which are very small and almost invisible to the naked eye. When the pest becomes more active, permethrin is applied weekly, up to two weeks before harvest. In Taiwan, cypermethrin, deltamethrin, carbofuran or fenthion during early fruit set is recommended to prevent damage later in the season. Moths can be excluded by enclosing the fruit panicles in nylon mesh bags, but this method is uneconomic in areas with high labour costs (Waite and Hwang, 2002).

Mealybugs (*Planococcus lilacinus* Cockerell, or *Planococcus citri*)

This insect mainly feeds externally on fruit of hosts in general causing formation of sooty mould on fruit (CAB International, 2002). *Planococcus* spp. in general feed on branches, inflorescences and fruits of rambutan. It occurs in India, Indonesia, Malaysia, Philippines, Seychelles, Sri Lanka, Taiwan, Thailand and Vietnam. This in general feeds on branches, inflorescences and fruits of rambutan. Infested fruits are of low quality and are non-marketable.

Control

In India, monocrotophos is used to control the pest and also soil application of dimethoate. The insecticides fenthion, phosphamidon, quinlphos, dimethoate, phosalone and endosulfan, in descending order of effectiveness, also gave good control.

Oriental Fruit Fly (*Bactrocera dorsalis* Hendel or *Dacus dorsalis* Henderk)

It is a very serious pest of a wide variety of fruits including rambutan and its damage level can be anything upto 100 per cent of unprotected fruit. The plant parts affected are mainly fruits. It occurs widely in mainland Asia. This insect attacks rambutan fruits prior to fruit ripening and fruits show symptoms of infestation like some necrosis around the puncture mark (sting). Premature fruit drop may take place.

Control

Wrapping of fruit bunch either in newspaper or paper bag is effective.

Mites (*Tetranychus cinnabarius* Boisd.)

The mites are widespread throughout the tropics. In this case of infestation, yellow spots develop on the upper surface of leaves and undersides are infested with minute green or red mites. The inflorescence may also be infested affecting fruit set. Leaves turn red, wither and fall down.

Control

Spraying of miticide along with removal of twigs showing leaf infestations is practiced.

Other Pests

Scale insects (*Coccus* sp.) and leaf eating caterpillar (*Dudusa nobilis*) also cause damage to rambutan plant parts, inflorescences and fruits. Of particular importance is fruit bats or flying foxes, which can devour rambutan fruits during fruiting seasons causing serious damage in Thailand and Australia. The only effective means of reducing damage is to erect a protective net around the perimeter of the orchard or over each tree or with low voltage electric current flowing wire fencing in the orchard.

Diseases

A few important diseases affecting rambutan fruits are as follows:

Powdery Mildew (*Odium nephelii*)

Powdery mildew is the most serious disease of rambutan in the field and occurs in most of the production areas. The disease affects mainly young and actively growing, leaves, shoots, inflorescence and fruit. On very young fruit, white powdery areas of mycelial growth appear on the surface, later becoming brown. Fruit surface spinterns are also stunted and turn brown in case of infestation. Conidia of the fungus are spread by wind and all stages of growth can be infected but young shoots are more susceptible.It appears as white colour on fruit. Later fruits wither, turn black and drop prematurely.

Control

Orchard hygiene is an important aspect to control powdery mildew. The orchard should be in weed free condition. Timely and effective pruning to increase the canopy ventilation can reduce the infestation. The spraying of fungicide (chemical formulation of sulphur) at several times during.the early stage of fruit development is also very effective (Tindal, 1994).

Phytophthora Leaf Blight and Fruit Rot (*Phytophthora botryosa*)

The disease occurs widespread in most of the rambutan production areas and infests young shoots, panicle, root, leaf, fruit (Tindal, 1994). The symptoms in rambutan include a dark necrosis of young shoots, brown blight on the panicles, flower drop, irregular brown lesions on fruit and premature fruit drop. During wet weather, the fruit cracks and lesions will be covered by white sporangia and sporangiophores (Coates *et al.*, 2003). *Phytophthora* leaf blight and fruit rot are serious diseases of rambutan where fruits are induced to set during the off-season.

Control

It can be strategically controlled through good orchard sanitation, with destruction of diseased plant materials to reduce the source of inoculum. Fungicide (copper based or Mancozeb) applications may be necessary during disease conducive periods.

Pink Disease (*Erythricium salmonicolor*)

Pink disease affects a wide range of economically important fruit trees including rambutan, particularly in humid tropical areas. The disease is severe in rambutan trees having dense canopy. Symptoms start with white fungal (mycelia) patches on twigs and young branches which later become light pink. Affected bark may crack and severely infested branches may die.

Control

It can be strategically controlled with timely orchard cultural practices which help in reducing the source of inoculum. Spraying of fungicide (copper based) can control effectively (Bhavakul, 1977).

Stem Canker (*Fomes lignosus/Dolabra nepheliae*)

These trees are afflicted with a stem canker disease associated with the ascomycete *Dolabra nepheliae*. Previously reported in Asia and Australia, this fungus was recently spotted from Hawaii and Puerto Rico, causing the stem canker in the Philippines and is fatal for rambutan trees if not controlled (Moton, 1987a) and it is a soil borne disease..

Control

Good orchard husbandry and spraying of systemic fungicides and also soil drenching with systemic fungicides control the disease at initial stage of infestation.

Some causal organisms like *Botryodiplodia theobromae, Gliocephalotrichum bulbilium* and *Colletotrichum* spp., also cause the major postharvest diseases in rambutans. Postharvest storage of fruit in the dark, with low temperatures, may discourage fruit rot.

Harvesting and Fruit Yield

Rambutan trees bear twice annually, once in late fall and early winter with a shorter season in late spring and early summer. The fragile nutritious fruit must ripen on the tree, then they are harvested. Fruits are the best harvested when the body of the fruit is fully coloured and spinterns still retain some green colour. Main horticultural maturity indices for rambutan fruits are attributed by skin and spine coloration. Fruits having green skin and greenish-red spines are sour. Fruits have both skin and spines red or yellow, depending upon variety are very sweet. Rambutan growing countries have their own maturity standards. Rambutan fruits are borne in clusters; Attainment of red color is the main harvest index. A minimum soluble solids content of 16 per cent may also be used; cultivars vary in their maximum soluble solids content from 17 to 21 per cent.

At fruit maturity stages, sugar content increases about 20 per cent, and acid levels are half that at the green stage (Mendoza *et al.*, 1982). The acceptable stage is 16 to 28 days after color break, at which time skin and spines are brightly colored (O'Hare, 1992). Over-ripe fruits have a watery texture (Somboon, 1984) which may be a senescence-induced tissue breakdown. The fresh fruits are easily bruised and have a limited shelf life. An average tree produces 5,000-6,000 or more fruits (60–70 kg/ tree). Yields begin at 1.2 tons/ha in young orchards and may reach 20 tons/per ha on mature trees. In Hawaii, 24 of 38 cultivated ha. were harvested producing 120 tons of fruit in 1997. Yields could be increased via improved orchard management, including pollination, and by planting high yielding compact cultivars. Yield varies from year to year. Individual trees 8 years old or older have borne as much as 200 kg one season and only 60 kg the next. In the Philippines, the average production/tree of 21 selections was 120 kg over a 4-year period, while the general average is only 48 kg. It is necessary to plant male trees with this cultivar-Seenjonja and Maharlika flowers are mostly hermaphrodite functioning female with a very few hermaphrodite functioning as males in the same panicles, always suggested for planting of male trees with these cultivars to improve production.

Individual fruits or whole panicles of fruits are picked depending on flower synchrony and market prices. Fruits are harvested several times according to ripeness of fruits. Picking should be restricted to early hours of the morning when cool to ensure fruits are fully turgid (hydrated) and avoid the field heat. Fruits harvested after sun-shine may deteriorate quickly by losing the attractive colour. Harvested fruits, rapidly loose water from many stomata which surround spinterns (Landrigan, 1996). Postharvest shelf-life of fruits markedly improves under low temperature (cool) and high humidity conditions. In Thailand, the rambutan produces fruits twice a year, the first, main crop is in June and the lesser one is in December. The general harvesting schedule for rambutan in Thailand is between 90-120 days after full bloom, with the greatest increase in fruit weight during the later stage of growth. For most of the rambutan varieties, the appropriate age for harvesting is 16-17 weeks after blooming. This is the time when fruits attain maximum sweetness. The best quality rambutan is generally that which is harvested still attached to the branch. This rambutan is less susceptible to rot, damage, and pests, and remains fresh for a much longer time than rambutan which is picked from the branch. Rambutan that is sold on a full branch is of much higher quality, maintains its true flavour, and is better fit for consumption than that which is sold as individually picked fruits. Another indicator of quality is the ease of detachment of the flesh from the seed. An easily detachable flesh normally will have bits of the woody seed coating.

Rambutan requires approximately 107 to 111 days from fruit set to harvest, with the greatest increase in fruit weight during the later stage of growth. In 'SeeChompoo', coloring occurs on rind and spinterns about 12 weeks after fruit set, total soluble solids increase from 16 per cent at 10 days after color break to 21 per cent at 31 days, while titratable acids fall from 0.26 to 0.16 per cent. The harvesting time of rambutan is based on pericarp color and by counting the number of days from fruit set or from color break. Rambutan fruit has the best rind and spinterns appearance and color if harvested 16 to 28 days after color break. Fruits are frequently harvested as early as 10 days after color break to capture the higher market price. The early fruits lack the sweetness and quality of mature fruit. The general harvesting schedule for rambutan in Thailand is between 90 to 120 days after full bloom; in Malaysia 100 to 130 days, and in Indonesia 90 to 100 days. In some varieties, it may take over 30 days between the first and the last fruit to mature, so harvesting involves several pickings. Fruits harvested 28 days after color break are over-mature, they usually have a darker color, lower sugar and acid content, flat flavor and are puffier to the touch. The flesh of over-ripe fruits are drier and firmer than the normal fruits, and has a more cloudy appearance. Rambutan is harvested in clusters and marketed as loose fruits or in fruit bouquets. Rambutan bouquets are made by arranging and tying clusters of fruits together along a stalk, usually in units of a kilogram or by counts of 20, 50 and 100 fruits.

Postharvest Maintenance of Quality Fruit

Rambutans are harvested when the fruits reached optimum visual and organoleptic quality. The rambutan fruit is highly perishable and non-climacteric. Fruits should be cut carefully from the clusters to maintain quality and suitable storage conditions are required. Organisation of storage at low temperatures (5–10°C) and quick packing in controlled conditions (perforated polyethylene bags) are necessary to avoid transpiration and excessive dehydration, and to prevent decolouration and darkening of the spinterns. The fruits are non-climacteric with little change in total soluble solids or titratable acidity after harvest. Rambutans rapidly deteriorate unless proper handling techniques are employed. Visual appearance can be maintained if moisture loss (largely from the spinterns) is minimised (95 per cent relative humidity) and the fruits are refrigerated (7–10 °C, depending on cultivar). Browning which detracts from visual quality, is directly associated with water loss and physical damage.

Sorting, Grading and Packaging

Grades, Sizes and Packaging: Harvested fruits are immediately transported to pack house facilities where fruits are sorted and graded based on size and degree of ripeness. Then they are washed and dried before packing. There are no U.S. or International grade standards. Quality criteria for grading and packaging include fruit size, shape and weight; bright skin and spine color; uniformity; absence of defects and freedom from disease and insects. High SSC and low TA are desirable (Ketsa and Klaewkasetkorn, 1992).

Figure 10.5: Rambutan fruits.

Fruit are sold in 2.25 kg and 4.5 kg, one-piece, fiberboard cartons. Sometimes fruit are pre-packed in punnets. In Southeast Asia, clusters of fruit are sold in bunches still attached to the stem. For the national and international trade, the fruits are transferred in plastic crates to the packing area, where they are selected by colour, sweetness, pulp loosening from seed, and by size and then graded on the basis of weight and quality. Management system should be in place to ensure that picked fruits are kept moist and rapidly transferred back to shed where they can be wet down with cool water prior to grading and sorting. Grading and handling systems vary with size of the enterprise and management preferences. Size grading is important to assist with packaging, particularly when punnet systems are used. The minimum acceptable fruit size is 34g, but most growers are packing fruits in the 40 to 50 g range. Size grading systems vary from simple diverging belt to mechanical and electronic weight systems. All packagings are designed to minimize the moisture loss from the fruit. The draft standards developed for categorizing rambutan fruits are Extra-class, Class-I and Class-II and only extra class fruits with the highest quality are exported. The fruits >30g weight and meeting the other characteristics are sent to the international packing area. The other fruits are destined for the national, regional and local markets. Agrochemicals are not used and the fruits are not washed. Fruits for the international market are packed in cardboard boxes, formed by a bottom and a cover. The box measures different

Figure 10.6: Quality fruits of Rambutan.

sizes with a net capacity of 2, 3 and 5 kg of rambutan fruits/box. For carton box packing, ideal size of the pack is 60cm × 28cm × 28cm. The packing is properly realised by ensuring proper ventilation and cushioning. The rambutans are inspected visually to avoid transporting deteriorated or damaged fruits.

Transportation and Marketing

The most common modes of transport used for rambutan are three-wheelers and lorries in most of the growing areas for local markets. Most intermediaries believe the cost of transport to be negligible relative to the total value of their cargo. Rambutan fruits are properly packed and with the maintenance of cool chain are transported by refrigerated van or air cargo for distant markets and exports.

The market chain is made up of growers, marketing groups, agents and Asian Wholesalers, boutique fruit retailers and supermarkets. There are potential export markets of United States and European Union. The significant marketing of rambutan in China is organized by Hainan Baoting Rambutan Co. Ltd., which was set up in 1995. Because rambutan production is low in China, demand in domestic market is far beyond the supply and fruits are sold at much higher price (Yang, 1995). The export of rambutan is very limited in Indonesia and only done by air. The crop has good export potential in Srilanka and is presently exported in very limited amounts to the Gulf Region, France, Germany and UK. Consumers prefer to buy rambutans having bright red skin. A little orange or yellow on the skins in addition to red is also okay, but green skins mean the rambutans are unripe. Exports to domestic markets and Japan are channelled via several marketing groups and their coordinators in Australia. They do not buy rambutans which have turned black or have a lot of black "hairs", as this indicates the over-ripe condition. Quarantine issue related to rambutan is fruit fly and the available treatments are irradiation and heat treatment. Heat treatment leads to rapid loss of skin color. Mealy bugs are often found on the fruit, but no damage to the flesh occurs (Ketsa and Klaewkasetkorn, 1992). Roughly about 50 per cent of the harvested produce was sold in 'distant markets', meaning major cities of the growing provinces in that particular country. The other half was resold at locations in the vicinity of the producer's farmgate/local markets. None of the small-scale producers sell to collectors sub-contracted by exporters. In case of semi to large producers, 10 per cent of their produce did reach exporters. The rest of their produce was resold across the growing areas. Intermediaries, when reselling their goods in the domestic markets, often acted on an informal basis by setting up street vendors alongside major routes and streets. Other important sales outlets are found next to campuses, at markets and bus/train stations where sellers tend to hold selling permits and are not intermediaries. In summary, three major marketing channels were identified for rambutan: (1) medium to large sized growers selling rambutan directly to consumers across the growing areas/ provinces; (2) small-scale producers reselling directly to local markets; (3) growers selling via intermediaries, who are reselling rambutan either to local markets or distant markets (which include exporters). Among the producing countries, only Malaysia and Thailand export rambutan, while in other countries, production is for the domestic markets. Malaysia exports rambutan to Singapore. Thailand exports fresh and canned rambutan to Asian and European countries. Processed rambutan products fetch more than 10 times the value of fresh rambutan.

Storage

Rambutan is non-climacteric, the postharvest ethylene production and high respiration rate are results of fruit deterioration associated with moisture loss. A rambutan fruit contains about 300-400 spinterns on the pericarp, each with many stomata. The large fruit surface area is the major cause of moisture loss during storage. Rambutan stored at ambient temperature for 5 to 8 days could lose 19 to 25 per cent of its weight, mostly as dehydration from the peel and spinterns (Landrigan *et al.*, 1996a). The drying causes a darkening of the peel and the soft hairs, which gives the fruit an awful appearance. The same fruits, however, are totally edible, there is little deterioration in fruit qualities in the form of flesh weight, flavour and total soluble solids. The browning and weight loss of rambutan can be reduced by using low temperature storage and moisture retentive materials. The recommended storage for rambutan is 10°C and in sealed plastic bags. The combination of low temperature and the self-generated carbon dioxide level of 7.5 to 9.2 per cent in a sealed plastic bag keep rambutan fruit marketable for 12 days. Rambutan fruits stored at a low temperature acquire a firmer texture and more translucent appearance. Sucrose content is increased from 51 mg/g at harvest to 76 mg/g after storage at 12°C for 6 to 11 days. Storage temperatures lower than 7°C caused chilling injuries to the peel and spinterns. O'Hare and Underhill (1995) found that the process of colour loss at chilling temperature was associated with significant anthocynin degradation (rather than discolouration). Browning of tissue also occurred and consequently the same colour transition seen with chilling injury occurred. Atmosphere control reduces the colour loss. For fresh consumption, keeping the fruits moist and shaded prolongs their shelf-life considerably. Fruits packed in shrink-wrapped containers allow fruit to arrive at the market in excellent condition. The studies on Optimum Storage Conditions(OSC) as to store at 8 to 15 °C with 90 to 95 per cent RH to achieve a storage-life of 14 to 16 days showed that there may be changes in the skin and spine coloration after storage, but the flesh is unaffected. Temperature recommendations vary for different cultivars (O'Hare, 1992). Fruits held at 20 °C with 60 per cent RH last about 3 to 5 days. For enhancing shelf-life of harvested rambutan fruits, Controlled Atmospheres (CA) condition of 7 to 12 per cent CO_2 + 3 to 5 per cent O_2 at 10 °C (50 °F) is recommended (Kader, 1993). At 9 to 12 per cent CO_2 colour loss is reduced and shelf-life extended by 4 to 5 days, while low O_2 (3 per cent) has little affect (O'Hare *et al.*, 1994; O'Hare, 1995). CO_2 levels >12 per cent have no additional effect, and decay begins after a few weeks storage. The MA/CA effect appears to be more via CO_2 elevation and minimizing water loss than through effect of low O_2. Storage in sealed polyethylene film bags or plastic containers is effective in reducing water loss (Mendoza *et al.*, 1972; Ketsa and Klaewkasetkorn, 1995; Mohamed and Othman, 1988), while wax coatings are less effective (Mendoza *et al.*, 1972; Brown and Wilson, 1988; Lam and Ng, 1982). With MA/CA efforts,a shelf-life of 14 to 21 days can be expected.

Commercial Processing of Rambutan Fruits

Most R and D efforts are geared towards main crop improvement and protection but little on fruit processing and off seasonal production of rambutan. The developing of proper processing and freezing technology (such as low oxygen transportation) for rambutan is seen as a major issue. The most popular form of processed rambutan is canning. There are two kinds of canned rambutan, one is plain de-seeded rambutan in syrup and the other is stuffed with chunk of pineapple, also in syrup, both of which are popular export commodities. A method of drying rambutan fruits in the Philippines by soaking peeled fruits in a 40°Brix for three days, then drying fruits to the moisture content upto 19 per cent are promising with 9 months of storability under normal room condition. There are a few other forms of processed rambutan products, namely rambutan pickle, dried rambutan, dried sugar-saturated rambutan, sugar granulated rambutan, dried taste-adjusted rambutan, taste adjusted agitated rambutan flesh, instant powdered rambutan drink etc.

Conclusion

Rambutan has good prospects as the yield can be increased significantly with improved cultural techniques, better cultivars and postharvest treatment. The rambutan fruits have immense potential but require concerted efforts towards R&D on development of suitable cultivars, production techniques and improved management practices at pre- and postharvest level in humid and sub-humid tropics. For fresh consumption, keeping the fruits moist and shaded will prolong their shelflife considerably and requires skill in handling and precision in treatment application. Fruits packed in shrink-wrapped containers allow fruits to arrive at the market in excellent condition. Cold storage at 12°C and fungicidal treatment can also extend the shelflife. Processing offers good prospects, especially for export, for example, as canned rambutan, either singly or in combination with pineapple (rambutan stuffed with a pineapple chunk).

References

Almeyda, N., Malo, S.E. and Martin, F.W., 1979. *Cultivation of Neglected Tropical Fruits with Promise, Part 6: The Rambutan*. Agricultural Research and Education Administration Division, Department of Agriculture, USA.

Anon., 2003. *Sapindaceae*. Seminar series. Unit of Tropical Crops, University of Bonn.

Aradhya, M.K., Zee, F.T. and Manshardt, M.T., 1998. *Journal of Horticulture Science*, 71: 847–857.

Baertels, A., 1990. Farbatlas Tropenpflanzen–Zier– und Nutzpflanzen. 2., verbesserte Auflage, Stuttgart: Verlag Eugen Ulmer, p. 243 and p. 250.c.f.

Barry, I.N., 2007. A value-chain analyses for the Sri Lankan rambutan subsector.–*Research Report No. 2, The International Centre for Underutilised Crops* (ICUC), Colombo, Sri Lanka, 55 pp.

Bhavakul, K., 1977. Diseases of rambutan. *Kasikorn*, 50: 3–7.

Brown, B.I. and Wilson, P.R., 1988. Exploratory study of postharvest treatments on rambutan (*Nephelium lappaceum*) 1986/1987 season. *Rare Fruit Counc. Austral. Newsletter*, 48: 16–18.

Buisson, D., 1986. *Fruits*, 41: 477–498.

CAB International, 2002. *Crop Protection Compendium* (2002 edition). CAB International, Wallingford, UK.

Coates, L.M., Sangchote, S., Johnson, G.I., and Sittigul, C., 2003. Diseases of lychee, longan and rambutan. pp. 307–325. In: Ploetz, R.C. (Ed.) *Diseases of Tropical Fruit Crops*. CABI Publishing, Wallingford, UK. 527pp.

Diczbalis, Y., 1997. Environmental Factors influencing the growth and yield of rambutan and cupuacu. (Final Report Project DNT–10A). Rural Industries Research and Development Corporation, Canberra, Australia.

Doasl, 2007. *Rambutan*. Department of Agriculture, Sri Lanka.

http: //www.agridept.gov.lk/more.php?pagelink=Rambutan&morelink=Introduction (accessed March 18th, 2007)

Franke, W., 1994. Nutzpflanzenkunde–nutzbare Gewächse der gemäßigten Breiten, Subtropen und Tropen. Stuttgart: Georg Thieme Verlag, p. 253 and 299–300.

He, D.P., 2001. An overview of integrated management of insect pests in litchi orchards of Guangdong. In: Huang, H. and Menzel, C. (eds). *Proceedings of the First International Symposium on Litchi and Longan*. Guangzhou, China, 16–19 July 2000. *Acta Horticulturae*, 558: 401–405.

Hwang, J.S. and Hung, C.C., 1996. Gracillariid insect pests attacking litchi and longan in Taiwan. *Plant Protection Bulletin Taipei*, 38(1): 75–78.

IPGRI, 2003. Descriptors for Rambutan (*Nephelium lappaceum*). International Plant Genetic Resources Institute, Rome: IPGRI.

Ito, P.I. and Hamilton, R.A., 1990. Fruits and nut for tropics with potential for improvement and increased importance. *Acta Hort.*, 269: 113–117.

Kalyasiri, P., 1996. Survey of seed oils for use as diesel fuels. *J. Amer. Oil Chem. Soc.*, 73: 471–474.

Ketsa, S. and Klaewkasetkorn, O., 1992. Postharvest quality and losses of 'Rongrein' rambutan fruits in wholesale markets. *Acta Hort.*, 321: 771–777.

Ketsa, S. and Klaewkasetkoan, O., 1995. Effect of modified atmosphere on chilling injury and storage-life of rambutan. *Acta Hort.*, 398: 223–231.

Kosiyachinda, S. and Salma, I., 1987. Changes in rambutan during growth and development. In: *Rambutan: Fruit Development, Postharvest Physiology, and Marketing in ASEAN*. Kuala Lumpur, Malaysia, p. 16–27.

Lam, P.F. and Ng, K.H., 1982. Storage of waxed and unwaxed rambutan in perforated and sealed polyethylene bags. MARDI Food Technol. Div. Rpt. No. 251, 23 pp.

Landrigan, M., Morris, S.C., Eamus, D. and McGlasson, W.B.., 1996. Postharvest water relationships and tissue browning of rambutan fruits. *Sci. Hort.*, 66: 201–208.

Landrigan, M., Morris, S.C., Eamus, D. and McGlasson, W.B., 1996. Postharvest browning of rambutan is a consequence of water loss. *Journal of American Society of Horticulture Science*, 121: 730–734.

Laxmi, L.D.S., Lam, P.F., Mendoza, Jr. D.B., Kosiyachinda, S and Leong, P.C., 1987. Status of rambutan industry in ASEAN. In: *Rambutan: Fruit Development, Postharvest Physiology and Marketing in ASEAN*, (Eds.) P.F. Lam and S. Kosiyachinda. ASEAN Food Handling Bureau, Kulalumpur, Malaysia.

Lim, A.L., 1984. The reproductive biology of rambutan, *Nephelium lappaceum* L. (Sapindaceae). *Gardens Bullettin*, Singapore, 37(2): 181–192.

Lim, T.K., 1992. Rambutan industry in the Northern Territory: Current status, research and development emphasis. *Acta Horticulturae*, 62–70: 321.

Lim, T.K., and Diczbalis, Y., 1995. Rambutans. In: *Horticulture Australia*. Morescope Publishing.

Li, Y.C., Davenport, T.L., Rao, R. and Zheng, Q., 2001. Nitrogen flowering and production of Lychee in Florida. *Acta Hort.*, 558: 221–224.

Mendoza, D.B., Pantastico, E.B. and Javier, F.B., 1972. Storage and handling of rambutan (*Nephelium lappaceum* L.). *The Philippines Agriculturist*, 55: 322–332.

Mendoza, D.B., Ramos, P.T., del Mundo, G.R., Garcia, N.L. and Bantoc, G.G., 1982. Maturity and ripening guide for rambutan cultivars. *ASEAN–PHTRC Tech. Bull.* No. 5. Univ. Philippines Los Banos.

Menzel, C.M., 2002. *The Lychee Crop in Asia and the Pacific*. Food and Agriculture Organization of the United Nations. Regional Office for Asia and the Pacific. Bangkok, Thailand. RAP Publication: 2000/16. http: //www.fao.org/DOCREP/005/AC681E/ac681e00.htm#Contents

Menzel, C.M., Olesan, T., McChonchie, C., Wiltshire, N., Diczabilis, Y. and Wicks, C., 2000. Lychee, longan and rambutan–Optimizing canopy management. A report for the Rural Industries Research and Development Corpotation (RIRDC), RIRDC Publication No. 00/29, RIRDC Canberra.

Mohamed, S. and Othman, E., 1988. Effect of packaging and modified atmosphere on the shelf– life of rambutan (*Nephelium lappaceum*). *Pertanika*, 11: 217–228.

Mollah, W., 1986. Rainfall variability in the Katherine–Darwin Region of the Northern Territory and some implications for cropping. *The Journal of the Australian Institute of Agricultural Science*, 52: 28–36.

Morton, J.F., 1987. Rambutan. In: *Fruits of Warm Climates*, last update: 4/2/99, Miami: USA, p. 262–265. http: //www.hort.purdue.edu (accessed May 1st, 2006)

Nagao, M.A., Leite, H.M., Kawabata, A.M. and Terada, A.Y., 2002. Abstract: Update of tropical fruit research in Hawaii. In: *Proceedings of the 12th Annual International Tropical Fruit Conference*. Keauhou Kona, Hawaii, p. 8.

Nakasone, H.Y. and Paull, R.E., 1998. *Tropical fruits*. CAB Intl., Wallingford, U.K., 445 pp.

Nazaruddin and Muchlisah, F., 1994. Commercial Fruits. Penebar Swadaya (in Indonesian). Cf. Poerwanto, R., 2005. Rambutan and longan production in Indonesia. *Proc. 2nd International Symposium on Litchi, Longan, Rambutan and other Sapindaceae Plants. ISHS Acta Horticulturae*, 665: 81–86

Nicholas, M., and Christie, B., 1993. Less well known tropical fruits. *Agribusiness Worldwide*, 15(4): 6–12.

Ngo, H.T. and Baker, I.W., 1990. Rambutan production in the NT Top End: An agronomic and economic evaluation. NT Department of Primary Industry and Fisheries, *Technical Bulletin* No. 152.

Nieuwolt, S., Zaki Ghazalli, M. and Gopinath, B., 1982. Agro-ecological regions in Peninsular Malaysia. Malaysia Agricultural Research and Development Institute, Serdang, Malaysia.

O'Hare, T.J., 1992. Rambutan: Postharvest physiology and storage. *Trop. Fruit News*, 26: 4–6.

O'Hare, T.J., 1995. Postharvest physiology and storage of rambutan. *Postharvest Biol. Technol.*, 6: 189–199.

O'Hare, T.J., Prasad, A. and Cooke, A.W., 1994. Low temperature and controlled atmosphere storage of rambutan. *Postharvest Biology and Technology*, 4: 147–157.

Ong, P.K.C., Acree, T.E. and Lavin, E.H., 1998. Characterization of volatiles in Rambutan fruit (*Nephelium lappaceum* L.). *Journal for Agricultural Food Chemistry*, 46(2): 611–615.

O'Hare, T.J. and Underhill, S.J., 1995. *Proc. Australlian Postharvest Conference*, Melbourne, Australia, September, 18–22, 1995. p 109–113.

Pohlan, J., Vanderlinden, Eva J.M. and Janssens, Marc J.J., 2008. *Stewart Postharvest Reviews*, 2: 11.

Ramirez, D.A., 1961. Cytology of Philippine Plants. VII. *Nephelium lappaceum* Lin. *Philippines Agriculture*, 45: 340–342.

Rehm, S. and Espig, G., 1984. Die Kulturpflanzen der Tropen und Subtropen–Anbau, wirtschaftliche Bedeutung, Verwertung. 2. Auflage, Stuttgart: Verlag Eugen Ulmer, p. 183–184 and p. 186–188.

Ruan, L., 1989. A report of preliminary selection of some elite rambutan lines. *Trop. Crop Sci. Technol.*, 4: 27–33.

Salakpetch, S., 2005. Rambutan production in Thailand, p. 67–72. In: (Eds.) N. Chomchalow and N. Sukhvibul. *Acta Hort 665: Proceedings of the Second International Symposium on Lychee, Longan, Rambutan and other Sapindaceae Plants*. ISHS, Leuven, Belgium.

Salma, I., 1983. Clonal variation and reproductive biology of *Nephelium lappaceum* Lin. in Peninsular Malaysia. University of Malaya, Kaulalumpur, Malaysia.

Schuetz, P., Sauerborn, J., Martin, K. and Hengsawad, V., 2002. Consequences of pesticide use and weed management to arthropod communities in litchi orchards in northern Thailand. *International Symposium: Sustaining Food Security and Managing Natural Resources in Southeast Asia–Challenges for the 21st Century*. Chiang Mai, Thailand, 8–11 January 2002.

Sivakumar, D., Wilson Wijeratnam, R.S., Wijesundera, R.L.C. and Abeyesekere, M., 1997. Postharvest diseases of rambutan (*Nephelium lappaceum*) in the western province. *Journal of National Science Council, Sri Lanka*, 25: 225–229.

Somboon, Y., 1984. Effect of temperature and maturity stages on biochemical changes during storage of rambutan (*Nephelium lappaceum* Linn.) cv. Seechompoo and cv. Rongrien. *M.S. Thesis*, Kasetsart Univ., Bangkok, Thailand, 64 pp. (in Thai)

Tatt, Ong Hean, 1976. Climate change in water balance and their effects on tropical flowering in rambutan. *Planter*, Kaulalumpur, Malaysia.

Tindall, H. D., 1994. *Rambutan Cultivation*. Food and Agriculture Organization of the United Nations. ISBN 9789251033258. http: / /books.google.com/books?id=Ag–1Fzs0bxMC.

Tindall, H.D., Menini, U.G. and Hodder, A.J., 1994. *Rambutan Cultivation*. Food and Agriculture Organization of the United Nations, Rome, Italy.

Valmayor, R.V., Palencia, C.O., H.B. Aycardo, and D.B. Mendoza., 1970. Growth and flowering habits, floral biology and yield of rambutan (*Nephelium lappaceum* Linn.). *Philippine Agriculturalist* 54(7/8): 359–374.

Van Welzen, P.C., Lamb, A. and Wong, W.W. W., 1988. Edible Sapindaceae in Sabah. *Nature Malaysiana* 13 #1. p.10–25. Malaysia.

Van Welzen, P.C. and Verrheiji, E.W.M., 1991. Nephelium. p. 233–239. In: *Plant Resources of South East Asia–No. 2. Edible fruits and nuts*. Wageningen, Netherlands.

Verheij, E.W.M. and Coronel, R.E. (Eds.), 1991. *Plant Resources of South East Asia No. 2: Edible Fruits and Nuts*. Backhuys Publishers, Leiden.

Waite, G.K. and Hwang, J.S., 2002. Pests of litchi and longan. In: *Tropical Fruit Pests and Pollinators: Biology, Economic Importance, Natural Enemies and Control*, (Eds.) J.E. Peña, J.L. Sharp and M. Wysoki. CABI Publishing, Wallingford, UK, pp. 331–359.

Walter, T.E., 1976. *Nephelium lappaceum* Lin.–Rambutan. In: The propagation of tropical fruit trees. R.J. Garner and Saeed Ahmed Chauhri. FAO., *CAB Horticultural Review* No. 4, U.K.

Watson, B.J., Cunningham, I.C., Walduck, G.D., Wait, A.J. and Goebel, R.L., 1988. *Rambutan Culture in North Queensland*. Queensland Dept. of Primary Industries.

Watson, B.J., 1984. In: *Tropical Tree Fruit for Australia*. Queensland Department of Primary Industries. (Ed.) P.E.Page. Horticulture Branch, p. 198–203.

Watson, B.J., 1988. Rambutan cultivars in north Queensland. *Queensland Agricultural Journal*, Jan–Feb.: p. 37–41.

Yaacob, O. and Subhadrabanhu, S., 1995. *The Production of Economic Fruits in South East Asia*. Oxford University Press, Kuala Lumur, Malaysia.

Zee, F., 1993. Rambutan and Pili Nuts: Potential Crops for Hawaii. In: *New Crops*, (Eds.) J. Janick and J.E. Simon. John Wiley and Sons Inc., p. 461–465.

Zee, F.T., 1998. *Rambutan*, 1998. (New Crop FactSHEET). Purdue University, Center for New Crops and Plant Products. Internet, Indiana, USA.

Zee, F.T.P., Chan, Jr. H.T. and Yen, C.R., 1998. Lychee, longan, rambutan and pulasan. In: *Tropical and subtropical Fruits.*, Ag. Science Inc. Auburndale, Florida, USA, p. 290–335.

Zhang, B.C., 1994. *Index of Economically Important Lepidoptera*. CAB International, U.K.

2011, Future Crops Vol. 1
Editor: Professor K.V. Peter
Published by: DAYA PUBLISHING HOUSE, NEW DELHI

Pages 203–208

Chapter 11
SAFFRON
(*Crocus sativus* L.)

Mohammad Ali Behdani
Saffron Research Group, Faculty of Agriculture,
The University of Birjand, P.O. Box: 97175/331, Birjand, Iran
E-mail: mabehdani@yahoo.com

Saffron (*Crocus sativus* L.) has considerable place as spice for the last 3500 years (Plessner *et al.*, 1989). It is predominantly used to give colour, flavor and aroma to food and specific chemical constituents have been identified (Koocheki, 2004). Recently use of saffron in medicinal industry as a cancer curing agent has brought more attention to this crop (Abdullaev, 2002). Increasing world interest in the use of fragrance, colour, flavour and medicinal sources with plant origin, has broadened horizon for production and consumption of saffron.

Origin of Saffron

Saffron belongs to Iridaceae family and is geographically distributed in Mediterranean climates, East Asia (latitudes 30-50° N and longitudes 10° E to 80° W), as well as Irano-Touranian regions with low annual rain fall, cold winters and hot summers. While some species are grown for their beautiful ornamental flowers, cultivated species of saffron have great economic importance. (Kafi *et al.*, 2006).

There are different theories about the origin of saffron. However, Iran and Spain are two producing and exporting countries. In addition, Kashmir and Afghanistan are regions where saffron is grown. It seems that saffron was introduced to Kashmir by the Zoroastrians who came to India when Arabs entered Iran (Abrishami, 1997). A few scientists believe that the saffron plant is native to the orient, others to Greece (Raina *et al.*, 1996; Sampathu, 2000; Winterhalter and Straunbinger, 2000).

Botany

Among the 85 species belonging to the *Crocus* genus (Iridaceae), *Crocus sativus* L. (Saffron) is the most fascinating and intriguing species. Saffron is an autumnal flowering geophyte with corms which are covered by a tunic, dormant during summer, sprouting in autumn, and producing 1-4 flowers in a cataphyll with linear leaves. The flower has an underground ovary, a style and three red trumpet-like stigmas (2.5 cm long) that once dried to form the commercial spice saffron. Flowering spans from late autumn until December according to climatic conditions (Wendelbo, 1966). Flowers of saffron appear before any vegetative development, its growth starts in autumn and ends in spring, its seeds are sterile despite production of many flowers, and its flowers should be harvested early morning before sunrise. (Rashed- Mohassel, 1993, 1994).

Saffron Production

Saffron is grown on a commercial scale in Iran, Spain, India, Greece, and a little in Italy, Morocco and France. Iran is the major producer accounting to 90 per cent of the world's production. In recent years, Spanish output is in constant decline. This is attributed to the increasing cost of labour and to the migration of younger people to better paid jobs in industrial areas (Koocheki *et al.*, 2006).

Cultivation of saffron in Iran is confined to the state of khorasan Razavi and Southern khorasan. Although this crop has a long history of production in the Southern Khorasan, not much progress is made in developing production and processing technologies (Behdani *et al.*, 2008).

Soil

Saffron can be grown in a wide range of soils, with moderate structure and well infiltration, siliceous, ferrugenous and gypsyferous soils are preferred. Adequate surface drainage is necessary for corm development. Since with soils in calcium, high amount of organic matter is facilitated, this type of soil is recommended. (Behdani *et al.*, 2005). Soil is ploughed in autumn or winter and animal manure of 20-40 tons/ha is applied. In Spain,a deep ploughing of 25-35 cm is practiced in winter and the fields are harrowed to eliminate weeds. In Kashmir, soil preparation is practiced in February and March with 3 to 4 tillage operations and application of 5 to 6 tonnes/ha of animal manure (Kafi *et al.*, 2006).

Planting

Planting of corm is done from August to early October, but in some case early planting in June and July may be carried out. In Spain,planting is done from the middle of May to the beginning of June, and in Kashmir from mid July to the end of August (Molina *et al.*, 2004)

Propagating Materials

Saffron is propagated by corms and therefore selection of proper corms is crucial for production. Mother corms reproduce annually giving rise to new cormlets. The young corm is onion shaped, a little compressed, white and fleshy. A series of fibrous sheaths cover the solid bulbus part. The cormlet develops and flowers only once in the following year, and also gives rise to new cormletes (Rashed-Mohassel, 2007). It seems, age of the field is an important factor in site of the corms. Therefore, it is recommended to reduce age of the field which is normally 8-10 years in traditional system in Iran to 4-5 years to avoid reduction of size of corms which have a negative impact on flowering potential (Sadeghi, 1988).

Ecology of Saffron

Saffron grows in climatically diverse regions. But vegetative growth coincides with cold weather and freezing conditions. Although early autumn is the expected time for saffron flowering in the Southern Khorasan, there is considerable variation in flower emergence across the region

(Behdani *et al.*, 2004). Matching this variation calls for quantitative understanding of flowering response of saffron to unexpected environmental variability. Development is an irreversible process of change in the state of a plant, which generally progresses according to a move of less fixed and species specific pattern (Atkinson and Porter 1996). Many developmental stages are crop specific such as silking in maize, double ridge in wheat, or tuber formation in potatoes. While all these stages occur after a substantial vegetative growth, flowering in saffron is a unique process that starts before regular vegetative events like leaf appearance and growth (Kafi *et al.*, 2002). Therefore, known approaches for quantitative prediction of developmental stages in field crops *e.g.*, growth degree days (GDD) or photothermal units (PTU) cannot easily be applied to saffron.

Flower emergence in saffron is influenced by factors like radiation, nutrients and water availability. However, previous studies show that it is principally controlled by temperature (Kafi *et al.*, 2002; Molina *et al.*, 2004) and therefore temperature would be the main criteria for estimating the time of flower emergence in this plant. Flower initiation in saffron starts in the late spring, however, flowers appear in the early autumn.

Saffron production areas of Khorasan province are extended over a wide range of climate with considerable differences in environmental condition, mainly temperature, and flowering emergence of saffron varies over this temperature gradient. Behdani *et al.* (2008) showed that flowering time of saffron could be predicted from mean September temperature using simple linear model to more complicated Beta model with promising results. Climatic parameters like temperature, rainfall and probability of freezing occurrence in the months of October and November are well documented (Vafabakhsh *et al.*, 2002). Since, flower development mainly proceeds under soil conditions, soil temperature plays an important role in flower production.

Irrigation

Saffron production does not require much water but timeliness of irrigation particularly the first irrigation is very important for flower emergence and length of flowering period (Kafi *et al.*, 2006). There is limited information about effects of amount and intervals of irrigation on yield components of saffron. In this regard, three aspects are of great importance: A- Effect of pre- flowering irrigation. B- Effect of irrigation intervals on saffron yield, C- Effect of after harvest irrigation on yield of the following year. Relationship between irrigation intervals and summer irrigation was investigated for their effects on yield of saffron in different age farms (Behdani *et al.*, 2008). Results showed that there was good correlation between irrigation intervals and yield of saffron. By reducing irrigation interval, yield was increased for different age farms. Higher saffron yield was obtained when irrigation was applied at 12 days intervals. Behdani *et al.* (2008), Kafi *et al.* (2002) and Molina *et al.* (2004) reported onset of saffron flowering happened between 2-3 weeks after first irrigation. Time of first irrigation is the most important factor in initiation of flower emergence in saffron. The timing of irrigation is so important with regard to labour intensive picking of saffron. In large- sized saffron farms, irrigation of different parts of farm should be split to stimulate gradual flower emergence in each segment to facilitate daily picking of flowers with available work force. Application of first irrigation in appropriate time, with regard to low mean temperatures, is so important for saffron yield, because if the first irrigation be applied earlier, when mean temperature is still high, vegetative organ (leaf) may also appear earlier than flowers, which is an unusual event in saffron growth.

Based on the physiological characteristics of saffron, (growth starts in early autumn with decreasing temperature of the area), first irrigation at that time stimulates emergence of flower. After flower emergence and harvesting which normally last 30 days, vegetative growth starts and the leaves emerge. During winter days, growth of leaves continues and by the end of May leaves are dried and go to dormancy. Flower initiation starts in early July (Farooq and Koul, 1983). Therefore, first irrigation is normally applied at the beginning of fall without any summer irrigation applied. However, in recent years one summer irrigation is practiced between end of July to middle of August. It is believed that application of one irrigation at time of early flowering helps this process and results in higher rate of flowering (Farooq and Koul, 1983; Sadeghi, 1998). Behdani *et al.* (2008) reported that mean yield of saffron for the farms with summer irrigation was higher than those with no summer irrigation (4.9 and 3.35 kg/ha, respectively) and as the whole, nearly 60 per cent of yield was obtained from these farms in which summer irrigation was practiced. Summer irrigation at flower differentiation stage will lead to yield increase. Traditionally saffron is irrigated four times during the growing season in Iran. However, to achieve higher yields, sufficient water is required from October until May. More experiments in different climates and with saffron fields at different ages are required to schedule proper irrigation regimes and summer irrigation for saffron.

Saffron and Organic Farming

Organic farming can be defined as a method for production, which puts the highest emphasis on environmental protection and diminish environmental pollution of agricultural production (Liebhardt, 2003). Organic agriculture is practiced in almost all countries of the world, and its share of agricultural land is increasing everywhere. Organic agriculture is the most dynamic sector within the whole countries and has grown at a rate of more than 20 per cent/year for the last eight years (Creamer, 2003).

Organic farming based on the standards of IFOAM is still rare, and there is no single certifying body at a national level in Iran. There is a growing tendency in Iran towards internal use of organic products and also for exports (Koocheki and Gorbani, 2005).

Production, processing, trade and consumption of saffron have evolved within the context of indigenous knowledge and from ecological points of view. This crop has occupied a specific niche where no other crops can be as environmentally friendly, economically feasible and socially justified in the saffron growing areas. Therefore, it is a crop of sustainable values. Bed preparation and cultivations are all carried out by labour or very simple machines. Saffron corms for planting in new fields are provided locally. Saffron soil fertility management is mostly based on cow manure composts. Farmers mostly use non- chemical methods like biological control, mechanical traps, smothering and steaming, flooding irrigation and chilled water during winter irrigations. Saffron is a family based crop and most of farming practices and particularly picking flower and separating the stigma are carried out by family members with community cooperation. Establishing national and local certifying organizations with affordable services for inspection, certification and improving facilities are necessary for organic growers of Iran (Behdani *et al.*, 2008; Koocheki and Gorbani, 2005).

A return to more sustainable production systems in a country like Iran with a long history of ecological agriculture could be one effective method to protect the fragile local environments.

Medicinal Applications

Saffron is used as a spice but also as a drug to treat different human diseases and in the recent years, many hopes are placed in the spices potentiality to cure or to prevent dangerous diseases like cancer (Abdullaev, 2002; Abdullaev, 2004). In view of its wide range of medical uses, the saffron has undergone extension phytochemical and biochemical studies and a variety of biologically active ingredients are isolated.

Characteristic components of saffron are crocin- responsible for the colour, picrocrocin- responsible for the bitter taste and saffranal- responsible for odour and aroma. During the last decade, different laboratories in the world reported many effects of saffron–saffron on coronary artery diseases; on learning behavior; on ocular blood flow and retinal function; on blood pressure; on contraceptive, anti-inflammatory and antiatherosclerosis activity; antigenotoxic and cytotoxic activities (Abdullaev; 2007).

References

Abdullaev, F. I., 2002. Biomedical properties of Saffron and is potential use in cancer therapy and chemoprevention trials. *Cancer Det. Preven.*, 28(6): 426–432.

Abdullaev, F. I., 2004. Caner chemo preventive and tumoroicidal properties of saffron (*Crocus sativus* L.). *Exp. Biol. Med.*, 227: 20–25.

Abdullaev, F. I., 2007. Biological properties and Medicinal use of saffron (*Crocus sativus* L.). *Acta Hort.*, 739: 339–345.

Behdani, M.A., Koocheki, A. Rezvani, P. and Jamial-Ahmadi, 2008 Agro-ecological zoning and potential yield of saffron in khorasan–Iran. *Journal of Biological Sciences*, 8(2): 298–305.

Behdani, M.A., Nassiri, M. and Koocheki, A., 2008. Models to predict flowering time in the main saffron production regions of khorasan province. *Journal of Applied Science*, 8(5): 907–909.

Behdani, M.A., Nassiri Mahallati, M. and Koocheki, A., 2008. Evaluation of irrigation management of saffron at agroecosystem scale in dry region of Iran. *Asian Journal of Plant sciences*, 7(1): 22–25.

Behdani, M., Koocheki, A., Nassiri, M. and Rezvani, P., 2005. Evaluation of quantity relations between yield and nutrient consumption in saffron: on farm study. *Journal of Iranian Field Crop Research*, 3(1).

Creamer, N.G., 2003. *Organic Farming in Center for Environmental Farming Systems*. Box 7609. Raleigh, NC 27695.

Farooq, S. and Koul, K., 1983. Changes in gibberellins like activity in corms of saffron plant (*Crocus sativus* L.) during dormancy and sprouting. *J. Plant Biochem.*, 178: 685–691.

Kafi, M., Koocheki, A., Rashed, M.H. and Nassiri, M., 2006. *Saffron (Crocus sativus) Production and Processing*. Science Publishers, New Hamshire, USA, pp. 244.

Koocheki, A., 2004. Indigenous knowledge in agriculture with particular reference to saffron production in Iran. *ISHS Acta Hort.* 650: International symposium on saffron Biology and Biotechnology, 175–182.

Koocheki, A. and Gorbani, R., 2005. Traditional agriculture in Iran and development challenges for organic agriculture. *Intl. J. Biod. Sci. Mgt.*, 11: 1–7.

Koocheki, A., Behdani. M.A. and Nassiri, M., 2006. Agronomic attributes of saffron yield at agro ecosystems scale in Iran. *Journal of Applied Horticulture*, 8(2): 121–124.

Liebhardt, B., 2003. What is organic agriculture? What I Learned from my transition? In Organic agriculture: Sustainability, Markets and Policies Ed. By OECD. CABI,UK.

Molina, R.V., Valero, M., Navarro, Y., Garcia-Luisand, A. and Gardiola, J.L., 2004. Temperature effects on flower formation in saffron (*Crocus sativus* L.). *Scientia Horticulturae*, 103(b): 361–379.

Plessner, O., Negbi, M., Ziv, M. and Basker, O., 1989. Effect of temperature on the flowering of the saffron crocus (*Crocus sativus* L.): Induction of hystheranthy. *Israel J. Bot.*, 38: 1–7.

Raina, B.L., Agarwal, S.G., Bhatia, A.K. and Guare G.S., 1996. Changes in pigments and volatiles of saffron during processing and storage. *J. Sci. Food Agric.*, 71: 27–32.

Rashed-Mohassel, M.H., 1993. *Khorasan Vegetations*, Vol. 1. Ferdowsi University of Mashhad Publication, Mashhad, Iran.

Rashed-Mohassel, M.H., 1994. *Khorasan Vegetations*, Vol. 2. Ferdowsi University of Mashhad Publication, Mashhad, Iran.

Rashed-Mohassel, M.H., 2007. Saffron from wild to the field. *Acta Hort.*, 739: 187–194.

Sadeghi, B., 1998. *Effect of Summer Irrigation on Saffron Yield*. IROST Mashhad Center, Iran.

Sampathu, S.R., Shivaskar, S. and Lewis, Y.S., 2000. Saffron (*Crocus sativus* L.) cultivation, processing, chemistry and standardization. *CRC Crit. Rev. Food Sci. Nut.*, 20: 123–157.

Vafabakhsh, G., Ahmadian, H., Shibani, D. and Badagh Jamali, G., 2002. Potential climatic zone for saffron growing in Iran. *Proceedings of the second National Symposium on Saffron and Medicinal Plants*.

Wendelbo, P., 1966. *Tulips and Irises of Iran and their Relatives*. Botanical Institute of Iran, Botanical Garden, Tehran, Iran.

Winterhalter, P. and Straunbinger, M., 2000. Saffron: Renewed interest in an ancient spice. *Food Rev. Int.*, 16: 39–59.

2011, Future Crops Vol. 1 ***Pages 209–217***
***Editor:* Professor K.V. Peter**
***Published by:* DAYA PUBLISHING HOUSE, NEW DELHI**

Chapter 12
SPEARMINT
(*Mentha spicata* L.)

A. Mishra
Plant Physiology and Biochemistry
Central Institute of Medicinal and Aromatic Plants, P.O. CIMAP, Lucknow – 226 015
E-mail: amisracimap@yahoo.co.in

Spearmint (*Mentha spicata* L.) is an indigenous plant of England. Being a temperate plant, its cultivation started in almost every part of the world as an aromatic crop. Hydrodistillation of the leaves yields an essential oil, commonly called spearmint oil. Due to pleasant fragrance of the oil, it is widely used in confectionaries, tooth paste, pickles, spices and luxury soap preparations. In pharmaceutical industries, it is used as a carminative and stimulating agent. Spearmint oil is a natural source of carvone. Due to its pleasant fragrance, it is highly used in alcoholic beverages for the seasoning. Being temperate in nature, extensive research has to be carried out to study its acclimatization, cultivation and popularization in tropical and sub-tropical conditions.

Spearmint is the popular name of *Mentha spicata* L. indigenous to Northern England. A variety of *Mentha spicata, M. spicata* var. *viridis,* called 'Pudina' in Hindi, is grown in kitchen gardens in India as a spice. But the plant is grown practically all over the world (Husain *et al.,* 1988).

Mints are a group of plants belonging to family Labiateae which is mainly grown for its essential oil. The essential oil is obtained by steam or hydrodistillation of mint leaves, which finds wide application in spice, perfumery and food flavour industries (Robert and Chastain, 1944).

Spearmint cultivation started only during the modern period. Spearmint was first cultivated in convent gardens during the 9th century. Spearmint is cultivated in USA in the States of Indiana, Michigan and Oregon. Although these plants were used by mankind since ancient times but the large scale cultivation came up only recently. In India, the spearmint cultivation has also been taken up recently by farmers of Badann and Barabanki districts of Uttar Pradesh.

Historical Background

The genus *Mentha* which includes true mints originated in the Mediterranean basin and spread to the rest of the world by both natural and artificial means. According to Greek legend, the first mint plant was the bodily transformation of the nymph Mintho, who had angered proserpine, wife of Pluto by casting covetous eyes on the Plulanodering God of the underworld. Proserpine gained her revenge by artfully turning Mintho into a herb, as humble mint plant. The nymph lost some of her beauty, yet continues to attract men of her freshness.

Distillation of mint oil was mentioned even in 410 AD in an account written by Synesius of Alexandria, the Bishop of Ptolemais. The leaves and oil of mint were widely used in culinary and medicine during this period.

Charlemagne recommended three mints for cultivation in 812 AD and Abbess included 4 mints in her subtilitatum in 1160. Five mints were described in the oldest German literature written by Hieronymus Brunschweig in1500.

The utility of mints made the plant quite popular as herb and from the convent gardens these plants reached almost every kitchen gardens. By 1892, mint oil was considered as an outstanding stomachic stimulant, antispasmodic and carminative.

Distillation of mint herb to extract oil originated in England in1750. Spearmint was cultivated commercially in the medicinal herb garden near Mitcham in County Surrey of England. Later on commercial mint cultivation moved from England to the European continent. It first came to Utrecht in the Netherlands during 1770 and soon spread to Germany, France and to other European countries.

Table 11.1: Indian export of spearmint oil.

Year	*Quantity (tonnes)*	*Value (Rs.lakhs)*
1990-91	0.5	2
1991-92	2.2	84.8
1992-93	66.2	167.1
1993-94	111.1	635

The plant was introduced to USA in 1790. However commercial production started only in 1812 in Massachusetts. Later on it moved to the states of Indiana, Michigan and Oregon and developed into one of the largest plants based industries in the world.

Uses

Due to pleasant fragrance of the spearmint oil, it is highly used in luxury soap preparations in the United States. It is mostly used as a flavouring agent in confectionaries, tooth paste, pickles and spices (De Vincenzi *et al.*, 1996). Spearmint oil is also used as carminative and stimulant in pharmaceutical preparations. Spearmint oil is widely used in pharmaceutical preparations as a non-bitter intensely sweet flavouring agent *i.e.*, in sucrolose and chewing preparations made with high concentration of sucrolose (1000 ppm). Spearmint oil showed considerable antimicrobial activity against *Staphylococcus aureus, Escherichia coli* and *Candida* (Morris *et al.*, 1979).

Economics

In this fast changing global scenario and materialistic world, people are very cautious regarding the economics of their trade they are involved in or wish to do it. Farmers also prefer to grow more profitable cash crops. Cultivation of spearmint crop is quite economical to farmers and traders which is evident from its constant increasing production in the world.

World Production

Spearmint oil is used as a source of carvone. It is present to the extent of 57-80 per cent in the oil (Bounosheta *et al.*, 1993). The total production of Spearmint oil in world is estimated 1500 tonnes. The major producing countries in the world are USA, China and India.

Production in India

India has not only become self sufficient in meeting its demand for spearmint oil but it is exporting spearmint oil to the international market. The total production of spearmint oil in India is about 15 tonnes.

The production of spearmint oil in India has increased tremendously during the period 1990 to 1994 bringing a considerable amount of foreign currency to our country. We are sure that with the high yielding varieties developed by CIMAP, India will be able to produce and export a huge amount of spearmint oil in the near future.

Area Under Cultivation in India

According to 1996 survey on spearmint cultivation, it was observed that spearmint cultivation is rapidly increasing in India. At present the total area under cultivation of spearmint in India is approximately 2000 ha.

International Prices of Spearmint Oil

In International market, the price of spearmint oil depends on the carvone content. A view on the variation of spearmint oil prices of different countries in New York market is given in Table 11.2.

Table 11.2: New York market price (Rs/kg).

Chinese (80 per cent)	–	735
Far West, Native	–	1151
Far East Scotch	–	1508

In India spearmint oil is sold at an average price of Rs.400/kg

Botany

Mints are a group of plants belonging to the family labiateae and the true mints are included in the genus mentha. *Mentha spicata* is a perennial herb. Its propagation is done by Stolons from which 30-60 cm erect ascending branches arise.

Morphological Characters

The leaves of *Mentha spicata* are sessile or nearly so, smooth lanceolate sharply serrate, smooth above and glandular below. Flowers are sharply pointed long and narrow. Calyx teeth are hirsute or glabrous and corolla is about 3 mm long, stem is always shorter and strands more erect.

Genetics

The interspecific somatic chromosome variation in F1 hybrid progeny of a cross between *M. spicata* and *M. pipenta* is reported. The chromosome number in the F1 hybrid progeny ranged 36 to 115. The normally expected chromosome number (2n = 96) was the most frequent. Only one hybrid with 2n = 36 resembling the female parent (*M. spicata* 2n = 72), suggested its origin either through female parthenogenesis or normal fertilization followed by selective chromosome elimination.

Agronomical Practices

Texture of the soil and proper irrigation are the two most important factors for the successful and profitable cultivation of mint crops.

Soil

Spearmint grows very well in deep loamy soils, which are rich in inorganic matters with pH of 6.0-7.5. The well drained soils with good water holding capacity are better suited for mint cultivation. On the other hand, sandy and heavy clay soils do not support good mint.

Irrigation

The water requirement of spearmint depends on texture of the soil. But in general 8-12 irrigations are required before the onset of monsoon. The crop requires three more irrigations after the monsoon in September, October and November. Water logging should be avoided and adequate drainage provided. Spearmint grown in temperate regions, requires 3-4 irrigations during July to October (Katoch *et al.*, 1978).

Fertilizers

Spearmint responds very well in nitrogenous as well as phosphorus fertilizers. Increase in herbage yield by application of potassium is reported and application of 120-150 kg of nitrogen and 50-60 kg of P_2O_5/ha is recommended to get two good harvests of mints. Besides above, application of K_2O/ha(10kg) is also recommended to prevent depletion of soil status. The entire quantity of phosphorus and potassium along with 30-40 kg of N_2/ha is applied as a basal dose at the time of planting.

Weeds and Herbicides

Weeds are big menace to mint growers as it effects both herb and oil yield to a great extent. The first thirty to seventy five days are critical for the crop from weed competition point of view. Hence proper control of weeds is very essential during this period. The weeds can be controlled either by weeding or by use of herbicides. The recommended herbicides and their doses for spearmint are given below.

1. Terbacil (1.5 kg ai/ha)
2. Oxyfluofen (0.5 kg a.i/ha)

Table 12.3: Mean performance of the selected mutant clones of *Mentha spicata* at Lucknow and Pantnagar.

Clones	*Herb Yield (q/ha)*		*Oil Yield (kg/ha)*		*Carvone Content (per cent)*	
MSKA17 LKO	153.89	191.00	101.99	103.41	78.8	72.1
MSMA35 PTN	138.56	200.50	86.93	133.27	75.9	71.3
Arka content	135.31	207.50	87.37	111.98	77.2	72.6
CD (5 per cent)	29.66	33.19	24.69	20.85		

High Yielding New Varieties

Mentha spicata L. (Syn. *M.viridis* L. *M. crispa* L., *M.crispata* Schradus, *M. longifolia*) are common or true mint. CIMAP,Lucknow is actively engaged in developing high oil yielding varieties of spearmint and good success is achieved in this area. A few of the high oil yielding varieties developed by CIMAP are given below.

- ☆ MSS-5: A selection from MSS-1 yielding 690 q/ha fresh herbage (348 kg/ha oil) from 3 cuttings consisting of 70 per cent carvone in the oil.
- ☆ MSS-1: A selection from cultivars of spearmint introduced from USA producing fresh herbage of 535 q/ha (oil 253 kg/ha) from 3 cuttings consisting of 60 per cent carvone.
- ☆ Arka (MASS-98): Developed by clonal selection of MASSA1. This is an early maturing variety with erect growth habit. It yields herbage of 332 q/ha and oil 177 kg/ha.
- ☆ Punjab Spearmint: Evolved through clonal selection, producing 200 q/ha herbage and 119.29 kg/ha oil containing 69 per cent of carvone in the oil.
- ☆ Neera: Developed by radiation of variety MSSA1. It has spreading type of growth habit, narrow leaves and compact inflorescence bearing flowers with pinkish twig. It produces about 310 q/ha fresh herbage and 113 kg/ha oil containing 58 per cent of carvone. On account of its unique odour and flavour, the oil has potential for its diversified use in quality products.

Two selected mutant genotypes MSM-17 and MSM-35 of *Mentha spicata* along with variety Arka as check were evaluated for their yielding potential and essential oil quality in replicated field trial, one each at Lucknow and Pantnagar. The data on yield and essential oil quality revealed that mutant clone MSH-17 at Lucknow and mutant clone MSM-35 at Pantnagar, proved superior by yielding 17 per cent and 19 per cent more oil respectively, over check. However, the quality of the essential oil in terms of carvone content of these mutants remained at par with var. Arka.

It is evident that in Pantnagar which is in the *tarai* belt of Uttarkhand and soil is clay loam, the herbage and oil yield are better than in CIMAP Lucknow which is in the plain and the soil is sandy loam (Singh and Duhan, 1979). Regarding the quality of oil, it is observed that the spearmint oil produced at Lucknow is slightly better than that of Pantnagar due to its higher carvone content.

Pathology

Like other cultivated crops, spearmint is also attacked by several diseases and insect pests. A brief on the diseases and insect pests which heavily effect the spearmint crops are given below.

Disease of Mints

There are a number of diseases in mint crops in different parts of the world. There are three major diseases affecting adversely the mint crop under Indian conditions.

Rust

Powdery Mildew and Stolon Rot

The various organisms responsible for these diseases and the methods of their control are given below.

Name of Disease	*Causal Organism*	*Control of Disease*
Mint Rust	*Puccinia menthae pers*	Karathane WD Morestan
Powdery Mildew	*Erysiphe cichoracearum*	Karathane WD Morestan
Stolon Rot	*Macrophomina fhareoli* Hob Ashy	Karathane WD Morestan

Mint Rust

The symptoms of this disease appear during April-May. These consist of yellowish lemon coloured blisters on the upper surface of leaves, which are covered with powdery substances consisting of the uredospores of the fungus. In case of severe infection, the leaves turn yellow and fall off. In the later stages, lemon coloured spots are replaced by compact brownish black spots consisting of the teliospores. Due to this brownish black colour of this disease, it is known as mint rust.

Powdery Mildew

Symptoms of this disease consist of powdery white patches on the upper and lower surface of the leaves and stems. The patches consist of mycelium and spores of the fungus. The disease can be controlled successfully by spraying sulphur formulations or organic fungicides.

This is the disease which directly influences the essential oil yield through the destruction of foliar oil gland and defoliation.

Aerial Blight in Mints

Mentha spicata is found badly damaged (25-30 per cent) by the aerial blight. The symptoms of this disease appear like water soaked faded patches which start from the base and extend upto the tip of the leaf and stem. The causal organism of this disease is *Rhizoctonia solani*.

Insect Pests

A large number of defoliating insects attack spearmint crop. Although these insects cause very little damage to the crop, but in localized condition, heavy damages are also observed. The important pests which attack the mint crops are:

1. Mint Leaf Roller (*Syngamia abruptalis* Walk)
2. Pyralid (*Laphygma exigua* Hub)
3. Hairy caterpillar (*Diacrisia oblique* Walk)

Chemical Composition of Mint Oil

The spearmint oil is used as a source of natural carvone which is present to the extent of 57-80 per cent in the oil (Lal and Lal, 1966).

Chemical Composition of Spearmint Oil of Different Countries

Spearmint oil has the various percentage of chemical constituents on the basis of its distribution in different countries (Table 12.4).

Chemical Constituents of Spearmint Oil

Spearmint oil is used as a source of carvone. It is present to the extent of 57-80 per cent in the oils. The other constituents of the oils are phellandrene, limonene, pinene, 1,8-cmeol, dihydrocarvone, diosmetin, diosmin and dipentene (Chopra *et al.*, 1972).

Carvone: p-mentha-6,8-dren-2-one

d-l-carvone is found in caraway and dill seed oils while l-carvone is found in spearmint and Kuromoji oils. On the other hand, d-l- carvone is found in ginger grass oil (Alonso *et al.*, 1992). Carvone is the major constituent of spearmint oil present to an extent of 57-80 per cent. The carvone(per cent) of spearmint oil is determined by GLC by potentiometric titration method. Carvone is also converted into various useful compounds like:

Table 12.4: Chemical composition of Egyptian, Indian, American and Hungarian spearmint oils.

Egyptian	*Indian*	*Italian*	*American*	*Hungarian*
The percentage of oilcontent of *M. spicata* increased with growth attaining 3.8% at full bloom and subsequently declining to 0.5% at senescence	Carvone (66.9%) Limonene (10.84%)	Carvone (71%) Limonene (11.00%)	Carvone (66%) Limonene	Carvone (61.72%) This oil has a higher specific gravity than the American or German oils
The quality compound of the oil samples were similar except for some juvenile herbage		The amount of monoterpene found (32.51%) was higher than in any of the other carvone rich mints		They are soluble in 80% alcohol and do not grow clearly on further addition of solvents like the American and German oils
Carvone is the dominant constituent, reaching the highest concentration (73.3%) in juvenile plant				

1. On catalytic hydrogenation (H2), carvone gave l-carvotanacetone
2. Reduction of carvone with Zn gave d-dihydrocarvone

Limonene: p-mentha-1,8-diene; cinene

Limonene occurs in various ethereal oils, particularly in the oils of lemon, orange, caraway, dill and bergamot limoneneis also present in optically inactive 9dl) form. It possesses pleasant lemon like odour. Practically limonene is insoluble in water but miscible with alcohol.

Diosmetin

3,5,7 A trihydroxyA4 Amethoxy flavone. This was isolated from the leaves of *Mentha spicata* (Chopra *et al.*, 1956).

Diosmin

It is a flavone glycoside and isolated from the leaves of *Mentha spicata* (Chopra *et al.*, 1956).

Spearmint (*M. spicata* L.) is a commercially viable temperate plant. Due to higher carvone content in the essential oil, spearmint is now grown in the tropical and sub-tropical countries of the world. Carvone is highly used for seasoning of spices, alcoholic beverages and flavouring of edible items.

References

Alonso, William, R., Rajaonarivony, Jean I.M., Gershenzon, J. and Croteau, Rodney, 1992. Purification of 4S Limonene synthase: A monoterpene cyclase from the glandular trichome of Peppermint and Spearmint. *J. Biol. Chem.*, 11: 267.

Bounoshita, Masao, Hibi Kiyokatsu and Nakamura Hiroshi, 1993. Determination of enantiomer ratios of d–l carvone. *Anal. Science*, 9: 425–428.

Bugaenko, L.A. and Raznikova, S.A., 1984. Production of highly heterotic hybrids of mint suitable for a two cut cultivation regime. *Tsitol. Genti*, 19: 92–7. *Plant Breeding Abstract*, 56: 2383, (1986).

Chakravarti, K.K. and Bhattacharya, S.C., 1954. Chemical examination of Indian spearmint oil. *Perfumery Essential Oil Record*, 45: 217–218.

Cherukuri, Subraman Rao, Wong, L., Lee, F. and Steven, M., 1990. Pleasant taste masking composition and methods for preparing them. Eur. Pat. Appl. EP 371,584 (Cl–A–23 LI/236)06 Jan, 1990, US Appl. 264–281, 28 Oct, 1988,15 pp. (Chem. Abs. 113, 190093).

Chopra, Nayar and Chopra, 1956. *Glossary of Indian Medicinal Plant*. Pub. by Publication and Information Directorate (CSIR), New Delhi, India p. 65 and Diosmetin and Diosmin isolated from leaves. Phytochemistry, 11: 452 (1972).

De Vincenzi, M., Mancini, E. and Dessi, M.R., 1996. *Monograph on Botanical flavouring substances used in foods. Part V. Fitoterapia* 67, 241.

Ellis, N.K. and Stevension, E.C., 1950. Domestic production of spearmint oil. *Econ. Botany*, 4: 139–149.

Fabian, F.W., 1942. Chemical action of Spice oil in pickles. *Western Canner and Packer* 34, 29A31 (*Chem. Abs*. 31; 1795).

Hiraga, Y., Shi, W., Ito, D.I., Ohta, S. and Suga, T., 1993. Biosynthetic generation of the species specific chirality of limonene in *M. spicata*. *J. Chem. Soc. Chem. Commun.*, 17: 1370–132.

Husain, A., Virmani, O.P., Singh, D.V., Singh, A. and Singh, K., 1988. *Bulletine on Mint farming in India*, Revised Edition, Central Institute of Medicinal and Aromatic Plants, Lucknow, India.

Katoch, P.C., Bhardwaj, S.D., Kausal, A.N. and Raina, M., 1978. Optimum time of planting and row spacing on herb yield and oil content in spearmint. *Indian Perfumer*, 23: 91–94.

Kugler, Kovats, 1963. λ-carvone is found in spearmint and Kuromojii oils. *Helv. Chim. Acta*, 46: 480.

Lal, R.N. and Lal, J.B., 1966. Chemical examination of Indian spearmint oil. *J. Proc. Inst. Chemist (India)*, 38: 127–133.

Lal, R.N. and Sen, T.K., 1965. Estimation of carvone in Indian spearmint oil by potentiometric titration. *Perfumery Essential Oil Record*, 56: 731–732.

Laverne, D. Small, James, E., Dusenberry and Walter, T. Gloor, Jr., 1952. Determination of carvone oil of spearmint. J. Am. Pharm. Association. 41: 280–5.

Maffei, Massimo, Codignola, A. and Fieschi, Maurizio., 1986. Spearmint cultivated in Italy. *Flavour Fragrance J*. 1: 105–9.

Melnikov, V.N., Hendriks, H., Bos, R. and Bugaenko, L.A., 1984. Chemical content of essential oil of *Mentha spicata*. *Rastl. Resur.*, 20: 131–136.

Misra, L.N., Tyagi, B.R. and Thakur, R.S., 1989. Chaemotypic variation in Indian spearmint. *Planta Med.*, 55: 5756.

Morris, J.A., Khettry, A. and Seitz, E.W., 1979. Antimicrobial activity of aroma chemical and essential oil. *J. Am. Oil Chem. Soc.*, 56: 595–603.

Nelson, E.K., 1944. Chemical Investigation of American Spearmint Oil. *Chem. Abs*. 6, 1041.

Padmavati, K., Udipi, S.A. and Rao, M.1992. Effect of different cooking method on carotene content of vegetables. *J. Food Sci. Tech.*, 29: 137–140.

Raghavan, B., Abraham, K.O., Jagmohan, Rao and Shankaranarayan, M.L., 1994. Effect of drying of flavour quality of Indian Spearmint. *J. Spices Aromatic Crops*, 3: 142–151.

Rastogi, R.P. and Mehrotra, B.N., 1984. Detection of carvone, limonene, 1, 8–cineole and dihydrocarvone essential oil by GLC in Compendium of Indian medicinal plant Vol.3, 456 Pub. by Publication and Information Directorate (CSIR), New Delhi, India.

Robert, O. Brown and Chastain, G. Harrel, 1944. Synthetic Spices Comprising Spice Oils and Edible Carmers. Pillsbury Flour Mill (Co.) U.S. 2, 356, 501, Aug. 22. (Chem. Abs. 39, 136).

Schweizer, J., 1841. Prakt. Chem. 24, 257 and Gladstone, Isolation of δ-carvone from mandarin pect oil, *J. Chem. Soc..*, 25: 1 (1872).

Shikimaka, A.P., Vorobiova, E.A., Kubrak, M.N. and Buga, T.V., 1993. Terpenoid composition of essential oil of *Mentha spicata. J. Biol. Chim.* 5: 75–6.

Sieves, A.F., Lowman, M.S. and Mabel, I. Ruttle., 1945. The yield and quality of the oil of some hybrid and tetraploid mint. *J. Am. Farm. Association.*, 34: 225–231.

Singaram, B. and Verghese, J., 1977. Chemical structures. The name of carvone. *Perfumary Flavours*, 2: 47.

Singh, A.K. and Naqvi, A., 1996. Postharvest storage to boost essential oil recovery in Cymbopogon and mint species. *Indian Perfumer*, 40: 26–30.

Singh, V.P. and Duhan, S.P.S., 1979. Response of spearmint (*M. spicata* L.) region of Nanital. *Indian J. Pharm. Sci..*, 41: 47–48.

Sticher, O. and Flueek, H., 1968. Composition of genuine, extracted and distilled essential oils of some *Mentha spicata. Pharm. Acta. Helv.*, 43: 411–446.

Snri, R.K. and Jain, P.P., 1981. Study of oil of *Mentha spicata* grown at Dehraian, India. *Indian Perfumer*, 25: 127–128.

Walker, J.T. and Melin, J.B., 1996. *Mentha spicata* and effect of their essential oil of Meloidogyne in soil. *J. Menatology*, 28: 629.

Warren, J.E., 1933. Some observation on the constant of oil of spearmint. J. *Am. Pharm. Asso..*, 22: 294–297.

Wilson, John B., 1950. Expert of flavour and non-alcoholic beverages. *J. Assoc. Offic Agr. Chemists*, 33: 302–10.

2011, Future Crops Vol. 1 ***Pages 218–242***
Editor: **Professor K.V. Peter**
Published by: **DAYA PUBLISHING HOUSE, NEW DELHI**

Chapter 13

Threatened Medicinal and Aromatic Plants of Mizoram

T.K. Hazarika and B.P. Nautiyal
Department of Horticulture, Aromatic and Medicinal Plants,
Mizoram University, Aizawl – 796 009
E-mail: tridip28@gmail.com

The North-Eastern Region (NER) of India lies between Longitudes 89°42′ and 97°12′ E and Latitudes between 22°19′ and 30° N and comprises of eight states *viz.*, Assam, Arunachal Pradesh, Manipur, Meghalaya, Mizoram, Nagaland, Sikkim and Tripura. The NER is spread over to wide geographical area of 262,138 sq km. The unique combination of varied topography and climate of almost tropical to alpine type provides a repository of rich vegetation wealth. Out of 9 important vegetation types of India, 6 are found in NER and according to cumulative estimate,the region harbours approximately 8000 flowering plant species including 825 species of orchids, 80 species of rhododendrons, 17 species of citrus with 52 variants, 16 species of banana, 60 species of bamboo and 25 species of canes. Besides 40 gymnosperms, 500 pteridophytes and rich genetic diversity of many other species are also reported from the region (Bhattacharya and Dutta, 1956, Hegde 2000, Anon, 2003). Owing to rich biodiversity, the region has been under the focus for its biodiversity and all eight states of the region are now included under Indo-Burma hotspot for mega biodiversity. However, the species endemism and other threats in the region are also leading into species rarity and endangerment. According to Indian Red Data Book (Nayar and Shastri, 2002), nearly 800 species of the region are already facing severe threats. Importantly, many species of the region are not assessed for threatened status as exploration is continuing. Therefore, it is imperative to document various species which are facing severe threats including over and illegal exploitation. Medicinal and Aromatic plants (MAPs) of the region in particular need immediate attention. This chapter highlights threatened MAPs of Mizoram state of NER with their medicinal uses in traditional medical knowledge (folk medicine) and threat status for future conservation and sustainable utilization.

Physiography of Mizoram

Mizoram state is an expanse of 7 North-south transverse parallel mountainous ranges of tertiary origin locally known as 'Tlang' situated in the extreme of the Himalayan ranges of India. It is located in between 21º58´ N and 24 º35´N latitudes and 92º16´E and 93 º29´E longitudes and bounded by Cachar district of Assam in the North, Manipur in the North- east, Myanmar in the East and South and Bangladesh and the state of Tripura in the West. It covers an area of 21,081 sq. km comprising mostly hilly terrains with deep gorges in between the hill ranges run in North- South direction with an average height of 920 m above msl. The mountains are intercepted by narrow deep valleys and criss-crossed by innumerable small hillocks.

Mizoram enjoys a pleasant and moderate climate. The climate is characterised by short winter and long summer with heavy rainfall. The hillock has a sub-tropical humid climate and the temperature

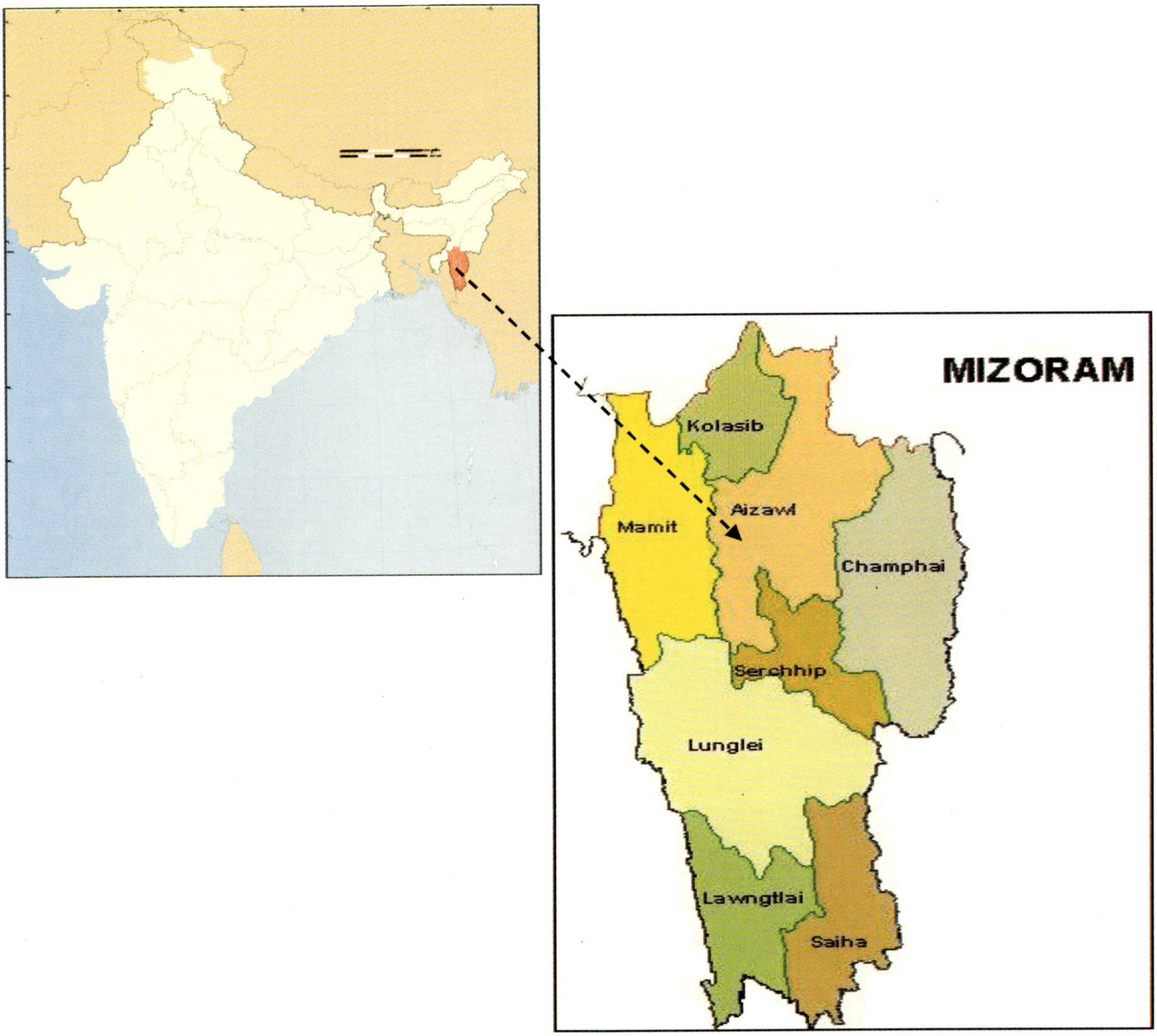

Figure 13.1: Map of Mizoram.

varies from 11ºC-30ºC. The entire state is under the direct influence of maritime tropical air mass brought in by South-Qwest monsoon. Mizoram receives heavy rainfall which generally starts in May and prevails upto September-October with an average rainfall of 2500 mm per annum. July- August receives highest rainfall. The soil type of Mizoram is generally acidic in nature and the texture varies from sandy-rocky to sandy-loam. The soils are generally very fertile and responsive to the vigorous growth of vegetation as well as arable crops.

In Mizoram, 88 per cent of population comprises of Mizo having 18 clans including Lushai, Khawlhring, Khiangte, Ralte, Renthlei, Chhangte, Paihte, Pawi, Hmar, Lakher, Thado, Chawngthu, Ngente, Tlau, Pautu, Rawite, Jongte and Vangchhia having more or less common culture and traditions of the Mizos speaking the common language recently called the Mizo. However, they have diverse traditional knowledge on natural resources with significant knowledge on wild plants including medicinal plants.

Floristic Diversity

The vegetation of Mizoram is tropical evergreen and semi-evergreen forest in the lower altitude hills; sub-tropical to montane sub-tropical in the high hills. A major portion of Mizoram's forests are therefore tropical evergreen and semi-evergreen. The floristic exploration of the state is incomplete including. Lalramnghinglova and Jha, 1996: Singh, 1997 and studies are still continued. Recently, Singh (1997) reported 2358 species of plants from the state of which 2141 species in 905 genera and 176 families belonged to flowering plants, 6 species of gymnosperms and 211 species of pteridophytes. The publication of Flora of Mizoram (in 3 volumes) is underway by Botanical Survey of India and Volume I already published (Singh *et al.*, 2002). The volume contents nearly 884 taxa belonging to 403 genera and 90 families of flowering plants. A large part of these vegetations are occupied by medicinal and aromatic plants. Earlier, Mahanti (1994) described ethno-botany of nearly 434 species from the state of which mainly used as medicine in traditional health care system by various ethnic groups. Likewise, Lalramnghinghlova (2003) also described nearly 236 medicinal plants of Mizoram with their ethno-botanical importance. However, the montane sub-tropical hill Forests in the eastern part of Mizoram are yet to be fully covered ethno-botanically. Moreover, the existence of the estimated 90 per cent of India's medicinal plants diversity in the forests and 10 per cent in the non-forest habitats suits the Mizoram condition and therefore provides an excellent opportunity to people for MAPs based entrepreneurship. Therefore, in-depth study and complete inventorization of medicinal plants available in rich biodiversity area is one of the 'Green' prospects of research field in Mizoram.

Status of MAPs in Mizoram

The socio-economic life of the rural people depends on ambient vegetation from where they derive all their material requirements *viz.* timber, food, fuel wood, medicinal plants etc. Nature has provided an excellent storehouse of remedies to cure all the ailments of mankind. The importance of botanical, chemical and pharmacological evaluation of plant derived agents used in the treatment of human ailments has been increasingly recognized in the last decades. We are now, witnessing a resurgence of interest in herbal medicines for treatment of various ailments. The progress achieved in botanical studies during 19th century had a revolutionary influence on utilizing plants for new drug discovery and traditional knowledge provides an excellent tool for new drug discovery. About 95 per cent of the interior population depends on herbal medicines and nearly 98 per cent of raw materials are harvested from the wild plant resources without replenishing the growing stocks. Love of the tribal people for forests has been maintaining their greenery in the past because they considered themselves as an integral part of the forests. However, with the rise of consumerism their attitudes is gradually changing and they have started looking at the forest as a source of wealth to sustain their increasing daily needs and luxuries. They have led to depletion of forest cover and a disturbance in the ecological balance. The village herbal preparations include uprooting of the plants that are detrimental to conservation of sub populations. As a result, commonly used and effective herbal plants become rare and endangered on the verge of extinction unless crash conservation measures are taken up for their revival.

The jhuming/shifting cultivation is the mainstay of the people as 80 per cent population is totally relay on agriculture for livelihood. Clearance of land for agriculture, uncontrolled fire, heavy indiscriminate destruction of forests for fuel wood and timber, agricultural demands, hunting for food and harvest for edible and medicinal plants in the forests in the past have contributed for depletion of medicinal and aromatic plant (MAP) habitats. The greatest threat to forests degradation and species endangerment, steep decline in soil fertility *vis-a-vis* crop production is slash-and-burn method of shifting agriculture in Mizoram. Moreover, decrease in jhum cycle (2-4 years compared to 10-15 years in past), replacement of virgin forests with bamboo forests resulted into forest degradation, habitat fragmentation and habitat lose. As a result, many valuable non-wood forests produce including MAPs are rapidly being lost. Moreover, 27 species are endemic to the state and 23 species are cited as threatened species (*www.bsienvis.nic.in*). Besides, many other species are facing severe threats especially MAPs used in traditional system of health care. These species along with the species enlisted in Red Data Book are given with brief account on habits, flowering period, part used for medicinal purpose and threat status.

The agro-ecosystems as well as natural ecosystems in Mizoram harbour immense genetic potential (Lalramnghinghlova 1999a; 1999b; 2002). Over 60 per cent people living in the interior parts of the state depends upon herbal medicine. Lalramnghinghlova and Jha (1998,1999) described more than 200 ethno-medicinal plants for their efficiency to cure diseases like bleeding from nose, fever, malarial fever, asthma, tuberculosis, calculi of kidney and gall-bladder, urinary troubles, hypertension, diabetes, stomachache, stomach ulcer, dysentery, diarrhoea, jaundice, hepatomegaly, fracture of bone, gyneic disorder and snake bite.

Based on available literature of medicinal uses of plants in Mizoram (Mahanti, 1994; Singh, 1997; Lalramnghinghlova and Jha, 1998, 1999; Lalramnghinghlova, 2003 and Singh *et al.*, 2002), threatened MAPs are observed for medicinal properties as well. The therapeutic uses and formulations prepared by using threatened MAPs in Mizoram by various ethnic communities are given below. It is evident from the table that the utility of plants in preparation of medicines is either alone or in combination with other plants. However, most of the plants are used singly for curing one or many diseases. Most of the time formulation in the form of infusion, decoction or milled powder or paste of plant is administered. The other uses like paste formation etc. is usually employed for external application, massaging or dabbing on to the effected part of the body. Fresh water is mostly used as solvent in the preparation. Sometimes animal fats, minerals and oils are also added in the mixture. A combination of two or even more ingredients is used for single disease while a single plant can be used for different diseases.

In traditional medical knowledge, beside primary health care, formulations prepared from these threatened MAPs for some chronic diseases is also mentioned. Nearly 14 species are used for urinary disorders and gynaecological purpose (leucorrhoea; contraceptive and labour pain); 8 species for orthopaedic purpose including rheumatism and inflammation); 6 species in liver related ailments; 6 species in respiratory problems including asthma and tuberculosis; 10 species for the treatment of various ulcer, tumour and cancers; 9 as febrifuge and 21 for gastrointestinal disorders (purgative,

Table 13.1: Threatened MAPs of Mizoram*

Sl.No.	*Scientific Name*	*Vernacular Name*	*Family*	*Habit*	*Flowering Time*	*Fruiting Time*	*Parts Used*	*Status*
1.	*Aegle marmelos* (L) Corr.	Belthei	Rutaceae	Tree	October-December	December-April	Pulp of fruit	EN
2.	*Aquilaria malaccensis* Lamk.	Thingrai	Thymelaeaceae	Tree	April-May	May-August	Agar/resinous substance, wood	CR
3.	*Ardisia polycephala* Wall.ex A.DC.	Sialtuai, Borsoilia	Myrsinaceae	Tree	April-May	November-April	Roots	VU
4.	*Averrhoa carambola* Linn.	Theiherawt	Averahoaceae	Tree	Mid March-October-November	July-Early February	Fruit	R
5.	*Begonia inflata* Cl.	Sekhupthur-hmul	Begoniaceae	Herb	April-May	June-October	Whole plant	R
6.	*Bergenia ciliata* (Haw.) Sternb.	Kham-damdawi, Pan-damdawi, Thual-thuap	Saxifragaceae	Herb	March-April	April-May	Roots	EN/R
7.	*Blumea lanceolaria* (Roxb.) Druce	Buarze	Asteraceae	Shrub	April	May	Leaves	VU
8.	*Bombax ceiba* L.	Phuanchawng	Bombaceae	Tree	February-March	April-May	Root bark, stem bark, leaves and cotton	VU
9.	*Cassia alata* L.	Reng-an	Caesalpiniaceae	Shrub	November-December	December-March	Leaves	VU
10.	*Cassia hirsuta* L.	Sabdaru and Rulchuk-damdawi	Caesalpiniaceae	Shrub	October-November	December-January	Roots and leaves	CR
11.	*Cautleya gracillis* (Sm.) Dandy	Pa-le	Zingiberaceae	Herb	June-July	August-September	Rhizome	R (cultivated)
12.	*Chonemorpha fragrans* (Monn.) Als.	Phungtheikelki; Theiklkisuak	Apocynaceae	Climber	June-July	November-January	Roots	VU
13.	*Clerodendrum wallichii* Merr.	Tratuba; Phuihnamria	Verbenaceae	Shrub	November-January	February-March	Roots	EN
14.	*Curcuma zedoaria* (Christ) Rosc.	*Ai-dizung*	Zingiberaceae	Herb	May-August	–	Rhizome	CR
15.	*Cyclea peltata* (Lamk.) Hook. f. & Thoms	Khauchhim; Bithikakhama	Menispermaceae	Climber	August-September	November-January	Roots	EN
16.	*Dalbergia pinnata var.acaciaefolia* (Dalz.) Thoth.	Hruitengtere	Fabaceae	Shrub	March-April	May-October	Root-bark	CR

Contd...

Table 13.1–Contd...

Sl.No.	Scientific Name	Vernacular Name	Family	Habit	Flowering Time	Fruiting Time	Parts Used	Status
17.	*Desmos longiflorus* (Roxb.) Saf.	Chi-ri-pi	Annonaceae	Small tree	November-January	March-April	Root & leaf.	R
18.	*Elaeagnus pyriformis* Hook.f.	Ramsarzuk, Zut	Elaeagnaceae	Shrub	November-January	January-April	Roots	VU/R
19.	*Elsholtzia blanda* (Benth.) Benth.	Nauhri	Lamiaceae	Shrub	September-October	October-January	Aerial part	R
20.	*Flemingia macrophylla* (Willd.) Prain.	Tuisithing	Fabaceae	Shrub	January-March	March-April	Roots	VU/R
21.	*Garcinia lanceaefolia* (G.Don) Roxb.	Pelh/pelh-te	Clusiaceae	Shrub	February-March	June-July	Leaves and fruits	EN/R
22.	*Garcinia pedunculata* G.	Theipumlian	Clusiaceae	Tree	January-March	April-June	Fruits	R
23.	*Garcinia sopsopia* Ham.	Thensaker	Clusiaceae	Tree	August-September	November-March	Bark, Fruit and wood	R
24.	*Gardenia coronaria* Ham.	Rul-hluah Lalran	Rubiaceae	Tree	August-September	December	Root & leaf	R
25.	*Gelsemium elegans* Benth.	Hnamtur, Chamai	Lo-ganiaceae	Shrub	April-May	June- August	Leaves and fruits	EN
26.	*Gynocardia odorata* R. Br.	Saithei	Flacourtiaceae	Tree	March-April	January-March next year	Seed- oil	E N / R
27.	*Hedychium coronarium* Koenig.	Ai-lalnu	Zingiberaceae	Herb	June-July	–	Rhizome	R (cultivated in home gardens)
28.	*Hydronocarpus kurzii* (King) Warp.	Khawitur	Flacourtiaceae	Tree	April	February-March next year	Barks and Fruits	EN
29.	*Ixora nigricans* R.Br.	Thainurual	Rubiaceae	Shrub/small tree	February-March	April-June	Leaf	R
30.	*Kalanchoe pinnata* (Lamk.) Pers.	Zihor	Crassulaceae	Herb	November-December	January-April	Leaf	R (introduced & naturalized)
31.	*Lepionurus sylvestris* Blume	Anpangthuam	Opiliaceae	Tree	February-March	May-August	Leaves	R

Contd...

Table 13.1–Contd...

Sl.No.	Scientific Name	Vernacular Name	Family	Habit	Flowering Time	Fruiting Time	Parts Used	Status
32.	*Lindernia ruelloides* (Colsm.) Penn.	Thasuih	*Scorphulariaceae*	Herb	November-December	January-March	Whole plant	EN
33.	*Lobelia angulata* Forst.	Choakthi	Lobeliaceae	Herb	April-May	July-August	Whole plant	EN
34.	*Lobelia nictianaefolia* Heyne.	Berawchal	Lobeliaceae	Herb	April-June	July-September	Leaves and Roots	R
35.	*Morus australis* Poir.	Lungli	Moraceae	Shrubs, trees	March-April	June-August	Bark, Leaf, Roots	R
36.	*Musa glauca* Roxb.	Saisu	Musaceae	Herb	March-July	July-January	Seeds and fibre	VU
37.	*Ocimum tenuiflorum* L.	Runhmui-dum	Lamiaceae	Shrub	November-December	December-January	Whole plant	R (Cultivated in home gardens
38.	*Pajanelia longifolia* (Willd.) K. Schum.	Ram-archangkawm	Bignoniaceae	Tree	November-February	February-March	Leaf & stem	R
39.	*Pentapetes phoenicea* L.	Parsenbial	Sterculiaceae	Herb	August-November	November-January	Leaf	VR
40.	*Phyllanthus airy-shawii* Brunal & Roux.	Mawsai	Euphorbiaceae	Shrub	April-May	July-January	Leaves	EN
41.	*Phyllanthus fraternus* Webs.	Mitthi-sunhlu, Lai-keng-thei-chah	Euphorbiaceae	Herb	August-October	November-March	Whole plant	VU
42.	*Rauvolfia serpentina* Benth.	Rulturzung, Chandoma	Apocynaceae	Undershrub	June	October-December	Roots	CR
43.	*Rhaphidophora hookeri* Schott.	Thiallawn	Araceae	Climber	June-August	Ausgust-November	Stem	R
44.	*Ricinus communis* Linn.	Mutih	Euphorbiaceae	Shrub or small tree	July-September	August-November	Leaf & stalk, seeds, roots	R
45.	*Rubia cordifolia* L.	Rawngsen/ Saphut	Rubiaceae	Climber or creeping Herb	June-August	September-December	Roots and leaves	EN/R
46.	*Ruellia suffruticosa* Roxb.	Sa-vang-ma	Acanthaceae	Herb	May-September	October-December	Roots	CR
47.	*Saraca asoca* (Roxb.) de Wilde	Mulhawih	Caesalpiniaceae	Tree	April-May	October-February	Bark and root bark	EN

Contd...

Table 13.1–Contd...

Sl.No.	Scientific Name	Vernacular Name	Family	Habit	Flowering Time	Fruiting Time	Parts Used	Status
48.	*Senecio scandens* Buch-Ham. Ex D. Don	Sai-ek-hlo, Si-chhia-va	Asteraceae	Climber	February-April	April-May	Leaves or aerial parts	CR
49.	*Smilax glabra* Roxb.	Tluang-ngil	Smilaceae	Climber	June-July	November-January	Roots/Tuber	EN
50.	*Solanum khasianum var. Chatterjeeanum* Sen Gupta	Athlo, Rulpuk	Solanaceae	Undershrub	July-August	November-February	Fruits/seeds	VU
51.	*Stemona tuberosa* Lour.	Kaimam	Stemonaceae	Twiner	March-April	June-July	Roots/Tuber	VU
52.	*Stephania japonica var discolor* (Miq.) Forman.	Hruifei	Menispermaceae	Climber	March-April	May-June	Root- stock	EN
53.	*Taraktegenos kurzii* (King) Warb	Khawitur	Bilaceae	Tree	July-October	November-February	Seed	R
54.	*Thunbergia grandiflora* Roxb.	Vako/ Zawngafian	Acanthaceae	Climber	April-May	September-November	Leaf	R
55.	*Tinospora cordifolia* (DC.) Miers. Ex Hook. f. & Thoms.	Theisawntlung	Menisper-maceae	Climbing shrub	March-April	April-May	Bark and leaves	CR
56.	*Trapa natans. var. Bispinosa* (Roxb.) Makino	Dil-hlo	Trapaceae	Aquatic Herb	August-September	November-December	Aerial parts	CR
57.	*Woodfordia fructicosa* Kurz.	Ainawn	Lythraceae	Shrub	March-April	April-May	Flowers	VU
58.	*Zanonia indica* L.	Lalruanga-dawibur	Cucurbitaceae	Climber	June-August	November-December	Fruits, seeds and leaves	CR
59.	*Zanthoxylum armatum* DC.	Arhrikreh	Rutaceae	Tree	April-May	October-December	Leaves and seeds	EN
60.	*Zingiber purpureum* Rosc.	Pale	Zingiberaceae	Herb	August-September	October-November	Rhizome	VU: Cultivated

*Source: Lalramnghinghlova (2003); Lalfakzuala *et al.* (2007); Rai and Lalramnghinglova, (2010).

Abbreviations used: VU: Vulnerable; EN: Endangered; R: Rare; CR: Critically Endangered.

Table 13.2: Brief description of the ethno-medicinal uses and formulations for therapeutics uses of Threatened MAPs.

Plants	*Therapeutic Uses*	*Medicinal Formulations*
Aegle marmelos (L) Corr.	Gastro-intestinal, Febrifuge	The orange coloured pulp is pounded to paste and mixed with water. The decoction is taken for the remedy of dysentery, stomachache & digestive problems. The bark decoction of about 20 ml is taken thrice daily for one week against intermittent fever.
Aquilaria malaccensis Lamk.	Orthopaedic, anti-emetic	The resinous substance is used in Gout and rheumatism. The agar extracted from the wood is used against vomiting.
Ardisia polycephala Wall. ex A.DC.	Haemostatic	Roots are mixed with *Amaranthus caudata* on grindstone and the paste is collected in a cup of water. The water is drunk @ ½ cup (50 ml) thrice daily as an effective remedy against excessive bleeding.
Averrhoa carambola Linn.	Anti-scorbutic	Three or four fruit slices are taken daily for jaundices or juice of crushed fruit is taken orally @ ½ -1 cup (50-100 ml) 3 times daily. Infusion of leaves is taken against enlargement of liver once daily; A poultice of fruits is used in bleeding piles and also used against scurvy.
Begonia inflata Cl.	Ureno-genital	Food poisoningThe plant is boiled with bark of *Engelhardtia spicata* and the water is drunk against piles disorder and dysentery. The plants is eaten raw against food allergy *e.g.* Fish etc. The white roots are boiled in water and water is taken as an effective remedy against genitor-urinary problems.
Bergenia ciliata (Haw.) Sternb.	Gastro-intestinal	Infusion of root is taken orally for diarrhoea and dysentery @ ½ cup (50 ml) twice daily. Juice of stem is used for sores and ulcers by external application
Blumea lanceolaria (Roxb.) Druce	Gastro-intestinal; respiratory disorders; Anti-cancer; Hepatoprotective	Juice of leaves is applied on wounds and chronic ulcers. Infusion of leaves is taken against dysentery @ ½ cup (50 ml) once daily. The leaves are boiled and the water is strained through chase- cloth and taken orally against bronchitis, asthma, cancer and liver ailment@ ½ cup (50 ml) twice daily.
Bombax ceiba L.	Urinary disorder; Anti-diarrhoea	The root-bark is crushed and boiled with the roots of *Firmiana colorata* and *Dracaena spicata* as a medicine of strangury @ 1 teaspoonful of sugar is added and taken for strangury @ a tablespoonful twice daily. The bark is crushed with that of *Mangifera indica* and then boiled with water. The water is drunk ½ cup (50 ml) twice daily against dysentery. The resin powder is used against diarrhoea.
Cassia alata L.	Diuretic, bronchodilators, dermatological problems, digestive	The leaves are crushed and mixed with a pinch of salt and soot and made into a paste. The ringworm or related skin diseases are rubbed with half sliced brinjal and the paste is applied externally and then bandaged to cure ringworm or related skin diseases. Decoction of leaves and flowers are used internally in constipation, bronchitis and asthma. The roots of this plant are diuretic.
Cassia hirsuta L.	Anti-dote	The roots are rubbed on grindstone and the paste is collected in a cup of water and drunk against snake-bite. The leaves are also crushed and the juice is applied externally on the part bitten.

Contd...

Table 13.2–Contd...

Plants	*Therapeutic Uses*	*Medicinal Formulations*
Cautleya gracillis (Sm.) Dandy	Flatulence; Hepato-protective	Infusion of rhizome is taken for flatulence, colic & hepatomegaly @ tablespoonful (10 ml) twice daily, the rhizome is eaten raw to relieve colic & hotness in the stomach & also used against cough
Chonemorpha fragrans (Monn.) Als.	Gynaecological problems	Infusion of root is taken orally as an effective medicine for easy labour as well as for the removal of retained placenta @ ½ cup (50 ml) twice daily.
Clerodendrum wallichii Merr.	Anti-tumour	The paste made of roots of *Ardisia paniculata, Caoxylon khasianum* and *Phogacanthus thyrsiformis* is applied externally on abdominal tumor once every day for 7 days. Fresh paste is used every day.
Curcuma zedoaria (Christ) Rosc.	Haemostatic	The rhizome is crushed and mixed with water and taken orally for piles and epileptic seizure. The medicine is taken in dilute solution @ 1 teaspoonful once daily.
Cyclea peltata (Lamk.) Hook. f. & Thoms	Fabrifuge, anti-diarrhoea, paediatric	The decoction of bitter taste root is taken orally for colic, fever and diarrhoea @ 1 teaspoonful twice daily.
Dalbergia pinnata var. acaciaefolia (Dalz.) Thoth.	Anti- stomatitis; hepato-protective	The root- bark is scrapped off and boiled in water for 1-2 minutes. The solution is strained through a clean soft cloth or filter paper and used against stomatitis, toothache and gum-boil. The medicine is retained in mouth for sometime, twice or thrice daily.
		The root- bark is scrapped off and ground to powder. About 5 g powder is mixed with a cup of milk (100 ml) and ½ cup is taken twice daily for effective remedy against hepatitis and jaundice.
Desmos longiflorus (Roxb.) Saf.	Respiratory disorders; Anti-ulcerous; Against abscess	The leaves with those of *Morinda angustifolia, Ardisia paniculata* and roots of *Desmos chinensis* are ground and the paste is collected in a splitted bamboo- culm. The paste is made warm in the fire and gently applied externally on chronic-sore. Poultice of leaves is used against sore and at the same time, infusion of leaves is taken orally @ ½ cup (50 ml) once daily. It constitutes one ingredient for the treatment of chronic ulcer; decoction of leaves used for asthma
Elaeagnus pyriformis Hook.f.	Against appendicitis pain	The roots are grinded and the paste is collected and dipped into a cup of water. The water is drunk against appendicitis. A lightly scorched single leaf is also banded on the waist of the patient.
Elsholtzia blanda (Benth.) Benth.	Febrifuge; Anti-cholera; dermatological problems	Infusion of aerial part of plant is used for children's disease called 'Nauhri', a combination of fever, cholera, skin diseases & inflammation @ 10 ml twice daily. Poultice of leaves is also used for inflammatory glands
Flemingia macrophylla (Willd.) Prain.	General body ache	Decoction of the roots is used as external application in swellings and body pain.
Garcinia lanceaefolia (G.Don) Roxb.	Stomachic; Diuretic	Infusion of the leaves is taken as stomachic and diuretic.

Contd...

Table 13.2–Contd...

Plants	*Therapeutic Uses*	*Medicinal Formulations*
Garcinia pedunculata G.	Anti-diarrhoeal	Acidic pericarp is taken out and mixed thoroughly with water and strained through tea-strainer or clean cloth. The solution is mixed with sugar & is taken orally against dysentery & diarrhoea @ ½ cup (50 ml) twice daily.
Garcinia sopsopia Ham.	Antidote	The powdered bark is applied on Snake-Bite and a twig is applied as snack repellent.
Gardenia coronaria Ham.	Antidote	The infusion of roots & leaves are taken orally against snake bite @ 50 ml. (twice daily)
Gelsemium elegans Benth.	Veterinary	Used as veterinary medicine. The root- bark is dried and ground to powder and the powder is added to the pig fed for pig's cough, fever and mange. Juice of root bark is applied on tiger–bite.
Gynocardia odorata R. Br.	Dermatological problems; Anti-leprosy; Ant-rheumatic	Fruits are crushed and seeds are extracted manually. The extracted seed oil is used as lotion in leprosy and other skin diseases. Used both externally and internally in leprosy, psoriasis, scrofula, skin-diseases and chronic rheumatism.
Hedychium coronarium Koenig.	Anti-rheumatic; Anti-inflammatory; Febrifuge	Base of stem is used for swellings & rhizome is used as febrifuge & antirheumatic
Hydronocarpus kurzii (King) Warp.	Anti-leprosy; Febrifuge	The seed yield oil, certain preparation of oil are injected for the treatment of Leprosy. The bark contain tannins and is used to cure fever.
Ixora nigricans R.Br.	Anti-diarrhoeal; Paediatric	Infusion of the leaves is prescribed for dysentery & colic problems @ ½ cup twice daily.
Kalanchoe pinnata (Lamk.) Pers.	Anti-inflammatory; Dermatological problems; Urinary disorders	Juice of leaves alone or in combination with *yolj* is applied on skin eruptions and inflammatory glands. The leaves are boiled and water is taken for strengthening the function of uterus @ 10 ml 3 times per day and for diabetes @ ½ cup (50 ml) once per day. Decoction of Leaves is taken orally against pain of throat. The roots are used poultice for headache in children.
Lepionurus sylvestris Blume	Against Vertigo	Scorched leaf is applied on forehead in dizziness, giddiness.
Lindernia ruelloides (Colsm.) Penn.	Anti-spasmodic	Infusion of the plant is used as a lotion for local swellings and sprains and as antispasmodic.
Lobelia angulata Forst.	Anti-diarrhoeal; anti- ulceration; anti-ophthalmic	The leaves and fruits are crushed and the juice is taken as an effective remedy against diarrhoea, dysentery and stomach ulcer. The medicine is taken @ 1 tablespoonful (10 ml) twice daily. Two to three fresh fruits are eaten against tonsillitis. Juice of crushed plant is taken orally for opthalmia@1 tablespoonful (10 ml) thrice daily.
Lobelia nictianaefolia Heyne.	Respiratory disorder; Anti-inflammatory	Plant juice is applied in Boils and sores. Aerial parts of the plant is dried in shade and used to treat asthma and bronchitis.
Morus australis Poir.	Purgative; Astringent	Bark is used for anthelia, purgative; leaf decoction as gargle in inflammation of vocal cord; white roots as astringent.

Contd...

Table 13.2–Contd...

Plants	Therapeutic Uses	Medicinal Formulations
Musa glauca Roxb.	Anti-diarrhoeal; Febrifuge; Antidote; Anti-diabetes	The seeds are made into beads and used as necklace by children against convulsions associated with fever. The sap of cut stem is taken orally against dysentry and as coolant. The watery juice of the stem is applied externally on insect bites, snake-bites, wounds, sores and whitlow. The powdered seeds are used as an effective remedy against tapeworm and diabetes.
Ocimum tenuiflorum L.	Hepatoprotective; Bronchodilators; Gastro-intestinal	The plant is boiled & the steam is inhaled against hepatitis; infusion of the plant is used in cough, bronchitis, gastric disorders & as mosquito-repellent
Pajanelia longifolia (Willd.)	Orthopaedic purpose K. Schum.	The young stem or branch is taken from the lowest rachis and cut into pieces. They are crushed and ground into a paste. The paste is directly applied on the fracture and then bandaged. The paste should be changed every day for 3-4 days. Leaf paste is also used for the same purpose.
Pentapetes phoenicea L.	Anti-inflammatory; Expectorants	The leaves are boiled & the water is taken for inflammatory glands, cough & cold; Juice of leaves is applied on inflammatory glands.
Phyllanthus airy-shawii Brunal & Roux.	Dermatological problems; Anti-inflammatory	The leaves are crushed and the juice is applied externally on measles, skin eruptions and inflammatory glands. An infusion of leaves is taken orally for diphtheria @1/2 cup (50 ml) once daily.
Phyllanthus fraternus Webs.	Anti-inflammatory in orthopaedic purpose; anti-ulcer; hepato-protective; diuretic; uro-genetial disorders; Gynaecological purpose	The whole plant is crushed and boiled in water is drunk as an effective remedy against arthritis and jaundice. The medicine is taken @1/2 cup (50 ml) twice daily for 7 days. It is also taken for gastro-intestinal troubles. The crushed leaves are applied externally on scabies and the juice is used for aphthae. Decoction of plant is taken for hepatitis and cirrhosis. Infusion of the plant is used in diuretic and genito-urinary infections. The powdered plant is mixed with powdered bark of *Terminallia chebula* and the mixture is taken internally with or without water as an effective contraceptive @ tablespoonful (5 ml) twice daily for 1-2 months.
Rauvolfia serpentina Benth.	Anti-diarrhoeal	The roots are grinded and paste is collected in a cup of water. The water is drunk as effective remedy against dysentery and stomach-ache @ 1 teaspoonful(5 ml) twice daily for children; ¼ cup(25 ml) twice daily for adult.
Rhaphidophora hookeri Schott.	Gynaecological purpose	Mixture of crushed stem juice and fatty oil of *Buceros bicornis* and mixture is given to pregnant women for easy labour. The plant in combination with *Beaumontia grandiflora* and net-like pale-green lichen (*Langakthei*) in the form of a paste for bone-setting by external application.
Ricinus communis Linn.	CNS disorder; dermatological; urinary disorder; Anti-inflammatory; Purgative; Antihelmintic; Anti rheumatism	Young leaves after heating are used in ulcer, sciatica & paralysis while crushed leaves are applied as bandage against urinary problems, skin diseases, wounds, boils, rheumatism. Root is good for urinary trouble. The seed oil is used as laxative and vermicide and purgative and against skin diseases and can be used in cases of inflammation of the intestine or dysentery.

Contd...

Table 13.2–Contd...

Plants	*Therapeutic Uses*	*Medicinal Formulations*
Rubia cordifolia L.	Anti-ulcer; vermifuge; Anti- inflammatory; Dermatological problems	The roots are ground and paste is applied externally on ulcers, inflammation and other skin diseases. Decoction of leaves and stem is used as vermifuge.
Ruellia suffruticosa Roxb.	Hepato-protective; Urinary disorder	The roots are boiled and the water is taken for kidney trouble and liver ailments @tablespoonful (10 ml) twice daily.
Saraca asoca (Roxb.) de Wilde	Diurectic; Antiulcer; Gynaecological; Haemostatic	Decoction of coat of inner bark is taken orally for placental disorder, easy labour and diuretic. The medicine is drunk ½ -1 cup (50-100 ml) twice daily. Fresh and dried fibrous bark is boiled with the roots of *Citrus limetta* and the water is taken orally for stomach ulcer @ tablespoonful (10 ml) twice daily. Infusion of root-bark is taken orally to stop haemorrhage. Pounded flowers mixed with water are given for haemorrhagic dysentery. Seeds are used to cure urinary discharges. Dried bark is astringent and used to treat excessive menstruation as a uterine sedative.
Senecio scandens Buch-Ham. Ex D. Don	Anti-ulcer; anti-cancer	The leaves or aerial parts are boiled in water and left overnight. The water is strained through a clean cloth in a container and the solution is used for treatment of ulcerated cancer/ulcers @ ½ cup (50 ml) daily for 2-3 days.
Smilax glabra Roxb.	Gynaecological problems; stomachic	The knotty root or tuber is trimmed and boiled and the water is taken as tea against gynaecological problems and as stomachic.
Solanum khasianum var. Chatterjeeanum Sen Gupta	Vermifuge	The smoke of burnt fruits/seeds is sucked through a pipe and retained in the mouth is said to expel tooth-worms from the mouth.
Stemona tuberosa Lour.	Febrifuge; Respiratory disorders	Decoction of the tuber is taken against fever and tuberculosis @ 1 teaspoonful (5 ml) twice daily.
Stephania japonica var discolor (Miq.) Forman.	Anti-diarrhoeal; Febrifuge; Purgative	Infusion of root is taken orally against Diarrhoea, fever and dyspepsia @ tablespoonful (10 ml) twice daily.
Taraktegenos kurzii (King) Warb	Anti-leprous; Dermatological problems	Seed oil is used in Leprosy and skin diseases.
Thunbergia grandiflora Roxb.	Anti-ophthalmic; Anti-diabetes; Antiseptic	The cut end of the stem when blow produces a sap which is dropped into the eye for eye-sore and opthalmia. Juice of the leaves is used for diabetes, eye diseases & as an antiseptic in cuts & wounds
Tinospora cordifolia (DC.) Miers. Ex Hook. f. & Thoms.	Anti-rheumatic; Anti-ottorrhoea; Tonic, Aphrodisiac; Anti-diarrhoeal	Decoction of bark is taken orally for insect-bite @1/2 cup (50 ml) twice daily. The bark is boiled with the leaves of *Centralla asiatica* and drunk for insanity. Infusion of leaves is taken against rheumatic pains and swellings @tablespoonful (10 ml) twice daily. Extracted juice of leaves is made warm and drop into orifice (2-3 drops) to cure ear- ache/ottorrhoea. Dried stem and bark is tonic, antiperiodic and aphrodisiac. The starch of roots and stems are nutritious and cure diarrhoea.

Contd...

Table 13.2–Contd...

Plants	*Therapeutic Uses*	*Medicinal Formulations*
Trapa natans. var. Bispinosa (Roxb.) Makino	Anti-diarrhoeal; Febrifuge; bronchodilator	The fruits are sweetish and useful for chronic fevers, biliousness and bronchitis. Infusion of the aerial part is used in Diarrhoea.
Woodfordia fructicosa Kurz.	Anti-ulcer; Antihelmintic and febrifuge; Anti-leprous; Anti-diarrhoeal; Gynaecological problems	The powdered flower is used in sores and ulcers by external application. In snake-bite, few drops of fresh leaves juice is given internally into each nostril and are rubbed on the part bitten. Dried flowers are used astringent and used @ twice daily for one week against dysentery and liver trouble. They are stimulant, antibacterial, anthelmintic and febrifuge. They are also useful in leprosy, leucorrhoea, diarrhoea and dysentery. The leaf paste is used against skin diseases. The leaf juice is used in conjunctivitis @ thrice daily for one week.
Zanonia indica L.	Anti-colic; Febrifuge; stomachic	Warm water is put into the cavity of cylindrical fruits and shaken well. The water becomes bitter and is drunk against stomachic twice daily. The powder prepared from seeds after grinding is taken with water against colic. Infusion of leaves is taken against fever @ tablespoonful (10 ml) twice daily.
Zanthoxylum armatum DC.	Anti-stomatitis	The powdered seeds are used against stomatitis and for removal of foul smell from the mouth. The leaves and fruits are chewed against teeth enamel disease. The stem is used as tooth brush and mouth purifiers.
Zingiber purpureum Rosc.	Anti-diarrhoeal; stomachic; carminative, appetizer; Gynaecological purpose	The rhizomes are used to cure stomachic & diarrhoea. They are carminative, digestive, appetiser and useful in leucorrhoea.

carminative, stomachic, colic pain etc). Besides, 3 species are reported having potential as antidote especially against snake bite. A few species are reported for other ailments *viz.*, 3 species for diabetes and 1 species as anti-emetic. In general, available knowledge on medicinal properties of these species is well known among the local inhabitants in Mizoram. Some important threatened MAPs found in Mizoram are described below for their taxonomic features.

Aegle marmelos (L)

A medium size deciduous tree bearing strong axiliary throne.; bark brownish-grey, thorny; leaves alternate, 3–foliolate; leaflets ovate-lanceolate, acuminate, 1.53- 4 cm, usbsessile, crenate, terminal long petioled; flowers greenish-white; Fruits 8-20 cm in diameters, globosely green, colour is changed to greyish when fruit mature; seeds numerous, embedded in sweet sour pulp.The fruits are valuable chiefly for its mucilage and pectin.

Aquilaria malaccensis Lamk.

Small to moderate sized semi evergreen tree, leaves elliptic oblong, shortly caudate acuminate. Flowers greenish-white.0.7 cm across, axillary umbels arising from branchelets, fruits obovoid, 1.5-2.5 cm long, tomentose, 2 valved, seeds 2.

Ardisia polycephala Wall. ex A.DC.

A small tree; bark brownish grey, ewarty; leaves obovate-lanceolate, acute, 5-6 x 15-232 cm, base narrowed to the petiole; flowers light pink, in axillary corymbose racemes; fruits globose, depressed, red when ripe.

Averrhoa carambola Linn.

An attractive, small, slow-growing evergreen tree with a short-trunk or a shrub. The branches are drooping and the wood is white and turns reddish. The pinnate leaves have a single terminal leaflet and 5 to 11 nearly opposite leaflets, each leaf is 15-20 cm long, and the 3.8-9 cm long leaflets are ovate or ovate-oblong in shape. The bell shaped, perfect flowers, are produced in loose panicles that are much-branched with pedicellate flowers; each flower is around 6 mm wide, with 5 petals that have recurved ends. The fruits are showy with an oblong shape: they are longitudinally 5- to 6-angled and 6.35-15 cm long and up to 9 cm wide. The fruits have a thin, waxy skin that is orange-yellow coloured. The juicy fruits are yellow inside when ripe and have a crisp texture and when cut in cross-section are star shaped. The fruits have an oxalic acid odour, which varies between plants from strong to mild, the taste also varies from very sour to mildly sweetish. Each fruit may have up to twelve 6-12.5 mm long seeds, which are flat, thin and brown.

Begonia inflata Cl.

Erect, branched herb up to 1 m; stem sometimes streaked with red; leaves oblong-lanceolate, acuminate, base obliquely cordate, sinuate-dentate, glabrous; flowers in short axillary dichotomous cymes, pinkish- white; fruits trigonous or triguetrous, inflated; seeds ellipsoid or obovoid.

Bergenia ciliata (Haw.) Sternb.

A perennial herb with thick rootstock. Stem thick fleshy, procumbent; leaves broadly ovate or sub-orbicular,5-35 cm long, coarsely hairy, dotted on lower surface; margin ciliated; flowers white pink, purple, corymbose or slender peduncles; fruits sub-globose; seeds sub pyramidal, smooth.

Blumea lanceolaria (Roxb.) Druce

An erect, aromatic shrub; stem thick; leaves crowded at apex or districhous, oblanceolate or elliptic-lanceolate, acuminate, 2-6 x 10-24 cm, serrate or distantly serrulate; nerves many; base narrowed into a winged petiole; flower-heads yellowish or greenish-white to pale red, clustered on elongated tomentose panicle; bracts tomentose; achenes ribbed, hairy, pappus red.

***Bombax ceiba* L.**

A tall deciduous tree with straight bole; branches whorled, horizontally spreading; bark grey, with sharp conical prickles which disappears in mature trunk; leaves digitate, large, 5-7, lanceolate or oblanceolate, 6-10 x 15-24 cm, acuminate; stipules triangular, caduceus; flowers conspicuous, scarlet-red, clustered at or near the ends of branches; fruits oblong-ovoid, 6-angled; seeds numerous, obovate, packed in white cotton.

***Cassia alata* L.**

A shrub to small tree; stem greenish; rachis furrowed, nodded; leaves large; leaflets 2.5 -7.5 x 6.5-13 cm, 5-10 pairs, sessile, oblong-obtuse, oblanceolate, mucronate; base oblique; nerves 10-20 pairs, flowers bright yellow, terminal, in dense apikes; pods 15-20 cm long, compressed with crenulate broad-wing down the middle of each valve.

***Cassia hirsuta* L.**

An erect shrub up to 2 m height; all parts hirsute and pungent smell; grooved; leaf rachis 10-16 cm long, with solitary gland at the base, stipulate; leaflets 3-4 pairs, opposite, covered with soft hairs, flowers yellow, terminal, 5 cm across, fruits (pods) 7-15 cm long, slender, hirsute, seeds ovate.

***Cautleya gracillis* (Sm.) Dandy**

An attractive member of the ginger family with slender stems to about 2ft tall. These stems are leafy to about two thirds of their length green lanceolate, acuminate, 3-4 x 10-22 cm leaves with maroon undersides and spikes of soft yellow flowers from brick-red bracts. Flower buds appear from the apical portion of the stem. Brilliant yellow flowers are produced successively over a period of a month and more with each flower lasting about 3 to 5 days in bloom.Fruits globose, red; seeds conical, ridged.

***Chonemorpha fragrans* (Monn.) Als.**

A large climber; stem dark-grey, exuding white latex, lenticellate; leaves broadly-ovate, bi-forked, shortly acute, 18 x 26 cm, dark-green above, pale beneath, pubescent; base shallowly cordate, fimbriate; flowers white, fragrant, axillary peduncled cymes; follicles in pairs, up to 30 cm long; beak curved; seeds pale brown, flat, surrounded by parachute-like white silky coma, up to 8 cm long.

***Clerodendrum wallichii* Merr.**

A tall shrub to 3 m height; bark grey; leaves opposite, elliptic oblong or ovate-oblong, 5-6 x 15-25 cm, acuminate; flowers white, pendent; peduncles horizontal and slender; drupe succulent, deep purple; druplets 1-4.

***Curcuma zedoaria* (Christ) Rosc.**

A large Rhizomatious perennial herb; rhizomes with cylindric annulate tubers, brownish outside, pale yellow and faintly aromatic inside; leaves 4-6 cm with long petioles, 10-15 x 20-45 cm, finely acuminate and curled, upper side often blotched with purple flush on either side of midrib, lower with narrower purple; Pseudostems and leaf stalks green. Flowers not longer than bracts, pale yellow in spikes, calyx pinkish; corolla- lobes white with pinks; lip pale yellow, emarginate, 3 lobed.

***Cyclea peltata* (Lamk.) Hook. f. & Thoms**

A slender climber, branches longitudinally ribbed, all parts hirsutous; leaves broadly ovate or deltoid, mucronate, 8 x11 cm; petiole hairy, 5 cm long; flowers greenish, axillary, hispid panicles; fruits globose or ovoboid drupes.

***Dalbergia pinnata var. acaciaefolia* (Dalz.) Thoth.**

A shrub with climbing branches or a small tree that can reach up to 12 m tall, twining left. Roots yellowish-brown; stem greyish. The leaves imparipinnate, pubescent with (13-)21-35(-61) leaflets. Inflorescence is an axillary, sometimes terminal which is up to 6 cm long. The flowers are about 6 mm long and greenish-white, in corymbose panicle racemes. The sepal teeth are short and subequal. There are 9(-10) stamens. The 5-6 cm long fruit is linear-oblong.

***Desmos longiflorus* (Roxb.) Saf.**

Erect, glabrous shrubs to small trees, 5 m tall. Leaves oblong-lanceolate, 15-30 x 5-8 cm, base rounded or cuneate, acuminate at apex, entire at margins. Flowers solitary, axillary, drooping, 4-10 cm long, horn shaped, reddish brown or orange outside, yellow inside; sepals 3, broadly triangular, minutely pubescent, petals 3, spathulate, red fleshy; stamens many, closely packed, carpels many, ovoid. Ripe carpels 15-20 borne on 3-4 cm long fruiting stalks, each with 1-4 seeded segments, terminal apiculate.

***Elaeagnus pyriformis* Hook.f.**

A scandent of straggling shrub; branches spreading, brown; leaves elliptic, acute, 2-3 x 3-7 cm, shining above, silvery beneath; base cuneate; flowers pale or greenish- yellow in axillary clusters; fruits pyriform, narrowed at both ends, yellowish when ripe.

***Elsholtzia blanda* (Benth.) Benth.**

Erect herbs, 1-1.5 m tall. Stems and branches densely pubescent. Petiole 3-15 mm, densely pubescent; leaf blade elliptic to elliptic-lanceolate, 3-16 × 0.8-4.5 cm, adaxially puberulent, glandular, abaxially grey-green, glabrous, strigose on veins, base narrowly cuneate, margin serrate, apex acuminate. Spikes terminal or axillary, subsecund, 4-8(-20) cm, mostly pubescent; verticillasters 7-10-flowered, short pedunculate; bracts subulate to lanceolate-subulate, 1.5-3 mm. Pedicel less than 1 mm. Calyx cylindric, 2-2.5 mm, teeth lanceolate; fruiting calyx slightly dilated at base, ovoid. Corolla white, 3-4 mm, funnelfrom, throat to 2 mm wide; upper lip emarginated; middle lobe of lower lip subcircular, slightly concave; lateral lobes semicircular, margin entire. Anterior stamens exserted, posterior 2 slightly longer. Nutlets yellow-brown, oblong, 0.8 mm. Fruits ellipsoid.

***Flemingia macrophylla* (Willd.) Prain.**

An erect shrub to 2 m height; stem rough, lenticellate, straiate; branches angular, sulcate, leaves 3-foliolate; petiole 3-quetrous, winged, 5 cm long; leaflets ovate-acuminate, 3 x 8 cm, 3-nerved; stipiles large, hispid above, pubescent along nerves beneath; flowers white, streaked with pink, sessile, branched, terminal; pods oblong, apiculate, 2 cm long, finely pubescent; seeds round, black and small.

***Garcinia lanceaefolia* (G.Don) Roxb.**

A shrub or small tree, decussately branched; leaves lanceolate, cauminate, 2-3X5-10 cm, glossy above, pale beneath, flowers bright- red or radish yellow, xaillary, 2-3 together, fruits obovoid, orange yellow, up to 2 cm long.

***Garcinia pedunculata* G.**

Trees are about 20 m tall. Bark thick, corky; branchlets obtusely 4-angled or subterete, striate, lenticellate, glabrous. Petiole 2-2.5 cm; leaf blade oblong, obovate, or oblong-lanceolate, (12-)15-25(-28) × 7-12 cm, papery, base cuneate, margin conspicuously narrowly involute, apex usually rounded, rarely obtusely acuminate. Flowers 4-merous, peduncle 3-6 cm; pedicels robust, 3-7 cm; sepals broadly ovate or suborbicular, thick, fleshy, margin membranous; petals yellow, oblong-lanceolate, 7-8 mm; stamen fascicles connate in capitate ring. 3 mm high, anthers sessile, anthers 2-celled, cells longitudinally dehiscent; pistillode columnar-cuneate, slightly angular; stigma peltate, inconspicuously tuberculate. Female flowers usually in pairs or solitary at apex of branchlet; pedicels robust, slightly tetragonous, ovary subglobose, 8-10-loculed; stigma radiate, 8-10-lobed, papillate. Fruit yellow, large, oblate, concave on both ends when mature, 10-18 × 11-20 cm, smooth; fruiting pedicel 5-6 cm. Seeds 8-10, reniform.

***Garcinia sopsopia* Ham.**

Medium sized evergreen trees 15-20 m tall; drooping branchlets decussate; bark dark grey. Leaves oblanceolate or broadly elliptic or often obovate, 6-20 x 4-9 cm, base cuneate, shortly acuminate at apex, entire along margins. Male flowers in terminal, brachiate panicles, dull white, sweet scented, sepels 4, petels 4, stamens numerous. Female flowers in 5-8 cm long, terminal spikes; ovary sublgobose, tubercled. Berries axillary, globbose to ellipsoid, 2.5-3 cm across, green when young, yellow when ripe, succulent, usually 4 seeded; seeds arillate, aril pulpy.

***Gardenia coronaria* Ham.**

A deciduous unarmed tree; branches stout, woody, younger stems notched at the nodes, pubescent, buds resinous. Leaves opposite and decussate, simple; stipules interpetiolar; subsessile; laminae abovate, Inflorescences usually axillary cymes, sometimes terminal, solitary or 2 to3 flowered, frequently leaf-opposed or supra-axillary; bracts foliaceous. Flowers ebracteolate, subsessile, bisexual, actinomorphic, pentamerous, epigynous, large and showy, fragrant. Calyx synsepalous, tubular, obscurely and irregularly 5-toothed, acutely 5-angled, spathaceous, caducous. Corolla synpetalous, 5-lobed. Androecium polyandrous, stamens 5, epipetalous, the anthers dithecous, linear-oblongoid, dorsifixed, dehiscence longitudinal. Pistil 1, ovary ellipsoid, 2-carpelled, syncarpous, the style long, thick, persistent, the stigma 2-lobed, clavate fusiform. Fruits are hard drupe, ellipsoid, 5-ribbed, orange to reddish brown when ripe; seeds many, reniform, compressed.

***Gelsemium elegans* Benth.**

An evergreen scandent shrub; bark thick, warty and grey; leaves opposite, ovate- acuminate, 4-8 x 9-12 cm, thin, membranous, flowers golden yellow, in axillary trichotomous cymes; fruits inflated, 2 celled; seeds oblong, numerous

***Gynocardia odorata* R. Br.**

A small to medium sized tree, evergreen; branches flexous; bark ashy- grey, warty; leaves oblong, abruptly acuminate, 3-7 x 14-24 cm, flowers pale yellow, fragment, in large fascicles on trunk or few flowered in leaf axils; fruits globose but broader than long, ash- coloured, hard, lenticellate outside, 6-8 cm across; seeds many, obovoid, embedded in pulp.

***Hedychium coronarium* Koenig.**

A stout rhizomatous herb to 2 m height; leaves are oblong-lanceolate, sessile, appressed silky above and sharp-pointed, 8-24 in long and 2-5 in wide and arranged in 2 neat ranks that run the length of the stem. From midsummer through autumn the stalks are topped with 6-12 in long clusters of wonderfully fragrant white flowers that look like butterflies. Flowers conspicuous, pure white and mildly fragrant; corolla tube 3-5 cm long; lobes 0.5-3.5 cm; staminodes entire or notched; lip sub-orbicular, distintly clawed; stamens short; filaments white; anthers 1.5-2 cm long; capsule sub-globose, 2 cm in diameter.

Hydronocarpus kurzii (King) Warp.

Up to 20 m tall evergreen tree with hanging branches; young leaves and inflorescence tawny puberulous; bark greenish-grey; leaves ovate-oblong or elliptic-oblong, shortly acuminate, 6-16 x 11-22 cm; flowers pale yellow in axillary cymes, dioecious; fruits globour, chocolate-brown, orange sized, rather longer than broad tawny-velvety with stout beak; seeds many, grooved, embedded in pulp

Ixora nigricans R.Br.

A small shrub or small tree, up to 2 m tall, bark greyish,lenticellate; branchlets white; leaves 15 X 6.5 cm, membranous, lanceolate or elliptic- ovate, acute or shortly acuminate, oblong or obovate; base acute or rounded;petiole up to 3 cm long; flowers white, fragrant, on terminal brachiate cymes, 2 cm long, pure white, turn black when dried, fruits globose or didymous, green when young, red when ripe, black when dry. 1 X 0.2 cm with persistent stylar at apex.

Kalanchoe pinnata (Lamk.) Pers.

Succulent perennial plant with hollow stems growing 3–5 feet tall. It is distinctive for the profusion of miniature plantlets that form on the margins of its leaves. Leaves are thick, soft, juicy, basal leaves are simple but upper one become 3-5 foliate pinnate, margin slightly serrated. Flowers pinkish or reddish-purple, pendent, *Kalanchoe pinnata* (syn. *Bryophyllum calycinum*, *Bryophyllum pinnatum*, also known as the Air Plant, Life Plant, Miracle Leaf, Goethe Plant. Fruits enclosed in persistent calyx and corolla.

Lepionurus sylvestris Blume

A large shrub or small trees, 0.5-3 m tall, switchy branches, glabrous,with grey bark; leaves alternate, ovate or obovate,oblong, lanceolate to elliptic or obovate, 6-10 x 3-6.5 cm, base acute, acuminate at apex, entire, membranous, glabrous, pustulate; flowers greenish to pale yellow, 1-2 mm long, faintly scented bisexual in axillary racemes. Drupes ellipsoid, up to 1.5 cm long, orange, 1 seeded, glabrous.

Lindernia ruelloides (Colsm.) Penn.

A small prostate herb, rooting at the nodes; leaves obovate-oblong, serrate, obtuse, 1-2.5 x 2-4 cm; flowers bluish, subtended by green calyx, in axillary-terminal cymes; fruit small, slender, up to 2.5 cm long; seeds yellow.

Lobelia angulata Forst.

A small creeping herb, pubescent, rooting on lower stem; leaves cordate-ovate, repand or slightly notches; 1 x 2 cm, denticulate; flowers greenish-pink; berries ellipsoid with persistent style on tip, pinkish-black.

Lobelia nictianaefolia Heyne.

Tall, erect, much branched, somewhat hairy herb grows from 1.5 to 3 m in height. The leaves, resembling those of tobacco, are narrowly obovate- lanceolate, the lower ones being 30 cm in length and 5 cm in width, while the upper ones gradually become smaller; all are smooth or nearly so above and hairy or smooth beneath. The flowers are large, white, and borne in terminal compound inflorescences. The calyx-tube is smooth or hairy, with linear teeth about 1.2 cm long, and smooth or hairy. The capsules are 2- celled, somewhat rounded and about 1.5 cm in diameter. The seeds are numerous very small, ellipsoid, and compressed

***Morus australis* Poir.**

A moderate sized deciduous tree with straggling branches sometimes grows as shrub. Leaves ovate, caudate-acuminate, serrate and densely lobed, membranous, pubescent when young. Flowers monoeceous,appear with young leaves. Spikes lax flowered, hairy, Fruit ovoid or sub-globoose, greenish yellow when young, red at maturity and ultimately black when ripe.

***Musa glauca* Roxb.**

A non-stoloniferous tree like herb to 3 m tall; trunk below the leaves; pseudostem stout, cylindric, broadest at base and tapering towards the apex; leaves oblong-lanceolate; up to 2.5 m long, acute; spike drooping from the base; bracts many, ovate, green, many flowered; fruits pale yellow with thick skin, obovoid-oblong, up to 12 cm long; seeds numerous, sub-globose, blackish, 0.5-1 cm across.

***Ocimum tenuiflorum* L.**

An erect, much branched sub-shrub 30-60 cm tall with hairy stems and purple or mauve colour with simple, opposite greenor purple leaves that are strongly scented. Leaves peltate, elliptic-oblong, acute, ovate, 1.5-2 x 3-5 cm, usually slightly toothed, pubescent. Flowers purplish-pink in elongate racemes in close whorls, often panicled; fruits forming 4 reddish-brown nutlets, ellipsoid, reddish.There are two main morphotypes: green-leaved (Sri or Lakshmi tulsi) and purple-leaved (Krishna tulsi).

***Pajanelia longifolia* (Willd.) K. Schum.**

A medium- sized to large deciduous tree, bark warty, reddish-brown; leaves imparipinnate, up to 100 cm long; rachis ridged, leaflets 9-11 pairs, 4.5-10 x 10-25 cm, elliptic or ovate-lanceolate, acuminate; base very oblique; flowers dull purple, yellow inside the throat, in large panicles; fruits erect, flat, winged on each margin; seeds compressed.

***Pentapetes phoenicea* L.**

An erect, branched, half-woody annual herb, 0.5-1 m in height. The stems and foliage are mostly glabrous, with a few scattered stellate hairs. Leaves are petiolate, stipulate, alternate, linear-lanceolate, crenate or serrate, and variably hastate or sagittate, 6-10 cm in length, toothed at the margins, usually having the broad, pointed base, and tapering to a pointed tip. Petioles are short (1-2.5 cm long) in comparison to the blades which are 5-10 cm long and 1-2 cm wide, stipules are triangular, base of the blade is variably triangular, round or truncate, and the apex acuminate.Scarlet flowers are borne in the axils of the leaves with 5 large, deep rose, and show petals. Flowers, 2.5-3.5 cm across, fruit is 5-valved, rounded, hairy capsule, about 1 cm in diameter. The seeds occur 8-12 in two series in each cell.

***Phyllanthus airy-shawii* Brunal & Roux.**

A shrub to 1.5 m, stem reddish-brown; branches slender; leaves bifarous, bipinnate, distichous or alternate; leaflets sub-sessile, 6-9 pinnate, ovate or obovate, acute, 1 x 1.3 cm, dark-green above, glaucous beneath, lower ones smaller; nerves 3-5; stipels lanceolate; flowers greenish-white, shortly pedicelled, borne on rachis beneath; fruits small, ribbed.

***Phyllanthus fraternus* Webs.**

An annual erect herb to 60 cm high; branchlets spreading, slender; stem terrete; leaves distichous, pinnae 10-30 pairs, sub-sessile, elliptic oblong, obtuse or shortly mucronate; base rounded, oblique; flowers greenish-white or yellowish, axillary or borne on lower rachis, shortly pedicelled; fruit sub-globose, ribbed on dorsal side of rachis, 3-5 mm across

Rauvolfia serpentina Benth.

Undershrub; bark grey; leaves 3-4, whorled, ablong-lanceolate,acute,2.5-6x6.5-16 cm; base attenuate; flowers white, terminal or axillary corymbose cymes; fruits reddish, obliquely ovoid, purplish-black when ripe; seeds 1, ovoid.

Rhaphidophora hookeri Schott.

A profuse climber on tree trunk; stem stout and fleshy, 2.5-4 cm in diameter; leaves simple, ovate-oblong, caudate-acuminate, sub-falcate, 8.5-13 x 20-30 cm; base oblique; petiole and midrib channelled.

Ricinus communis Linn.

Coarse, erect, branched, smooth, somewhat woody bush, 1 to 4 meters high. The younger parts are glaucous, and the vegetative parts and inflorescences are green or purplish. Leaves smooth, alternate, palmately divided, and 20 to 60 cm in width, the lobes being oblong and toothed. The racemes are stout and erect. The male flowers are about 1 cm in diameter. The capsules are ovoid, 1 to 1.5 cm long, green or purplish, and covered with soft spine like processes.

Rubia cordifolia L.

Climber or creeping herb; roots long, flexous, stem 4 angled, hispid, slender; leaves 4-whorled, 3-nerved, scabrid, 2-3.6 cm x 3-14 cm, cordate or ovate-lanceolate; tips incurved; flowers small, white or greenish, panicled, terminal, trichotomous; petels lanceolate, tips incurved; fruits globose, 2-seeded.

Ruellia suffruticosa Roxb.

A small pubescent herb; stem gnarled, 15-30 cm high; roots slender with fusiform swelling, pellucid; leaves arising from apex of stem, elliptic-oblong or oblanceolate, acute, villous with hairs on both surfaces, more on nerves beneath; nerves-5, arcuate, distant, intramarginal; petiole channelled,6-8 cm long;tertiaries transverse; flowers terminal, solitary, white or pale purple; bracts foliaceous; fruits globose or oblong, reddish when ripe, 0.5 cm across, 1 seeded.

Saraca asoca (Roxb.) de Wilde

Small evergreen beautiful tree; crown umbrella shaped; bark dark or black, rough; leaves paripinnate, 15-25 cm long; leaflets 4-6 pairs, oblong-lanceolate, acute, 3-5x 10-20 cm, drooping when young, dark-green above; base slightly oblique; flowers orange-scarlet, fragrant, in dense axillary corymbs appearing with the leaves; fruits oblong 2.5-3.5 x 10-16 cm, veined, compressed, caute at both the ends; seeds 4-8, oblong, compressed.

Senecio scandens Buch-Ham. Ex D. Don

A slender climber; stem greenish, ribbed, terette, branches zig-zag, grooved or ribbed; young parts lsightly pubescent; leaves hastate while young, ovate-lanceolate when old, acuminate, 1-2.5 x 5-12 cm, crenate or distantly denticulate, coarsely hairy; midrid channelled above, raised beneath; nerves arcuate, anatomosing; base acute; petiole auricled; flowers heads yellow, on lax divaricate terminal corymbs; involucres bracts 10-15, linear-oblong, retrose; achenes ribbed or 4 angled, recurved or retrose; pappus white, 1 cm long.

Smilax glabra Roxb.

Slender climber with knotty roots; stem unarmed; leaves alternate, distant, elliptic-lanceolate, 3.5-4.5 x 5-18 cm, acuminate or subcaudate, glucous beneath, 3- costate; base rounded or cuneate; petiole and sheath up to 1.8 cm long; cirrhi very slender and coiled; flowers in axillary umbels, many flowered, white, minute; berries dark-blue.

Solanum khasianum var. _Chatterjeeanum_ Sen Gupta

Much-branched spinous undershrub to 1 m tall, whole part covered with straight and curved prickles; branches angled; leaves ovate or deltoid-acute, lobed; lobes triangular, hirsute and prickly on both surfaces, 5-12 x 6-18 cm; base sub-cordate; petiole to 4 cm long; flowers white or pale yellow in few flowered axillary racemes; fruits globose, greenish spotted while young, yellowish when ripe, 2 cm across; seeds many, compressed, smooth and brown.

Stemona tuberosa Lour.

Twiner or scandent with fasciculated roots; stem dark-green, smooth, nodes present; leaves ovate-cordate, 2-3 whorled or opposite, 7.5-11 x 15-20 cm size, acuminate, coriaceous; base 9-12, costate, petiole 7-10 cm long; bracts lanceolate; flowers 2-3, usually 2, often opposite, companulate, 6.5 cm long; capsule ovoid-oblong, 4-5 cm long; seeds 5-8.

Stephania japonica var discolor (Miq.) Forman.

A slender climber, branches straiate or ribbed; leaves triangular-ovate, broader than long, peltate, pubescent under surface; apex blunt or sub-acute; nerves 9-11 at the base; flowers yellow, umbellate on very slender axillary pedicels; fruit globose, red when ripe.

Taraktegenos kurzii (King) Warb

10-20 m tall evergreen trees, with narrow crown and hanging branches. Leaves bifarious, variable, 10-20 x 4-6.5 cm, oblong or oblong-lanceolate, elliptic oblong, base cuneate, obtusely short or abruptly acuminate at apex, entire along margins, glabrous; Inlforescence in axillary cymes. Male flowers 8-12 mm across, pale yellow; sepals and petals ovate-orbicular, stamens 20-30. Female flower in 2-5 in a compact fascicles. Berries globose, 5-8 cm across, stout beaked, velvety; seeds 12-18, 2-3 cm long, angular ovoid.

Thunbergia grandiflora Roxb.

Vigorous evergreen vine; leaves opposite along the stem and are choko-like; broad-based narrowing to a pointed tip, usually with deeply scalloped lobes towards the base, hastate-ovate, (8-)11-22 cm long, 3-15 cm wide, margins with widely spaced triangular, acute to rounded teeth. The trumpet shaped flowers have a short, broad tube, white on the outside, yellowish inside, which expands to five rounded pale lavender-blue petals, one larger than the others, flowers are up to 8 cm long and 6-8 cm across. Flowers often in pendent inflorescences, each one subtended by ovate bracts 25- 40 mm long; calyx reduced to a truncate, ring-like structure; corolla blue or white, 5-7.5 cm long. Seed pod is inconspicuous, cone shaped, 3-5 cm long, with a rounded base. The seed is flat, up to 1 cm long and covered with brown scales. It is catapulted several metres when the ripe pod splits.

Tinospora cordifolia (DC.) Miers. Ex Hook. f. & Thoms.

A large, glabrous, dioeceous, deciduous climbing shrub with succulent sparsely lenticellate, stems throwing long, filiform, fleshy, aerial roots; bark is warty, grey brown or creamy white, leaves broadly ovate-cordate, shortly acuminate, 5-10 cm long, flowers unisexual, small, greenish-yellow, in axillary and terminal racemes on leafless branches, Male flowers occur in clusters and female flowers are solitary.Fruits ovoid, succulent red when ripe, pea sized.

Trapa natans var. _Bispinosa_ (Roxb.) Makino

An aquatic herb; stem sub-merged, very slender, stoloniferous with fixing roots on mud and free floating on water; leaves whorled, crowded at apex, 2.5-4 cm long, rhomboidal or triangular, serrate-dentate, acute, glossy green above, densely hairy beneath; base truncate; petiole swollen, radish-purple or green; flowers white, axillary, solitary, opening above the surface of water; fruits obovoid, angular with short conical beak at apex and sharp sinous horn at either side; seed triangular.

***Woodfordia fructicosa* Kurz.**

A straggling shrub to 3-5 m high; bark reddish brown; young parts hairy; branches spreading, drooping; leaves opposite or in whorls of 3, sessile, ovate-lanceolate or oblong-lanceolate, acuminate, puberculous above, white with black dots beneath; base cordate; flowers numerous, brilliant-red in dense axillary clusters from the old wood, often completely covering the branches; fruits ellipsoid, irregularly dehiscent; seeds brown, obovoid, smooth, small

***Zanonia indica* L.**

A slender climber; stem green, straiate, blaze greyish; all parts bitter; leaves alternate, 3-foliolate; leaflets oblong-oblanceolate, acute, bright green, 4-8 x 6- 14 cm, 3- nerved at base; base slightly cordate; petiole channelled; tendrils lateral; flowers greenish-yellow, clustered on slender panicles; fruits cylindric-obconic, like a candle extinguisher, about 8 cm long, slightly tapering towards a rounded base, truncate and triangular at the apex, pale yellowish-brown; seeds laterally compressed, white to pale yellow.

***Zanthoxylum armatum* DC.**

A small, strongly aromatic tree; branches armed with prickles; twigs lenticellate; leaves imparipinnate; rachis with a foliaceous green wing; leaflets opposite, elliptic-lanceolate, caudate-acuminate, serrate and each serrature with punctuate, terminal up to 11 cm long, lateral smaller; flowers greenish-yellow, in dense pubescent lateral panicles; fruits red, globose; seeds solitary, globose, shining black.

***Zingiber purpureum* Rosc.**

A large, aromatic herb, 1-2 m in height, arising from rootstocks; leaves oblong-lanceolate, acuminate, sessile, 30-40 x 5-7.5 cm, pubescent along midrib beneath; spikes on long peduncles, up to 15 cm long, yellow; bracts greenish red turning to bright red at maturity; flowers yellow coloured, oblong-ellipsoid, capsules small, globose.

Traditional knowledge of herbal remedy to treat human diseases is fast declining in many parts of the world, including India. The tribal of Mizoram make wider use of a large variety of medicinal plants. Such activities on one hand contributed to our knowledge of various uses of biodiversity and on the other hand, resulted in rapid research on its collection, conservation and sustainable use. With the erosion of the tribal cultures, the traditional healers have become a threatened category. In addition, the genetic diversity in medicinal plants has diminished due to shifting cultivation and large scale destruction of the natural habitats. Therefore, there is an urgent need to record all ethnobotanical information among the diverse ethnic communities before the traditional culture is completely lost. The ethnomedicinal plants, particularly threatened ones, should be cultivated in herbal gardens, agroforestry systems and home gardens to encourage their sustainable utilization and hence conservation.

Destructive harvesting along with forest fire is a causative factor threatening species existence in the state. In addition, threatened MAPs described here are mostly herbs/shrubs (24 herbs/climbers and 20 shrubs/under shrub) by habits and for medicinal purpose, underground parts (root/rhizome etc) or entire plant (*Panchang*) of 30 species is used leading to destructive harvesting. Moreover, fruits, wood and bark of some trees are also used. The primary health care system in Mizoram is handled mostly by elderly people and they harvest the drug as and when required further hampers natural regeneration. Therefore, awareness regarding the importance of these plants; systematic and scientific harvesting and ensuring sustainable utilization are required with suitable conservation strategies at the earliest in the state.

References

Anonymous, 2003. *Forest Survey of India Reports*. 2003, Dehradun.

Bhattacharjee, S.K., 2004. *Handbook of Medicinal Plants*. Pointer Publishers, Jaipur, India.

Bhattacharya, S.C. and Dutta, S., 1956. *Classisication of Citrus Fruits of Assam*. ICAR Monograph No. 20

Biswas, K., 1956. *Common Medicinal Plants of Darjeeling and Sikkim Himalaya*. Bengal Government Press, West Bengal, Calcutta.

Chetna Bisht and Anoop Badoni, 2009. Distribution and indigenous Uses of Some Medicinal Plants in District Uttarkashi, Uttarakhand, India. *Researcher*, 1(6): 38–40.

Deorani, S.C. and Sharma, G.D., 2007. *Medicinal Plants of Nagaland*. Bishen Singh Mahendra Pal Singh, Dehradun.

Hegde, S.N., 2000. Conservation of North East Flora. *Arunachal Forest News.*, Vol 18 (1&2).

http: / /www.bsienvis.nic.in.

http: / /www.forest.mizoram.gov.in.

Jain, S. K. and Dam, N., 1979. Some ethnobotanical notes from northeastern India. *Economic Botany*, 33(1), 52–56.

Lalfakzuala, R., Lalramnghiinglova, H. and Kayang, H., 2007. Ethnobotanical usages of plants in Western Mizoram. *Indian Journal of Traditional Knowledge*, 6(3): 486–493.

Lalramnghinghlova, J. H. and Jha, L.K., 1996. Prominent Agroforestry Systems and important multipurpose trees in farming system of Mizora. *Indian For.*, 122 (7): 604–609.

Lalramnghinghlova, H. and Jha, L.K., 1997. Forest Resources: An Overview. In: *Natural Resource Management*, Vol. 1. (Eds.) L.K. Jha and S.B. Nangia A.P.H. Publishing Corporation, New Delhi, pp. 203–255.

Lalramnghinghlova, H. and Jha, L.K., 1998. Ethnomedicinal plants among the hill tribes of Mizoram. In: *Prospects of Medicinal Plants*, (Eds.) P.L. Gautam, R. Raina, U. Srivastava, S.P. Raychaudhari and B.B. Singh. Indian Society of Plant Genetic Resources, New Delhi, pp. 67–86.

Lalramnghinghlova, H. and Jha, L.K., 1999. New records of ethnomedicinal plants from Mizoram. *Ethnobotany*, 11: 57–64.

Lalramnghinghlova, H., 1999a. Prospects of Ethnomedicinal plants of Mizoram in the new millennium. Proc. Symp. *On Science & Technology for Mizoram in 21st Century*, 17-18 June, Aizawl, p. 119–126.

Lalramnghinghlova, H., 1999b. Ethnobotanical and agro-ecological studies on genetic resources of food plants in Mizoram state. *J. Econ, Taxon. Bot.*, 23(2): 637–644.

Lalramnghinghlova, H., (2000. Ethnomedicinal plants development in Mizoram. Proc. *Int. Workshop on Agroforestry and Forest Products*, Aizawl, November, 28-30 pp. 395–404.

Lalramnghinghlova, H., 2001. Ethnobotanical interpretations and future prospects of ethnobotany in the North-East India. *Science Vision*, 1: 24–31.

Lalramnghinghlova, H., 2002. Bioresources of Mizoram: An overview. *Journal of the North Eastern Council*, p. 56-64.

Lalramnghinghlova, H., 2003. *Ethnomedicinal plants of Mizoram*. Bishen Singh Mahendra Pal Singh, Dehradun.

Mahanti, N., 1994. *Tribal Ethnobotany of Mizoram*. Tribal studies of India, Series T 171. pp. 1-177. Inter India Pub., New Delhi

Nayar, M.P. and Sastry, A.R.K. (ed.), 2002. *Red Data Book of Indian Plants*, Vols. I-III. Botanical Survey of India, Calcutta.

Rai, L.K. & Sharma, E., 1994. *Medicinal Plants of Sikkim Himalayas: Status Uses and Potential*. Bishen Singh Mahendra Pal Singh, Dehradun.

Rai, P.K. and Lalramnghinglova, H., 2010. Ethnomedicinal Plant Resources of Mizoram, India: Implication of Traditional Knowledge in Health Care System. *Ethnobotanical Leaflets*, 14: 274-305.

Sharma, U.K., 2004. *The Medicinal Plants of Assam*. Bishen Singh Mahendra Pal Singh, Dehradun.

Singh. K.P., 1997. Mizoram. In: Mudgal, V. And Hazra, P.K. (eds.) *Floristic Diversity and Conservation Strategies in India*.III. pp. 1217–1256, B.S.I., Dehra Dun

Singh, N.P., Singh, K.P. and Singh, D.K. (Ed.), 2002. *Flora of Mizoram*, Vol. I. Botanical Survey of India, MOEF, GOI.

2011, Future Crops Vol. 1
Editor: Professor K.V. Peter
Published by: DAYA PUBLISHING HOUSE, NEW DELHI
Pages 243–259

Chapter 14
Underutilized Vegetables of Kashmir

N. Jabeen, S. Mufti and S.H. Khan
Division of Olericulture,
Sher-e-Kashmir University of Agricultural Sciences and Technology of Kashmir,
Shalimar, Srinagar – 191 121, J & K (India)
E-mail: nayeem_jabeen@yahoo.co.in

Jammu and Kashmir is situated 32°-17′ to 37°-05′ N latitude and 72°-40′ to 80°- 30′ E longitude. It consists of three regions: Jammu, Kashmir valley and Ladakh. Jammu and Kashmir is home to several valleys–the Kashmir Valley, Tawi Valley, Chenab Valley, Poonch Valley, Sindh Valley and Lidder Valley-. The total geographical area of the state is about 0.12 million sq km of which 10 per cent area (941.77 thousand ha) is under agricultural operations. More than 81 per cent holdings fall below one ha, 12 per cent under two ha, 5 per cent under four ha and only one per cent of holdings four ha and above.

The state is endowed with agriculture and horticulture potentials and has four distinct agro-climatic zones *viz.* Sub-tropical zone, Intermediate zone, Temperate zone and Arid zone. Due to diversity of agro-climatic conditions, state grows a number of vegetables and other horticultural crops. In Jammu and Kashmir, the area under vegetable crops is 60.55 thousand ha with a production of 1254.27 thousand tons and the productivity is 16.90 t/ha. There is huge gap between availability and requirement of vegetables. To meet this ever increasing gap, underutilized vegetables can be of great significance and would be of value in future.

Underutilized and Future Vegetables

Underutilized plants have under-exploited potential for contributing to food and nutrition security by combating 'hidden hunger' caused by micronutrient deficiencies. They often have medicinal properties and other multiple uses. They provide options for improved income to the poor and for environmental services to the global community.

These plants collectively receive little attention from research, extension services, farmers, policy and decision makers, donors, technology providers and consumers, yet they form intricate components for the wellbeing of millions of people. Many have untapped potential for processing into refreshments and snacks and for marketing in domestic and regional markets, thus providing poor people with new income opportunities.

Role of Underutilized Vegetables as Future Crops

Green leaf vegetables occupy an important place among the food crops as these provide adequate amounts of vitamins and minerals. They are rich sources of carotene, ascorbic acid, riboflavin, folic acid and minerals like calcium, iron and phosphorus. In nature, there are many underutilized greens of considerable nutritive value, which can nourish the ever increasing human population. Many of them are resilient, adaptive and tolerant to adverse climatic conditions. Although they can be raised comparatively at lower management costs even on poor marginal lands, they have remained underutilized due to lack of awareness and popularization of technologies for utilization. Now-a-days underutilized food is gaining importance as a means to increase the per capita availability of foods.

Jammu and Kashmir being hilly terrains with severe cold winter and from time immemorial, the people living in this region mostly depend on underutilized leaf vegetables for their source of food when other vegetables are in scarce to supplement their nutritional requirements. Such leaf vegetables are growing naturally in this region. Besides using leaves of such vegetables, petioles, seeds, roots of some species are also utilized. Most of these crops are easy to grow, resistant to pests and diseases, hardy and acceptable to the local taste. Such indigenous underutilized plants will be of considerable value as the cheapest source of nutrition and could be used for crop diversification and future use. Since they can be raised comparatively at lower management costs even on poor marginal lands, they can be used for increasing income of farmers and can be made available year round when other vegetables are in scarcity.

Underutilized vegetables are inexpensive native crops which could be cultivated by indigenous people in rural areas for reasons stated below:

1. Easily produced and harvested because they occur naturally.
2. Produce more yield for a longer period.
3. Highly tolerant to various diseases and stresses (biotic and abiotic).
4. Good sources of vitamins, minerals, proteins and dietary fiber, so can be used to combat 'hidden hunger'.
5. Can prove as a source of additional income to poor farmers.

Some of underutilized indigenous leaf vegetables of Jammu and Kashmir are discussed below:

Taraxacum (Dandelion)

Botanical name: *Taraxacum officinale*

Family: Asteraceae

Vernacular name: Handh

General Description

Taraxacum officinale has 24 or 40 pairs of chromosomes but some plants have 16 or 32 chromosomes.It grows from generally unbranched taproots and produces one to more than ten stems which are typically 5 to 40 cm long. The stems are tinted purplish, upright or lax, which produce flower heads held as tall or taller than the foliage. The foliage is upright growing or horizontally orientated, with leaves having narrowly winged petioles or they are unwinged. The stems can be glabrous or are sparsely covered with short hairs. Plants have milky sap and the leaves are all basal. The 5–45 cm long and 1–10 cm wide leaves are oblanceolate, oblong, or obovate in shape with the bases gradually narrowing to the petiole. The leaf margins are typically shallowly lobed to deeply lobed and often lacerate or toothed with sharp or dull teeth. The calyculi (the cup like bracts which

Flower Seed Parachute

Figure 14.1: *Taraxacum* (Dandelion).

hold the florets) are composed of 12 to 18 segments, each segment is reflexed and sometimes glaucous. The lanceolate shaped bractlets are in 2 series with the apices acuminate in shape. The 14 to 25 mm wide involucres are green to dark green or brownish green with the tips dark gray or purplish. Each flowering stem lacks bracts and has one single flower head. The yellow flower head lack receptacle bracts and all the flowers, which are called florets, are ligulate and bisexual. The florets number 40 to over 100/head, having corollas yellow or orange-yellow. The fruits, which are called cypselae, range from olive-green or olive-brown to straw-colored to grayish, they are oblanceoloid in shape and 2 to 3 mm long with slender beaks. The fruits have 4 to 12 ribs with sharp edges. The silky pappi, which form the parachutes, are white to silver-white in color and around 6 mm wide.

Habitat

It is a common weed of orchards, roadsides, vegetable fields or uphill's of valley. Germination of seeds is in large numbers and seeds germinate readily on suitable conditions. Most of seeds germinate in early March and October, optimum germination temperature is 15-20ºC. They flower in May-June or sometimes in November and seeds are disseminated by wind.

Uses

Dandelions are grown commercially on a small scale as a leaf vegetable. The leaves (called dandelion greens) can be eaten cooked. They are probably the closest in characters to mustard greens. Usually younger leaves are cooked. Raw leaves have a slightly bitter taste. The leaves are high in vitamin A, vitamin C and iron, carrying more iron and calcium than spinach (Dolland and Trower, 2000). A leaf decoction may be drunk to "purify the blood", for the treatment of anaemia, jaundice, and also for nervousness. Dandelion leaves have a diuretic effect as they increase salt and water excretion from the kidney (Anonymous, 2003). Dandelion contains luteolin, an antioxidant, and has demonstrated antioxidant properties without cytotoxicity (Chun and David,2004). In the valley, dandelion is usually fed to lactating women, being rich in iron and cooked as raw vegetable or dehydrated.

Mallow

Botanical name: *Malva sylvestris*

Family: Malvaceae

Vernacular name: Sonchal

General Description

M. sylvestris is a vigorously healthy plant with showy flowers of bright mauve–purple, with dark veins; a handsome plant, often standing 1 to 1.25m high and growing freely in fields, hedgerows and in fallow fields. It is a spreading herb, usually annual and perennial with a growth habit which can be straight or decumbent, branched and covered with fine soft hairs or none at all. The roots are mostly adventitious and remain in the superficial horizon of the soil. The adventitious roots arise from a thick stoloniferous sucker like root stock. It has thick, round and strong stem. The leaves borne upon the stem, are roundish and have three or five to seven or five to nine shallow lobes, each 2 to 4 cm long, 2 to 5 cm wide and 5 to 10 cm in diameter, downy, with hairs radiating from a common center and prominent veins on underside petiole either 2 to 6 cm or 2 to 13 cm long. Flowers are reddish purple, bright pinkish purple with dark stripes and bright mauve-purple, which appear in axillary clusters of 2 to 4 and form irregularly and elongated along the main stem with flowers at the base opening first. Nutlets strongly reticulate, 10-12 mericarps, usually without hair, with sharp angle between dorsal

Figure 14.2: Mallow.

and lateral surfaces, 5-6 mm in diameter. Seeds are brown to brownish green when ripe, about 2.5 mm long and wide,5 to 7 mm in diameter and are shaped like a cheese wheel. Cultivation is by sowing seeds directly outdoors in early spring. The seed is easy to collect, and they often disseminate by irrigation water.

Habitat

An occasional weed of vegetable fields and are indicators of moist nitrogenous alluvial soils.

Uses

Mild tasting young mallow leaves can be a substitute for lettuce, whereas older leaves are better cooked as a leafy green vegetable. The buds and flowers can be used in salads. The fruit is employed medicinally, as demulcents and emollients and the seeds are used internally in a decoction as a demulcent and diuretic and the leaves can be made into poultices as an emollient for external applications(Balfour,1863) The young leaves when boiled and fried are wholesome vegetables and eaten in several parts of Valley.

Portulaca

Botanical name: *Portulaca oleracea* L.

Chromosome Number 2n=54

Family: Portulacaceae

Vernacular name: Nunar

General Description

Common Purslane (*Portulaca oleracea* L.) is widely considered an edible plant, and in some areas an invasive type of weed. It has smooth, reddish, mostly prostrate stems and alternate leaves (2 cm diameter) clustered at stem joints and ends which are oval or oblong, sub succulent, lower ones opposite, middle alternate and upper whorl glabrous. The yellow flowers have five regular parts and

Figure 14.3: Portulaca.

are up to 6 mm wide. The flowers appear depending upon rainfall and occur year round. The flowers open singly at the center of leaf cluster for only a few hours on sunny mornings. Seeds are many, minute and are formed in a tiny pod, which open when the seeds are ready. Purslane has a taproot with fibrous secondary roots and tolerates poor, compacted soils and drought.

Habitat

This weed commonly grows in orchards, vegetable fields and fallow lands. It prefers moist and steady situation and is indicator of aerated clay soils. The seeds sprout in spring, flower in June and fruit in August.

Uses

Purslane contains more Omega-3 fatty acids (alpha-linolenic acid in particular) (Artemis, 2004) than any other leafy vegetable. Purslane has 0.01 mg/g of EPA. This is an extraordinary amount of EPA for land based vegetable sources. It also contains vitamins (mainly vitamin A, vitamin C, and some vitamin B and carotenoids), and dietary minerals, like magnesium, calcium, potassium and iron. Also present are two types of betalain alkaloid pigments, the reddish betacyanins (visible in the coloration of stems) and the yellow betaxanthins (noticeable in the flowers and in the slight yellowish cast of the leaves). Both of these pigment types are potent antioxidants having antimutagenic properties in laboratory studies(Artemis,2004).One hundred grams of fresh purslane leaves (about 1 cup) contain 300 to 400 mg of α-linolenic acid. One cup of cooked leaves contain 90 mg of calcium, 561 mg of potassium, and more than 2,000 IU of vitamin A. Purslane is used as a remedy for constipation and inflammation of the urinary system. Although purslane is a weed in the State, it is mostly eaten as a leaf vegetable. In Valley, it is cooked along with amaranths and an egg to form a delicious curry. The stems, leaves and flower buds are all good to eat. Purslane can be used fresh as a salad, stir-fried, or cooked like spinach, and because of its mucilaginous quality, it is also suitable for soups and stews.

Spiny Amaranth

Botanical name: *Amaranthus spinosus*

Family: Amaranthaceae

Vernacular name: Lissa

General Description

Amaranthus, collectively known as amaranth or pigweed, is a cosmopolitan genus of herbs. Although several species are often weeds, people consider amaranths as leaf vegetables, cereals, and ornamentals. In general, most of the species are monoecious. It is an annual herb, erect or trailing and rarely armed, scare to profuse branched, shallow to deep taproot system, stem green to purple or with mixed shades of these two colours, glabrous and succulent, leaf simple, alternate or opposite, with acute tip, green to purple or with different shades of these two colours, purple colouration prominent in young leaves and fades away at maturity. Inflorescence terminal and axillary, branched spikes differently coloured from green to deep purple and orange, and with different shades of these colours. Flowers are small, regular, mostly unisexual, and monoecious. Stamens 2-5, placed opposite to perianth parts, often reduced to staminodes, filaments either free or united below, anthers one or two celled, carpels 2-3, syncarpous, ovary superior, style single to 2 or 3 branched, fruit a utricle, indehiscent, enclosed by the persistent perianth parts, seed varying in colour, black, brown or white, compressed, smooth, shiny with floury endosperm.

Figure 14.4: Spiny Amaranth.

Habitat

A common weed of orchards, vegetable fields and fallow lands are indicators of dry claying loams.

Uses

Amaranthus species have a 30 per cent higher protein value than other cereals like rice, wheat flour, oats and rye. Amaranth seeds, like buckwheat and quinoa, contain protein that is unusually complete for plant sources. Several studies show that like oats, amaranth seed or oil may be of benefit for those with hypertension and cardiovascular disease. Regular consumption reduces blood pressure and cholesterol levels, while improving antioxidant status and some immune parameters (Czerwinski *et al.*, 2004). While the active ingredient in oats appears to be water-soluble fiber, amaranth appears to lower cholesterol via its content of plant stanols and squalene. Amaranth species are cultivated and consumed as a leaf vegetable in many parts of the state. Besides leaves, seeds of amaranth are taken with jaggery in the form of small pellets (Ganhar) commonly known in valley, which are good source of protein, dietary fiber and dietary minerals like iron, magnesium, copper and especially manganese.

Rhubarb

Botanical name: *Rheum rhabarbarum*

Family: Polygonaceae

Vernacular name: Pambh hawk

Figure 14.5: Rhubarb.

General Description

Rheum is a genus of perennial plants which grow from thick short rhizomes. The genus is in the family Polygonaceae and includes the vegetable rhubarb (*Rheum rhabarbarum* or *Rheum hybridum*.) The plants have large leaves, somewhat triangular shaped with long fleshy petioles. The flowers are small, greenish-white to rose-red, and grouped in large compound leafy inflorescences. *Rheum* species are herbaceous perennials with hermaphrodite flowers, consisting of a coloured perianth, composed of six to nine segments, arranged in two rows. The flowers have nine stamens inserted on the torus at the base of perianth. They are free or subconnatent at their base. The ovary is simple and triangular shaped with three styles. The fruits are a three-sided achene with winged sides, the seeds are albuminous and have straight embryos.

Habitat

Grows mostly under shady areas in moist soils.

Uses

A number of varieties of rhubarb are domesticated both as medicinal plants and for human consumption. The roots and stems are rich in anthraquinones like emodin and rhein. These substances are cathartic and laxative, which explain the sporadic use of rhubarb as a slimming agent. In Kashmir, Rhubarb is usually cooked with fish. Rhubarb is used as a strong laxative and for its astringent effect on the mucous membranes of mouth and nasal cavity. The roots are used as an aggressive laxative. It is mostly used in making pie in star hotels visited by foreign tourists.

Rumex

Botanical name: *Rumex obtusifolius*

Family: Polygonaceae

Vernacular name: Abuj

General Description

The genus *Rumex* L., is a genus of about 200 species of annual, biennial and perennial herbs in the buckwheat family Polygonaceae. They are erect plants, usually with long tap roots. The fleshy to

Figure 14.6: *Rumex.*

leathery leaves form a basal rosette at the root. The basal leaves may be different from those near the inflorescence. They may or may not have stipules. There are minor leaf veins. The leaf blade margins are entire or crenate. The usually inconspicuous flowers are carried above the leaves in clusters. The fertile flowers are mostly hermaphrodites, or they may be functionally male or female. The flowers and seeds grow on long clusters at the top of a stalk emerging from the basal rosette. In many species,the flowers are green, but in some (such as sheep's sorrel, *Rumex acetosella*) the flowers and stems may be brick-red. Each seed is a 3-sided achene, often with a round tubercle on one or all three sides.

Habitat

An occasional weed of fields and vegetable areas, which grows under shady places

Uses

Broad-leaved dock (*Rumex obtusifolius*) is called butter dock because its large leaves are used to wrap and conserve butter. The leaves of most species contain oxalic acid and tannin, and many have astringent and slightly purgative qualities. *Rumex acetosella*, common sorrel is grown as garden herb for their acidic taste (Lukasz, 2008). Dock leaves are a traditional remedy for the sting of nettles. Stings are cured by vigorously rubbing a dock leaf onto the sting.

Other Underutilized Vegetables of Future Value

Top Onion

Botanical name: *Allium cepa var. proliferum*

Family: Alliaceae

Vernacular name: Pran, Topset onions

Top onions are a strong-growing onion with a bunch of bulblets where normal onion would have flowers. They are members of the "proliferum" cultivar of the onion (*Allium cepa*), and are commonly called Top Onions or Topset Onions. In some varieties, these bulblets sprout and grow while still on the original stalk. This phenomenon of forming bulblets instead of flowers is also seen in garlic and other various wild species of *Allium*.

General Description

Tree onions produce up to 6-10 tender, juicy, mild onions at the base of plant; and a stiff stem with 6-8 bulbils, or little onions, at the top. It is the most cold-hardy of all the onions and survives frozen ground.

Uses

The top bulbils are used for pickling. The base bulb is a very important ingredient and is a traditionally used spice in the valley in wazwan for improving aroma and taste. It replaces onion in nost of the local dishes. Tree onions keep for up to 12 months. In valley, it is also an important ingredient in the spice cake (wari) made up of chilli powder, garlic, top onion and other minor spices like cumin.

Figure 14.7: Top onion.

Lotus Stem

Botanical name: *Nelumbo nucifera*

Family: Nelumbonaceae

Vernacular name: Nadru

General Description

The roots of *Nelumbo nucifera* are planted in the soil of the pond or river bottom, while the leaves float on top of water surface. The flowers are usually found on thick stems rising several centimeters

Figure 14.8: Dal lake.

Figure 14.9: Sliced lotus stems.

Figure 14.10: Seed cup "Phambach".

above the water. The plant normally grows up to a height of 150 cm and a horizontal spread of up to 3 m. The leaves may be as large as 60 cm in diameter, while the showy flowers can be up to 20 cm in diameter. In Kashmir, various species and varieties are found in famous lakes like Anchar, Dal and Mansbal. Lotus stem of Dal Lake is famous because of fresh water springs which give it a typical taste.

Habitat

Lotus grows in ponds, lakes and marshy lands. This plant is an aquatic perennial. Under favourable circumstances, lotus seeds remain viable for hundreds of years. It is commonly cultivated in water gardens, an entirely different plant having distinctive circular seed pod.

Uses

Fruit of *Nelumbo nucifera*; the dried seed cup is commonly used in flower arrangements. Boiled, sliced lotus roots are used in various Asian cuisines. The seeds, stems and "roots" (rhizomes) are all edible. The distinctive dried seed heads or nuts locally known as "Phambach" are quite versatile and are sold in valley and can be eaten raw or dried. Stem has a crunchy texture. In Kashmiri cuisine, it is popular with fish and curds. Lotus roots are rich in dietary fiber, vitamin C, potassium, thiamin, riboflavin, vitamin B_6, phosphorus, copper, and manganese, while very low in saturated fat. Most importantly it contains iodine and was used for curing goiter/hypothyroidism, while the seed heads are used for curing intestinal disorders.

Centaurea

Botanical name: *Centaurea iberica*

Chromosome Number 2n=16

Family: Asteraceae

Vernacular name: Kretch

Figure 14.11: Centaurea.

General Description

It is an annual or biannual herb stem erect, upto 45 cm, high green branched, angular, finally pubescent. Radical leaves forming a rosette, purified to pinnate with spreading lobes. Knapweeds are robust weedy plants. Their leaves, spiny in some species, are usually deeply divided into elongated lobes.The "flowers" (actually pseudanthium inflorescences) are diverse in colour, ranging from intense blues, reds and yellows to any mixture of these and lighter shades towards white. Often, the disk flowers are much darker or lighter than the ray flowers, which also differ in morphology and are sterile. Each pseudanthium sits atop a cup- or basket-like cluster of scaly bracts, hence the name "basket flowers". Many species, in particular those inhabiting more arid regions, have a long and strong taproot.

Habitat

A common weed of fallow lands, orchards, roadsides and grasslands. The seeds sprout in early spring and flowering and fruiting last from July-September.

Uses

It is a rich source of vitamin A usually consumed by goldsmith and local artists of Kashmir for improving vision.

References

Anonymous, 2003. University of Maryland. Complimentary medicine.

Artemis, P. Simopoulos, 2004. Evaluation of the Antimutagenic Activity of Different Vegetable Extracts Using an *in vitro* Screening Test.

Artemis, P. Simopoulos, 2004. Omega-3 Fatty Acids and antioxidants in Edible Wild Plants. *Biol Res.*, 37: 263–277.

Balfour, John Hutton 1863. "Products and Secretions of Plants". A manual of botany: being an introduction to the study of the structure, physiology, and classification of plants. *Edinburgh:* A & C Black.

Balfour, John Hutton 1863. *Malvaceae: A Manual of Botany: Being an Introduction to the Study of the Structure, Physiology, and Classification of Plants.* Edinburgh: A & C Black.

Chun, Hu and Kitts, David D., 2004. Food, Nutrition and Health, Faculty of Agricultural Sciences, University of British Columbia, Vancouver, BC, Canada. October 2004. Luteolin and luteolin-7-O-glucoside from dandelion flower suppress iNOS and COX-2 in RAW264.7 cells. *Springer Netherlands.* 245:1-2(107-113).

Czerwinski, J., Bartnikowska, E. and Leontowicz, H., *et al.*, 2004. "Oat (Avena sativa L.) and amaranth (*Amaranthus hypochondriacus*) meals positively affect plasma lipid profile in rats fed cholesterol containing diets". *J. Nutr. Biochem.*, 15(10): 622–629.

David Beaulieu. "Edible Landscaping With Purslane". *About.com.*

Dey, William Mair, Kanny Lall and William Mair 1896. *Indigenous Drugs of India.* The indigenous drugs of India: short descriptive notices of the principal medicinal products met with in British India. Thacker, Spink & Co., pp. 387.

Doll, J. and Trower, T., 2000 *Dandelion.*

Gonor, K.V., Pogozheva, A.V., Derbeneva, S.A., Mal'tsev, GIu, Trushina, E.N. and Mustafina, O.K., 2006. "[The influence of a diet with including amaranth oil on antioxidant and immune status in patients with ischemic heart disease and hyperlipoproteidemia]" (in Russian). *Vopr Pita,* 75(6): 30–33.

Hiley, John S., 1841. On the medical botany of the province of Halifax in Thomas Wakley. The Lancet, In Two Volumes (Volume The Second ed.), pp. 430.

Kaul, M.L. *Weed flora of Kashmir Valley*.

Lukasz, Luczaj, 2008. Archival data on wild food plants used in Poland in 1948. *J. Ethnobiol. Ethnomed.*, 4(4): 4.

Martirosyan, D.M., Miroshnichenko, L.A., Kulakova, S.N., Pogojeva, A.V. and Zoloedov, V.I., 2007. Amaranth oil application for coronary heart disease and hypertension. *Lipids Health Dis.*, 6: 1.

Uses of common sorrel. (*Rumex acetosa*). *About.com*.

2011, Future Crops Vol. 1 ***Pages 260–276***
***Editor:* Professor K.V. Peter**
***Published by:* DAYA PUBLISHING HOUSE, NEW DELHI**

Chapter 15

VEGETABLE PIGEONPEA [*Cajanus cajan* (L) Millspaugh]

C.L.L. Gowda, R.K. Srivastava, K.B. Saxena and R.V. Kumar
International Crops Research Institute for Semi Arid Tropics (ICRISAT), Patancheru – 502 324, A.P.
E-mail: c.gowda@cgiar.org

Majority of food proteins of plant origin derived from legumes in developing countries are generally grown under low input conditions in marginal environments. Currently, the protein availability among the rural poor in the developing world is less than one-third of its normal requirement; and with a growing population and stagnation of production, the availability of protein is projected to reduce further. Since the food production programmes in most countries are always in favour of staple cereals, the issue of protein availability assumes a greater significance for human nutrition. Among the legumes, pigeonpea or red gram [*Cajanus cajan* (L.) Millspaugh] occupies an important place in rainfed agriculture. Globally, it is cultivated on 4.92 million ha, with 3.56 million ha in India alone. Although, pigeonpea is reported to be grown in 22 countries, the remaining major producers are Myanmar, China, and Nepal in Asia. Kenya, Malawi, Uganda, Mozambique, and Tanzania also produce considerable amounts of pigeonpea. A few countries in Caribbean islands and South America also cultivate pigeonpea.

In India, de-hulled split cotyledons of dry pigeonpea seed are cooked to make *dal* or sambar for eating with chapati or rice; while in southern and eastern Africa, and South America, whole dry seeds are used for porridge, and its green seeds are used as fresh, frozen, or canned vegetable. Broken seeds, husks, and pod shells are fed to cattle; while the dry stems are used as household fuel wood. Pigeonpea is credited to be the most suitable crop for subsistence agriculture that needs minimum external inputs. Pigeonpea is known to produce reasonable grain yield even under unfavorable production conditions, mainly due to its abilities such as drought tolerance, nitrogen fixation, and deep root system. Its seeds contain 20–22 per cent protein and contain important amino acids such as glutamic acid (19.2 per cent), aspartic acid (9.8 per cent), phenylalanine (8.7 per cent), and leucine (7.6 per cent) (Duke, 1981).

Table 15.1: Global area, production and yield of pigeonpea in different countries in 2007.

Country	*Area ('000 ha)*	*Production ('000 tonnes)*	*Productivity (kg ha^{-1})*
India	3560	2310	649
Myanmar	570	540	947
Kenya	161.5	95.6	987
Malawi	154.5	159.3	619
Uganda	87.0	89.0	1023
Tanzania	67.5	48.5	719
Nepal	21.0	19.2	917
Dominican Republic*	20.4	19.4	952
Congo	9.9	5.8	582
Haiti*	6.2	2.5	403
Panama*	4.8	2.0	406
Venezuela*	2.5	2.0	797
Burundi	2.0	1.8	900
Bangladesh	1.5	1.4	961
Philippines	0.8	1.4	1636
Jamaica*	0.7	0.9	1197
Grenada	0.6	0.5	964
Trinidad and Tobago*	0.5	1.1	2444
Comoros	0.4	0.3	727
Puerto Rico*	0.3	0.2	807
Bahamas	0.2	0.1	675
Total/mean	4672.3	3300.8	706

Source: FAO 2008, and other resources.

*Vegetable pigeonpea.

Origin and Taxonomy

The presence of pigeonpea seeds in Egyptian tombs suggested that it was cultivated there during the 2000 BC. A perusal of early literature indicates that pigeonpea originated either in Africa (Zanzibar or Guinea) or India. The presence of high diversity among landraces made Vavilov (1939) to conclude that India is a primary center of origin of cultivated pigeonpea. De (1974) and van der Maesen (1980) also supported its Indian origin and is believed that pigeonpea was taken to Africa about 4000 years ago from India.

Linnaeus (1753) gave pigeonpea its first binomial nomenclature–*Cytisus cajan*. van der Maesen (1986) studied the taxonomy of genus *Cajanus* in detail and reported it to be closely related to genus *Atylosia*. Based on further taxonomic studies, he merged genus *Atylosia Cajanus* and hence, the genus *Cajanus* now contains a total of 32 species and among these the only cultivated species is *Cajanus cajan*.

According to De (1974) *Tuvari* and *Tuvar* are the oldest vernacular names of pigeonpea as these are mentioned in "Gathasaptasati" that was published during 300–400 AD. There are several local names of pigeonpea in different parts of the world (Saxena, 2008) but today's most popular name 'pigeonpea' was coined by Plukenet (1692) in Barbados, where the crop was grown in barren lands for feeding its grains to pigeons.

Important Attributes of Vegetable Pigeonpea

Growth and Development

The traditional pigeonpea cultivars and most landraces are of medium (160–180 days) to long (>250 days) maturity durations. However, some early maturing types have also been bred and the earliest maturing pigeonpea variety MN–5 flowers in 54–58 days (Saxena, 2008) and matures in 80–85 days at Patancheru, India (17º N). Between these extremes, there exists an almost a continuous variation for maturity which plays an important role in different cropping systems. Pigeonpea is a photo-period sensitive plant and its photo-period reaction is positively linked to maturity duration and biomass production (Wallis *et al.*, 1981), and so far no germplasm has been identified that is truly photo-insensitive. Besides photo-period, the flowering initiation and development is also strongly influenced by temperature and its interaction with photoperiod. Turnbull *et al.* (1981) showed that the time from floral initiation to flower opening (floral primordial development rate) varied from 40 days under an 8 hour photo-period at 24/16ºC to 22 days under extended photoperiod at 32/34ºC.

Growth Habit

There are two major plant growth habit groups in pigeonpea, and these are designated as determinate and non-determinate. The determinate (Figure 15.1a) type plants have pods in clusters at the top of their canopy and a somewhat corymbs–shaped inflorescence terminates the plant growth. Hence, the plant growth almost ceases after flowering and maturity of such plant is relatively uniform.

Figure 15.1(a): Determinate type of plant.

Figure 15.1(b): Non-determinate type of plant.

The determinate types are more susceptible to pod borer attack and both *Helicoverpa armigera* and *Maruca vitrata* can cause serious yield losses. On the contrary, in the non-determinate (Figure 15.1b) types, the floral racemes form a long terminal panicle and pods are borne in axillary clusters and the terminal buds are vegetative and continue to grow under conducive environment. The determinate types are shorter in height than the non-determinate types. In comparison to determinate types, the non-determinate genotypes tolerate major biotic and abiotic stresses much better due to their inherent capacity to regenerate under conducive environments. These types show good adaptation to a wide range of moisture regimes, soil types, and hence in subsistence agriculture.

Pod Colour

There is a large variation for fresh pod colour in pigeonpea. Saxena *et al.* (1983) studied the effect of pod colour on important organoleptic properties of vegetable pigeonpea, and found that fresh seeds from purple pods had poor texture, flavour and taste as compared to green seeded types; but such differences disappeared after cooking, suggesting that the pod colour may not play any important role in determining the organoleptic qualities of vegetable pigeonpea. In a survey conducted in Gujarat state of India, where green pigeonpea is consumed as a vegetable on a large scale, it was revealed that the rural consumers preferred pods which had green base with purple streaks on its surface. In contrast, the urban consumers preferred green coloured fresh pods. In a consumer survey conducted by Yadavendra and Patel (1983) in Gujarat State found that cultivar ICP 7979 was the best because of its green coloured pods, easy shelling, and good taste of seeds.

Pod and Seed Size

In pigeonpea, seed size and pod size are invariably correlated. The large podded types have large immature and dry seeds. In some vegetable type lines the immature seed is large but their size reduces gradually with maturity. On average, the number of seeds in a pod varies from 2 to 9 Saxena (2008) observed that all the ovules in the long podded genotypes did not develop to their full size due to ovule abortion (Figure 15.2). The exact reason for the abortion of ovules is not fully understood, but there appears to be blockage in the supply of carbohydrates and other vital nutrients to the growing ovules, causing their pre-mature cessation.

Quality Parameters

Quality Traits

In general, green pigeonpea seeds are considered superior to *dal* in nutrition. The observations recorded at ICRISAT, and some other laboratories show that pigeonpea *dal* is better than vegetable with respect to starch and protein. On the contrary, the green pigeonpea grains have higher crude fibre, fat, and protein digestibility. As far as mineral elements are concerned, the green peas are superior in phosphorus by 28.2 per cent, potassium by 17.1 per cent, zinc by 48.3 per cent, copper by 20.9 per cent, and iron by 14.7 per cent. The *dal* however, has 19.2 per cent more calcium and 10.8 per cent more manganese. Singh (1988) reported that the vegetable type pigeonpea had higher amount of polysaccharides and low crude fibre content than *dal*, irrespective of their seed sizes. They also reported that crude fibre content in vegetable pigeonpea was similar to that of garden pea (*Pisum sativum*). Trypsin inhibitor activity, however, was higher in pigeonpea than that of garden pea.

Like other legumes, pigeonpea seeds also contain some anti-nutritional factors. Pigeonpea seeds contain polyphenolic compounds which inhibit the normal activity of digestive enzymes like trypsin, chymotrypsin, and amylase. According to Kamath and Belavady (1980) dry pigeonpea seeds have unavailable carbohydrates which adversely affect bio-availability of certain vital nutrients. Some of

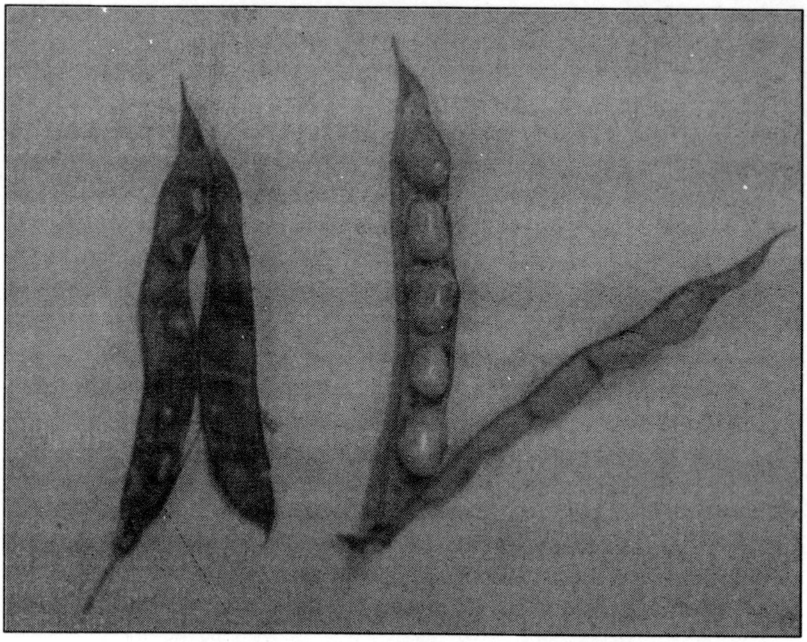

Figure 15.2: Ovule abortion (left) in a developing pod.

the anti-nutritional factors such as phytolectins are heat sensitive and are destroyed during cooking. Some of the flatulence causing oligosaccharides such as staychyose, raffinose, and verbascose are also present in dry pigeonpea seeds.

Table 15.2: Comparison of green pigeonpea seeds and *dal* for important quality parameters.

Constituent	*Green Seeds*	*Dal*
Starch content (per cent)	48.4	57.6
Protein (per cent)	21.0	24.6
Protein digestibility (per cent)	66.8	60.5
Soluble sugars (per cent)	5.1	5.2
Crude fibre (per cent)	8.2	1.2
Fat (per cent)	2.3	1.6

Morphological and Chemical Changes in Developing Seeds

Under normal growing conditions, pigeonpea plants produce profuse flowers and pods. In small seeded varieties pod load on an individual plant is much higher than those of large seeded varieties. Pod development takes about 45–50 days from pollination to seed maturity under sub-tropical growing conditions. During this period both pods and seeds pass through a number of physiological, morphological, and chemical changes. It has been observed that the floral petals wither completely and the ovary starts emerging three days after fertilization. A young pod of about one centimeter long is generally visible after one week. Such pods grow rapidly and reach their full size in about 25 days. During this period of pod growth, the young seeds (ovules) inside pods remain alive and intact but do not gain noticeable size and weight. Soon after achieving the potential pod size, a greater proportion of food reserves of the plant get diverted into the ovules, and soon rapid increases in seed size and weight are observed for the next 10–12 days. From nutritional and marketing view points, it is essential that the green pods are picked at a proper stage to harvest maximum grains with high nutritional quality.

Table 15.3: Trace and mineral elements (mg 100^{-1}g) identified in green seeds of vegetable variety ICP 7035 and *dal* of variety C11.

Element	*Green seeds (ICP 7035)*	*Dal (C 11)*	*SEm*	*Superiority of Vegetable Grains (per cent)*
Phosphorus	264.00*	206.0	± 3.95	28.2
Potassium	1498.00*	1279.0	± 12.74	17.1
Calcium	92.30	114.3*	± 1.98	(-) 19.2
Zinc	3.07*	2.07	± 0.01	48.3
Copper	1.39*	1.15	± 0.08	20.9
Iron	5.16*	4.50	± 0.06	14.7
Manganese	0.99	1.11*	± 0.02	(-) 10.8
Magnesium	108.30	108.5	± 10.86	–

**Source*: Singh *et al.* (1984)

Table 15.4: Major anti-nutritional factors and toxic substances identified in dry pigeonpea seed.

Constituent	*Range*	*Mean*
Protease inhibitors (units mg^{-1})		
Trypsin	8.1–12.10	09.9
Chymotrypsins	2.1–3.60	03.0
Amylase inhibitors (units g^{-1})	22.5–34.20	26.9
Oligosaccharides (100 g^{-1})		
Raffinose	0.24–1.05	0.47
Stachyose	0.35–0.86	0.49
Polyphenols (mg g^{-1})		
Total phenols	3.0–18.30	10.70
Tannins	0.0–0.20	0.03
Phytolectins (units g^{-1})	400	400

**Source*: Singh (1988)

Genetic Enhancement of Vegetable Pigeonpea

Breeding Objectives

Popular vegetable type pigeonpea varieties are characterized by their large pods and seeds. It has been generally observed that in most germplasm accessions these two traits are linked and such lines are invariably photo-sensitive, late maturing (>180 days at 17°N) and perennial in nature. These cultivars flower at the onset of short photoperiods and produce fresh vegetable pods for about 40–50 days, permitting 2–3 pod pickings. To ensure profitable processing and marketing, a regular supply of quality green pods for extended periods is essential. Besides this, the vegetable pigeonpea should have good appearance, taste, and other organoleptic properties. The breeding objectives in a vegetable pigeonpea breeding programme revolve around such traits. Therefore, the major objectives in an ideal vegetable pigeonpea breeding programme are:

- ✰ early podding,
- ✰ round-the-year flowering and pod production,
- ✰ dual purpose varieties,
- ✰ high multiple-harvest pod yield,
- ✰ long green pods with fully grown ovules,
- ✰ non-sticky pod surface with easy shelling,
- ✰ large attractive seeds with good taste, and
- ✰ long shelf-life of pods and seeds.

Vegetable Pigeonpea Germplasm

ICRISAT has a global responsibility for collection, characterization, maintenance, and distribution of pigeonpea germplasm. Currently the ICRISAT Genebank has 13,548 accessions from 76 countries that are available for use in breeding programmes globally. Since long pod size is the most important characteristic of vegetable pigeonpea, the accessions with more than 6 seeds pod^{-1} are considered in this group. A study of 231 long–podded germplasm accessions available in the gene bank showed that a considerable variation exists for important agronomic traits. In this material 50 per cent flowering varied from 80 to 229 days, and maturity duration varied from 133 to 270 days. The plant height ranged from 85 to 285 cm, while pod length varied from 3 to 12 cm. It was also observed that the majority of long-podded accessions originated from the African continent where traditionally large, white seeded cultivars and landraces are grown.

Table 15.5: Variation for some agronomic traits within vegetable type pigeonpea maintained at germplasm bank at ICRISAT, Patancheru, India.

Region of Origin	*No. of Accessions Available*	*No. of days to*		*Plant Height (cm)*	*Seeds Pod^{-1}*	*Pods Plant^{-1}*	*Pod Length (cm)*
		Flower	*Mature*				
Eastern Africa	106	117–229	166–270	130–270	5.4–6.7	26–406	5–12
Southern Africa	17	131–194	163–260	185–260	5.4–6.1	33–154	5–11
Central Africa	4	141–166	215–232	200–230	5.4–5.6	74–130	7–9
Western Africa	13	142–156	194–218	170–250	5.4–5.6	67–246	7–10
Central America	26	106–151	167–202	85–240	5.4–7.2	19–160	7–11
South America	16	132–158	182–230	100–285	5.4–6.1	27–420	5–11
South Asia	39	80–175	133–235	85–230	5.4–7.2	55–830	3–9
South-east Asia	8	134–201	190–264	140–210	5.4–5.9	24–119	5–9
Europe	2	156–174	222–237	210–260	5.4–5.8	137	9
Total	231	80–229	133–270	85–285	5.4–7.2	19–830	3–12

Inheritance of Traits Related to Vegetable Pigeonpea

A good understanding of the genetic systems controlling important qualitative and quantitative characters is essential for breeding of vegetable pigeonpea. Research on this aspect in pigeonpea is limited and fragmentary. For most qualitative characters, there are large differences in the genetic interpretations, and sometimes the observed phenotypic segregation ratios cannot be explained with clarity. Also in most cases, only a few crosses have been studied, which do not generate accurate information on allelic control of important traits. Also, the genetic interpretations are often restricted to the parental materials used in the study and, therefore, any generalization of genetic information is difficult. The presence of both additive and non-additive gene actions for yield and other characters have been reported in literature (Saxena and Sharma, 1990). Important traits for which inheritance should be revealed for vegetable pigeonpea are pod length, seeds pod^{-1}, pod colour, seed colour, seed size, and quality of cooked seeds.

Breeding Methods

Progeny Selection

Vegetable pigeonpea breeding programmes in most countries are predominantly based on selection and purification of native germplasm. The local landraces are generally well adapted to the area but natural out-crossing and mechanical mixtures have resulted in genetically impure varieties. With 25–30 per cent natural out-crossing, the pure lines have been contaminated and become heterogeneous. Breeders generally select individual plants of interest within such materials with due consideration to plant type, pod colour, seed colour etc. One or two branches of such plants are bagged using muslin cloth bags after removing fully opened flowers and young and old pods. An insecticide spray on the selected branches before selfing is always useful to inactivate insect eggs and larvae. At maturity, these selfed branches are harvested separately and their seed is used for evaluation in progeny rows in the subsequent season. Selection should be made among lines and in each selected progeny five plants be bagged again for raising their single plant progenies. In the subsequent generation again 4–5 plants are selfed in each selected progeny. The open-pollinated seeds can be used for evaluation, while the selfed seeds are used as nucleus seed for further multiplication.

Hybridization and Selection

Globally, very little work is being undertaken to breed specifically for vegetable type pigeonpea. However some efforts were made in the West Indies, Dominican Republic, and ICRISAT, Patancheru (India) to breed new varieties that produce vegetable pods early in the season and produce several flushes of flowers and pods. To achieve this, selection of parents for hybridization is the first step towards breeding. Attempts should be made to use pure lines in crosses. Early maturing varieties should be used as female parents. This will help in identifying the selfed plants present in an F_1 population. Selection in F_2 generation should be exercised for pod colour, seed colour, and their size and maturity. These plants can be handled further using classic pedigree selection method.

Choice of Varieties

The most popular vegetable pigeonpea cultivars have long pods and large seeds (weighing about 15 g 100^{-1} seeds when dry), with relatively easy shelling. In Gujarat state of India, the vegetable pigeonpea cultivars with dry seeds weight of about 10–12 g 100^{-1} seeds are popular. These cultivars are grown as a normal field crop, but green pods are harvested at an appropriate stage for use as vegetable. This practice is more prevalent around cities where green pods can readily be marketed at attractive prices. The crop is left for producing dry seeds after harvesting green pods. Such dual purpose varieties are very profitable for peri-urban farmers.

Asia

In India consumers prefer vegetable pigeonpea pods with green colour. Such varieties usually command better prices than the pods with other colours. However, studies at ICRISAT have shown that the differences in pod colour are not related to cooking time, taste, or nutritive value.

Figure 15.3: Seeds and pods of vegetable pigeonpea line ICP 7035.

Sweetness of fully grown green seed is also a preferred trait. Normal sugar content in green pigeonpea seeds are around 5.0 per cent; but researchers at ICRISAT have identified a line (ICP 7035) with a sugar content as high as 8.8 per cent. ICP 7035 is highly resistant to *Fusarium* wilt and sterility mosaic virus disease. Its flowers are dark red that produce purple coloured pods. The seeds of ICP 7035 are large, purple with a mottle pattern; while its pods are 7–8 cm long and, on average, contain six seeds (Figure 15.3). Another cultivar T 15–15, is widely grown in Gujarat state, India for both green and dry seed harvests. In southern India, the large-seeded lines such as HY–3C and TTB–6 are popular as vegetable. In hilly tribal areas of India many large-seeded landraces are traditionally grown for vegetable purpose. Scientists at ICRISAT have also bred an early maturing, determinate variety ICPL 87 which is also used for dual purpose. It produces pods for relatively longer time and allows 2–4 pickings within a crop season.

Africa

Considering the importance of pigeonpea in African agriculture, priority is given to breed high yielding wilt disease resistant late maturing varieties for deep soils, and medium maturing types for drought prone areas. The first early maturing variety ICPL 87091 was released in Kenya, Malawi, Uganda, and Tanzania for vegetable as well as dry seed production. In Eastern and Southern Africa, about 20 per cent of the farmers have adopted new medium maturing pigeonpea varieties such as ICEAP 00554 and 00557 both for grain as well as green vegetable purposes. In Tanzania, about 50 per cent of the farmers in Babati district have adopted new varieties and pigeonpea production area has now extended to the neighbouring districts of Karatu and Mbulu (SN Silim, personal communication). The adoption of a late maturing, *Fusarium* wilt resistant and consumer/market preferred variety ICEAP 00040 in Northern and Central Tanzania, Kenya, and Malawi has resulted in increased grain yields.

Southern and Central America and the Caribbean Regions

The Caribbean region constitutes a chain of island countries extending from Trinidad in the south to Jamaica in the north. In this region Dominican Republic is the longest pigeonpea growing country (17000 ha) with an average yield of 945 kg ha^{-1} (FAO, 2008). The other pigeonpea growing countries are Panama, Venezuela, Jamaica, Trinidad & Tobago, Puerto Rico, and Grenada. Pigeonpea in these countries is essentially a small farmers' enterprise but it is an important crop at national level. Pigeonpea is generally grown as intercrop for consumption as fresh peas.

The first vegetable type variety released in the West Indies was 'Prensado'. It was early in maturity and determinate in growth habit. This variety did not become popular among farmers. Subsequently, three more varieties ('Tobago', St.Augustine', and 'Lasiba') were released, which were similar to traditional types in their phenology and are still under cultivation. In the mid 60s selections from breeding populations were also made which provided varieties with good quality grains and high yield under intercrop situations (Ariyanayagam and Spence, 1978). The important breeding objective for the Caribbean region was to develop varieties which could provide year-round fresh pods for marketing. In Dominican Republic, pigeonpea is mainly grown by small farmers and about 80 per cent of the annual harvest is exported in the form of canned or frozen green peas. According to Mansfield (1981) information on varieties in Dominican Republic is unclear and farmers generally use mixture of varieties such as 'Kaki' and 'Saragateado'. In general four pigeonpea varieties are recognized in Dominican Republic. These are 'Kaki', 'Pinto Villalba', 'UASD', and 'Year-round'. All these varieties have long pods with large and white seeds.

According to Rivas and Rivas (1975) a cultivar called 'Panameno' was released in Venezuela, 1972. In Puerto Rico, the main breeding objectives are to develop high yielding varieties of different maturities and to breed dwarf lines suitable for mechanical harvesting (Abrams *et al.*, 1978). 'Kaki' is the most popular pigeonpea variety in Puerto Rico (Aponte, 1963) and 2B Bushy is another early maturing semi-dwarf variety. The pigeonpea breeding programmes in Puerto Rico and Venezuela have released a few vegetable type varieties such as 'Panameno', 'Amarillo', 'Kaki', 'Saragateado', and 'Totiempo' (Rivas and Rivas, 1975). 'Guerrero' and 'Cortada' were released in Puerto Rico in 2000 and 'Navideño' was released in the Dominican Republic in 2005.

Cultivation of Vegetable Pigeonpea

Pigeonpea is known to be sensitive to photoperiod and temperature which influence the development of plant phenology. Early maturing vegetable type varieties are short in stature and require high plant density (about 300000 plants ha^{-1}) in contrast to late maturing non-determinate types which require 40,000–50,000 plants ha^{-1} for optimum yields. Hence specific agronomic practices are required for different agro-ecological conditions. However, some general guidelines for pigeonpea cultivation are given below:

Backyard and Bund Cultivation

Many families grow pigeonpea plants in their backyards for domestic use (Figure 15.4). Such plants are maintained up to 4–5 years and they attain a height of over 3 m. The plants start flowering at the onset of short days and pods are picked for house-hold use as and when required. Under good soil moisture conditions, new flowers are produced for extended periods and one can see buds, flowers, young, mature, and harvestable pods on the same branch.

Relatively large populations are grown on field bunds, mainly around paddy fields for local market (Figure 15.5). In this system, generally 3–4 seeds are sown in a single hill and the plants produce a large number of branches on either side of the bunds. The green pods are picked manually and sold in market either as whole pods or shelled seeds.

Figure 15.4: A backyard crop of vegetable pigeonpea for domestic use.

Figure 15.5: Green pods are sold in retail market in India.

Peri-urban Commercial Crop

Field Preparation

The field should have a known history of good soil fertility. Since pigeonpea cannot withstand water-logging, low-lying fields should be avoided for vegetable pigeonpea production. Sowings on raised beds with appropriate slope is ideal for pigeonpea to avoid flooding. Application of 100 kg ha^{-1} of di-ammonium phosphate (DAP) and other soil amendments for the known soil deficiencies is advisable.

Sowing

Sowing should be undertaken at the onset of rains. This will ensure good plant growth and canopy development. For early maturing types, the row-to-row spacing of 50 cm at lower latitudes and 60 to 75 cm at higher latitudes, with plant-to-plant spacing of 25 to 30 cm is generally adopted. For medium and late maturing types, the seed should be sown to maintain row-to-row and plant-to-plant spacing of 100 and 50 cm, respectively.

Weed Control

Slow seedling growth of pigeonpea makes it prone to weed competition during the first six weeks of growth. In general, three hand weedings at appropriate intervals are sufficient to control most weeds. Alternatively, one spraying of pre-emergence herbicide such as *Basalin* or *Prometryn*, each @1.5 L ha^{-1}, followed by two hand weedings have been found very effective in managing weeds.

Irrigation

Irrigation is generally not recommended if the crop is grown for domestic consumption on deep Vertisols. However, if the crop is grown for seed purposes, either on light Vertisols or Alfisols, irrigation at the early podding stage is considered beneficial for pod filling and production of new flush of flowers and pods.

Chemical Control of Insects

Pod borers such as *Helicoverpa armigera* and *Maruca vitrata*, and pod fly (*Melanagromyza obtusa*) are major pigeonpea insects. These may cause severe pod damage and affect seed quality. To control the pod borers in pigeonpea, the following insecticides have been found effective at ICRISAT.

Helicoverpa armigera & Maruca vitrata

Indexacarb 70 SC @ *70 ml ai* ha^{-1}

Spinosad 45 SC @ *45 ml ai* ha^{-1}

Endosulfan 35 EC @ *350 ml ai* ha^{-1}

Monocrotophos 36 EC @ *360 ml ai* ha^{-1}

Pod fly (*Melanagromyza obtusa*)

Dimethoate 30 EC @ *300 ml ai* ha^{-1}

Imidacloprid 17.8 SC @ *20 ml ai* ha^{-1}

The first insecticide spray is generally recommended at flower initiation, and the second and third sprays should be done, if required, at 10–15 day intervals. If Knapsack sprayers are used, then 500 L of spray liquid is recommended to cover one hectare of a pigeonpea field. Since vegetable pigeonpea are consumed fresh, it is always advisable to consult a pest management expert before finalizing the spray schedule.

Disease Management

Fusarium wilt and sterility mosaic are two major diseases of vegetable pigeonpea. Wilt is caused by a soil-borne fungus *Fusarium udum* Butler. This pathogen can survive in field for three years or more. Using wilt resistant cultivars reduces risk of damage by wilt.

Sterility mosaic virus is transmitted through eriophide mite (*Aceria cajani*). These virus-carrying mites survive on a number of alternative hosts and pigeonpea stubbles left in fields after harvesting the main crop. The disease management options include: (*i*) grow sterility mosaic resistant cultivars. (*ii*) uproot and destroy infected plants at an early stage of disease development, and (*iii*) spray Metasystox @ 0.1 per cent to control the mite vectors in the early stages of plant growth.

Harvesting of Pods

Green pods are harvested for sale as fresh vegetable in nearby townships and cities. Since fully grown bright green seeds are preferred for vegetable purpose, the pods are harvested just before they start loosing their green colour.

Production and Maintenance of Quality Seed

Maintenance of genetic purity of elite genotypes is essential to ensure uniform crop. Considerable (20–30 per cent) natural out-crossing takes place (Saxena *et al.*, 1990) through insects in pigeonpea, making the seed maintenance more difficult than other pulses. A study conducted by Indian Council of Agricultural Research (1993) revealed that in most cases the farmer-saved seeds were found to be sub-standard with respect to genetic and physical purity. Pigeonpea, being a partially out-crossing crop, requires extra precautions to maintain variety purity.

Commercial Processing of Vegetable Pigeonpea

Commercial vegetable pigeonpea is commonly processed in to canned or frozen peas especially in the Caribbean countries and a few countries of South America. Dominican Republic stands first among countries exporting vegetable pigeonpea to the United States and other countries. The literature on various aspects of processing is scanty, but Mansfield (1981) gives details of vegetable pigeonpea processing technology. A brief review primarily taken from this publication are presented below:

The steps followed in canning and freezing procedures of vegetable pigeonpea are summarized in Figure 15.6.

Step 1. Vining

To maintain freshness of harvested green pods, they should be shelled as quickly as possible. Vining (shelling) of small lots of pods is usually done manually and the shelled peas are generally consumed in local market either as fresh or frozen peas. The bigger lots are used for commercial canning, and for this purpose, vining is performed mechanically.

Step 2. Cleaning of Shelled Peas

For local market the shelled peas are washed and cleaning operation is carried out to remove unwanted peas and inert materials. The mechanically vined peas are cleaned soon after shelling. The seed cleaning operation is performed by passing the shelled peas through an air blast which helps in removing small pieces of pods, dust, etc. The cleaned lot passes through a mesh screen that allows the peas to drop through it but retains large size peas and extraneous materials. Subsequently, the product passes through a fine mesh that retains shelled peas but removes fine dirt and splits. This dry cleaning operation is followed by washing for removing floating dirt, skins, split peas, etc. The washing is

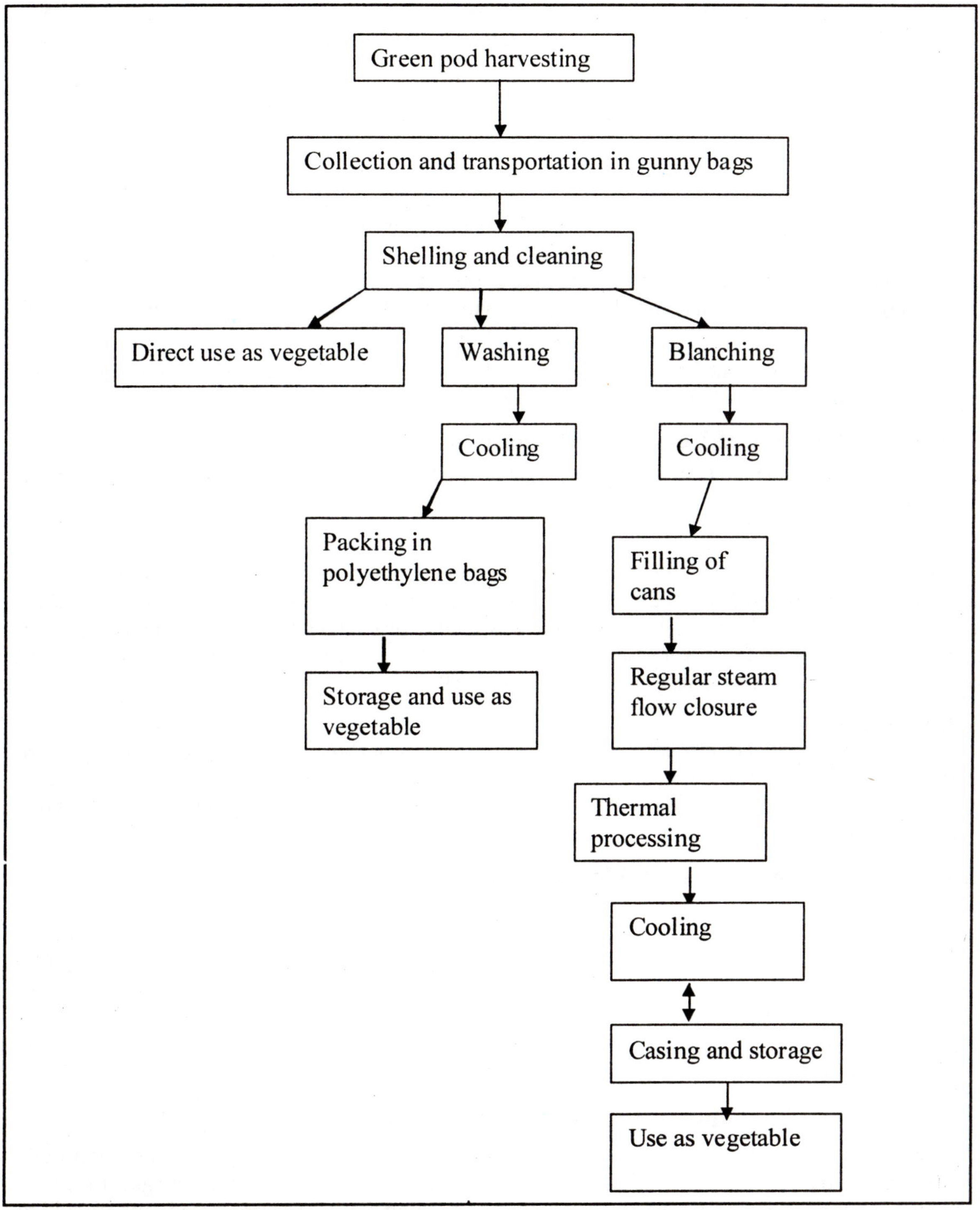

Figure 15.6: Processing flow chart for vegetable pigeonpea.

carried out more than once in various types of flotation washers with cold running water. After washing the shelled peas are forced to pass through rotary rod washers where splits, undersize, and mashed peas are separated. The washed peas fall on a belt where off-coloured remaining worm-damaged and broken peas are removed manually for further processing.

Step 3. Blanching

Heat treatment or blanching is an essential treatment for both freezing as well as canning. This helps in stabilizing colour and flavour besides improving the texture of seeds. According to Mansfield (1981) the blanching operation also helps in producing clear brine by discarding mucous substances, starch particles, and inter-cellular gases. The best blanching is done by heating the peas to 185° F for five minutes in hot water followed by cooling in cold (80°F) water (Sanchez Nieva *et al.*, 1961). Melmick *et al.* (1944) showed that steam is excellent in preserving nutrients of fresh peas but in most cases this process is not cost effective.

After the above mentioned series of treatments, the processed peas could be used either for canning or for freezing. These two follow-up treatments are summarized below:

Step 4(a). Freezing of Vegetable Pigeonpea

According to Mansfield (1981), the following two methods of freezing peas are used in Dominican Republic.

Automated System

In this system the peas are cooled in water at ambient temperature soon after blanching and then taken to fluidized bed freezer. In this freezer, operating between -10° F to -20° F, the peas are quick-frozen individually while moving inside a vibrating conveyor screen which receives a rapid moving current of cold air from the lower side (Mansfield, 1981). The frozen peas are then hand picked and kept in wax treated cartons. These cartons are stored at 0° F.

Batch Freezing System

By this system a blast freezer is used for small quantities of shelled peas. The blanched peas are dropped in cold water tanks and then the peas are hand picked in polyethylene bags and placed for freezing in a batch freezer between –2° F to –10° F for 4 to 10 hours. These packets are stored at 0° F (Mansfield, 1981).

Step 4(b). Canning of Vegetable Pigeonpea

For canning purpose the blanched peas are taken to volumetric filler through an elevator. Here the cans are filled with peas and 2 per cent brine at near-boiling (195–200° F) temperature. No additives are used for canning (Mansfield, 1981). For closing the cans, if near-boiling brine is maintained then the exhaust or steam closure is not adopted. This follows a thermal processing to check the growth of any thermophilic bacterium. After the thermal processing, the cans must be cooled immediately to reduce the thermal quality losses by putting the cans in cool water ponds to bring down their temperature to 90–105° F.

Marketing of Vegetable Pigeonpea

According to Mansfield (1981) the growth for pigeonpea cultivation in the Dominican Republic has been due to the impulse given by canning plants from Puerto Rico. From the farm gate the green pigeonpea pods are collected by the representatives of canning plants. The processed cans are sold to wholesalers for export to the United States, Puerto Rico and other Latin American countries. In India marketing of vegetable pigeonpea is not well established. Vegetable market local vendors (Fig. 5) buy the product from farmers for retail marketing.

Conclusions

The importance of vegetables in human diet can not be under emphasized. Vegetable pigeonpea can be good source of valuable proteins, vitamins, carbohydrates, and dietary fibre for humans. Vegetable pigeonpea compliments the nutritional profile of cereals, and is a good source of protein, vitamins (A, C and B complex) and minerals (Ca, Fe, Zn, Cu). Vegetable pigeonpea scores manifold advantages over green peas (*Pisum sativum*). It has more than five times beta carotene content, three times more Thiamine (Vitamin B_1), Riboflavin (Vitamin B_2), and Niacin. The ascorbic acid content is more than two times over peas. Similarly, it scores over peas in terms of minerals such as calcium and copper (more than two times higher), and magnesium. Besides all this, the shelling percentage of vegetable pigeonpea is 72 per cent compared to 53 per cent of green peas. All these factors render vegetable pigeonpea a highly nutritive potential crop for all ages. It can become one of the most nutritionally rich vegetables of the daily cuisine, especially for the poor in India, Nepal and Myanmar. It is already a vegetable of choice for Kenya, Tanzania, Malawi, Uganda, and the Caribbean.

References

Abrams, R., Morales, A. and Julia, F.J., 1978. Status of research on pigeonpea in Puerto Rico. *Tropical Grain Legume Bulletin*, 11 and 12: 17–21.

Aponte, A.F., 1963. El cultivo de gandulus en Puerto Rico. *Caribbean Agriculture*, 197: 7.

Ariyanayagam, R.P. and Spence, J.A., 1978. A possible gene source for early day-length neutral pigeonpeas, *Cajanus cajan*. *Euphytica*, 27: 505–509.

De, D.N., 1974. *Pigeonpea*. In: *Evolutionary Studies in World Crops: Diversity and Change in the Indian Subcontinent*, (Ed.) J. Hutchinson. Cambridge Press, London, pp. 79–87.

Duke, J.A., 1981. *Handbook of Legumes of World Economic Importance*. Plenum Press, New York.

FAO Stat., 2008. http: / /www.faostat.fao.org.

ICAR, 1993. Report of quinquennial review team on technology research and breeder seed production under the national seed project. New Delhi, India, Indian Council of Agricultural Research (ICAR).

Kamath, M.V. and Belavady, B., 1980. Unavailable carbohydrates of commonly consumed Indian foods. *Journal of Science of Food and Agriculture*, 31: 194–202.

Linnaeus, C., 1753. *Species Plantarum*. 739 (cited by van der Maesen 1986).

Mansfield, G.M., 1981. Processing and marketing of green pigeonpea: the case of the Dominican Republic. *In* Proceedings of the International Workshop on Pigeonpeas, volume 2, 15–19 December 1980, ICRISAT Center, India. Patancheru 502 324, Andhra Pradesh, India: International Crops Research Institute for the Semi–Arid Tropics 1: 344–350.

Melmick, E., Hachberg, M. and Oser, B.L., 1944. Comparative study of steam and hot water blanching. *Food Research*, 9: 148–153.

Plukenet, L., 1692. *Phytographia* 3, table 213, figure 3. (cited by van der Maesen 1986).

Rivas, N. and Rivas, E.G., 1975. Estudio de la calidad para enlatado de la variedad de quinchonchos (*Cajanus cajan* (L.) Millsp.) "Panameño", Rav Fac Agron (Maracay) 83(3): 77–81.

Sanchez Nieva, F., Rodrigues, A.J. and Benero, J.R., 1961. Improved methods of canning pigeonpeas. University of Puerto Rico Agricultural Experiment Station Bulletin 157, Mayaguez, Puerto Rico.

Saxena, K.B., 2008. Genetic improvement of pigeonpea: A review. *Tropical Plant Biol.*, 1: 159–178.

Saxena, K.B., Singh, L. and Gupta, M.D., 1990. Variation for natural out-crossing in pigeonpea. *Euphytica*, 46: 143-148.

Saxena, K.B., Singh, U. and Faris, D.G., 1983. Does pod colour affect the organoleptic qualities of vegetable pigeonpeas? International Pigeonpea Newsletter 2: 74–75.

Saxena, K.B. and Sharma, D., 1990. Pigeonpea Genetics. Pages 137–158 In: *The Pigeonpea*, (Eds.) Y.L. Nene, S.D. Hall and V.K. Sheila. CAB International, Wallingford, U.K.

Singh, U., Jain, K.C., Jambinathan, R. and Faris, D.G., 1984. Nutritional quality of vegetable pigeonpea (*Cajanus cajan*). Mineral and trace elements. *Journal of Food Science*, 49: 645–646.

Singh, U., 1988. Anti-nutritional factors of chickpea and pigeonpea and their removal by processing. *Plant and Foods Human Nutrition*, 38: 251–261.

Turnbull, L.V., Whiteman, P.C., and Byth, D.E., 1981. The influence of temperature and photoperiod on floral development of early flowering pigeonpea. In: *International Workshop on Pigeonpea*, Vol. 2, 15–19 December 1980, Patancheru, AP. India, pp 217–222.

van der Maesen, L.J.G., 1980. India is the native home of pigeonpea. Pages 257–262. In: *Libergratulatorius in Honorem*, (Eds.) H.C.D. de Wit, J.C. Arends, G. Boelema, C.T. de Groot and A.J.M. Leeuwenberg. Agricultural University. Miscellaneous Paper, 19, Wageningen, the Netherlands.

van der Maesen, L.J.G., 1986. *Cajanus DC.* and *Atylosia W.&A.* (*Leguminosae*). Agricultural University Wageningen Papers 85–4 (1985). Wageningen, the Netherlands: Agricultural University. 225 pp.

Vavilov, N.I., 1939. The new systematics of cultivated plants. Pages 549–566 In: *The New Systematics.* (Ed.) J. Huxley. Oxford University Press, London.

Wallis, E.S., Byth, D.E. and Saxena, K.B., 1981. Flowering responses of thirty–seven early maturing lines of pigeonpea. Pages 143–150 in International Workshop on Pigeonpeas, Vol 2, 15–19 December 1980, ICRISAT, Patancheru, Andhra Pradesh, India. Patancheru 502 324, Andhra Pradesh, India: International Crops Research Institute for the Semi–Arid Tropics.

Yadavendra, J.P. and Patel, A.R., 1983. Consumer preference of vegetable pigeonpea cultivars. *International Pigeonpea Newsletter* 2: 73–74.

Index

(+)-3, 3′-bisdemethyltanegool 157

1, 4-dihydroxy-2-methoxy-7-methylanthraquinone 157

2-methoxy-1, 3, 6-trihydroxy-anthraquinone 157

3, 3'-bisdemethylpinoresinol (xix) 158

β-carotene 44

A

A. caudatus 41

A. cruentus 41

A. hypochondriacus 41

Abuj 252

Acalles barbarus 73

Acrocercops cramerella 191

Aegle marmelos (L) Corr. 222, 226, 232

Aerial Blight in Mints 214

Aesculetin 156

Agronomical practices 212

AICRP on AZF 23

AIDS prevention 99

Albugo capparidis 73

Alectoria ochroleuca 95

Alexipharmac 69

Alliaceae 253

Allium cepa var. *proliferum* 253

Alternative approaches 102

Amarameal 46

Amaranth 44

Amaranth weevil 50

Amaranthaceae 41, 249

Amaranthus spinosus 249

Amaranthus spp. 40

Amarlac 46

American 215

Americanin A 156

Americanol A 158

Amino acids 154

Analgesic activity 162

Anaptychia ciliaris 95

Antarctic plants 101

Anthesis 59

Anti-bacterial activity 160

Anti-cancer Activity 164

Anti-fungal activity 160

Anti-helmintic Activity 161

Anti-herbivorous properties 93

Anti-inflammatory actions 99

Anti-inflammatory activity 162

Anti-obesity 161

Anti-oxidant activity 161

Anti-viral activity 161

Ants 60

Anxiolytic activity 162

Aphrodisiac 69

Applications and uses 46

Aquilaria malaccensis Lamk. 222, 232

Ardisia polycephala Wall. ex A.DC. 222, 226, 232

Area and distribution 79

Area and production 23

Area under cultivation in India 211

Arecanut + Noni system 154

Arka (MASS-98) 213

Aromatic plants 218

Aschochyta capparidis 73

Asexual/Vegetative Propagation 124
Asia 268
Asians 115
Aspergillus flavus 72
Aspergillus niger 129
Asperulosidic acid 158
Asphondylia capparis 73
Aspicilia esculenta 95, 97
Associated insects 72
Asteraceae 79, 244, 256
Asthma 69
Australia 109, 110, 124
Automated system 274
Averrhoa carambola Linn. 222, 226, 232
Aztec people 42

B

Bacillus subtilis 160
Backyard and bund cultivation 269
Bactrocera dorsalis Hendel 191
Baidum 123
Barhee 24, 26
Bark 57
Barley 44
Batch freezing system 274
Batu 123
Begonia inflata Cl. 222, 226, 232
Bergenia ciliata (Haw.) 226
Bergenia ciliata (Haw.) Sternb. 222, 232
Biew Khiew 114, 122, 136
Biochemical markers 9
Biodiversity conservation 71
Biological activities 160
Biology of lichens 100
Biophysical limits 56
Biopsesticides 143
Biotechnology 119, 183
Bird management 36
Blanching 274
Blumea lanceolaria (Roxb.) Druce 222, 226, 232
Bolting 86
Bombax ceiba L. 222, 226, 233
Botanical description 25
Botanical name 244, 246, 248, 249, 250, 252
Botany 80, 203, 211
Botany and reproductive biology 5
Botany and taxonomy 110, 173
Botany, taxonomy and physiology 40
Botany: Tree characteristics 57
Botrytis–gray mold 73
Bottom rot 87
BR-4 185
BR-7 186
Braim 24, 27
Breeding 47
Breeding and genetic improvement 117, 181
Breeding methods 7, 118, 182, 267
Breeding objectives 6, 117, 181, 265
Brown stain 86
Bryoria fremontii 93, 94
Bryoria spp. 95, 97
Buckwheat 44
Bud Sport Selection 119
Butterhead 81, 82

C

C. nivalis 96
C. pyxidata 96
C_4 plant 41
CABI 71
Cajanus cajan (L) Millspaugh 260
Calligonum polygonoides 64
Calocoris memoralis 73
Calories 136
Camarosporium suseganense 73
Cambodia 109, 114
Canning of vegetable pigeonpea 274
Canopy management 127, 189
Capparaceae 55
Capparis decidua (Forsk.) Edgew. 55, 62, 64, 72
Carbohydrates 136
Carbon tetrachloride (CCl_4) 161
Cardiovascular activity 163
Carminative 69
Cassia alata L. 222, 226, 233
Cassia hirsuta L. 222, 226, 233
Cautleya gracillis (Sm.) Dandy 222, 227, 233
CAZRI 37
Centaurea 256
Centaurea iberica 256
Central Institute for Arid Horticulture, Bikaner 30
Cercospora capparidis 73
Cetraria ericetorum 94
Cetraria islandica 94, 95, 96, 98
Chayote 1
Chemical changes in developing seeds 264
Chemical composition 3, 61, 214
Chemical control of insects 271
Chilling injury 128, 190
China 109, 120
Chinese 115

Chingko 122
Chip chap 24, 27
Choice of varieties 267
Chompoo 114, 122
Chonemorpha fragrans (Monn.) Als. 222, 233
Chromosome number 256
Chuliang 120
Cladina rangiferina 93, 95, 98
Cladina spp. 95, 98
Cladina subtenuis 101
Cladonia 99
Cladonia coccifera 96
Cladosporium sp. 129
Cleaning of shelled peas 272
Clerodendrum wallichii Merr. 222, 227, 233
Cletus sp. 50
Climate 28
Climate and soil 4, 112, 177
Cloeosporium hians 73
Cluster bean 12
Coccotrypes dactyliperda F. 35
Coccus sp. 192
Commercial processing of longan fruits 135
Commercial processing of rambutan fruits 198
Commercial processing of vegetable pigeonpea 272
Compositae 80
Composition and uses 115, 180
Conotrachelus seniculus 50
Constraints 73
Consumer unawareness 134
Control 129, 130, 131, 190, 191, 192, 193
Controlled atmosphere 134
Copper 136
Cough 69
COX-1 162
Crisphead 81, 82
Crocus sativus L. 203
Crop improvement 47
Crop protection 50
Cucumis sp. 62
Culinary uses 116
Cultivars 17, 26
Cultivars and varieties 47
Cultivation 49, 187
Cultivation and production 66
Cultivation of longan 125
Cultural practices 83
Curcuma zedoaria (Christ) Rosc. 222, 227, 233
Cuttings 125
Cyamopsis tetragonoloba L. 12
Cyclea peltata (Lamk.) Hook. f. & Thoms 222, 227, 234
Cycloartenol 156
Cydia capparidana 73
Cylindrocladiella disease (*Cylindrocladiella peruviana*) 131
Cystenine 44
Cytogenetics 6

D

Dacus dorsalis Henderk 191
Dalbergia pinnata var. *acaciaefolia* (Dalz.) Thoth. 222, 234
Damnacanthal 156, 164
Damping off 50
Dang 114, 123
Date palm 22
Date stone beetle 35
Daw 122, 136
Dehydrated longan nuts 136
Descriptor 8
Desmos longiflorus (Roxb.) Saf. 223, 234
Determinate type 262
Diabetes 70
Diacrisia oblique Walk 214
Dietary fiber 136
Dimocarpous longan Lour 109
Dimocarpus australianus 110
Dimocarpus crinita 172
Dimocarpus dentatus 110
Dimocarpus foveolatus 110
Dimocarpus fumatus 110
Dimocarpus gardneri 110
Dimocarpus longan 110
Diosmetin 215
Diosmin 215
Disease management 272
Disease of mints 213
Diseases 34, 87, 130, 192
Diseases and pests 72
Diversity and genetic resources 8
DNA polymerase activity 99
DNA sequences 104
Doctrine of signatures 99
Doka fruits 37
Dolabra nepheliae 193
Downy mildew 87
Dried fruit beetle 35
Drosophila species 159
Dudusa nobilis 192

E

E-Phytol 156
Ecology and distribution 56, 101
Ecology of lichens 101
Ecology of saffron 204
Economics 210

Ecosystem 71
Eczema 70
Egyptian 215
Elaeagnus pyriformis Hook.f. 223, 234
Elsholtzia blanda (Benth.) Benth. 223, 227, 234
Emmenagogue 69
Environmental adaptation of the crop 41
Erosion control 71
Erythricium salmonicolor 192
Escherichia coli 160
Estrogenic activity 163
Ethnobotanical expeditions 71
Euphoria longan 109
Euphoria nephalium 172
Eurydema ventralis 73
Evernia 99
Evernia prunastri 94, 96
Everniastrum cirrhata 95

F

Family 244, 246, 249, 250, 252
FAO 43
Fatty acids 45, 159
Febrifuge 116
Fengko 121
Fertilizer application 49
Fertilizers 212
Field preparation 49, 271
Flavocetraria cucullata 94
Flemingia macrophylla (Willd.) Prain. 223, 234
Florida 123
Floristic diversity 220
Flower 5
Flower panicle 175
Flowering and fruiting 58
Flowers 5, 57, 112, 174
Fodder 68
Foliation 62
Fomes lignosus 193
Food 68
Food value 180
Forket budding 125
Freezing of vegetable pigeonpea 274
Fruit and seed 6
Fruit blotch (*Phomopsis longanae*) 130
Fruit cracking 129
Fruit growth and development 33
Fruit ripening 34
Fruit rot 192
Fruit rot (*Phytophthora palmivora*) 130
Fruit set 176
Fruit thinning 34
Fruit yield 132
Fruiting 60
Fruits 57, 112, 177
Fuel 69
Fuhko2 114
Fungi 73
Fusarium sp. 129, 161
Future breeding needs 9
Future food 93
Future research areas 74
Fuyan 121

G

Garcinia lanceaefolia (G. Don) Roxb. 223, 234
Garcinia pedunculata G. 223, 235
Garcinia sopsopia Ham. 223, 235
Gardenia coronaria Ham. 223, 235
Garuda 184
GB Pant University of Agriculture and Technology, Ranichauri, 47
Gelsemium elegans Benth. 223, 235
General characteristics of selection 119
General description 245, 246, 248, 249, 251, 252, 253, 254, 257
Genetic enhancement of vegetable pigeonpea 265
Genetic resources and conservation 8
Genetic resources and crop improvement 25
Genetic variability 48
Genetics 211
Geographic distribution 56
Germplasm collection 47
Global and the Indian scenario 101
Glycine 44
Good fruit quality 127
Grading 194
Grafting 125
Grafting and budding 187
Grain amaranth 40
Graphiola leaf spot 34
Green leaf 82
Growth and development 262
Growth habit 262
Guar scenario in Western Rajasthan 13
Gynocardia odorata R. Br. 223, 235

H

Habitat 246, 250, 251, 253, 256, 258
Haew 114, 123

Hairy caterpillar 214
Halawy 24, 26
Halwa 46
Handh 244
Haptoncus luteolus Ericl. 35
Harvest and yield 50
Harvesting 36, 66, 85, 132
Harvesting and fruit yield 193
Harvesting and threshing 20
Harvesting of pods 272
Hedychium coronarium Koenig. 223, 235
Hendersonia rupestris 73
Hepatoprotective Activity 161
Heterodermia 93
Hiccough 69
High perishability 134
High yielding new varieties 212
Higher CO_2 atmosphere 134
Histidine 44
Historical background 210
History of cultivation 56
Hongko 122
Horticultural treatments 126, 189
Horticultural types 81
Hungarian 215
Hybridization 119, 183
Hydronocarpus kurzii (King) Warp. 223, 236
Hymenis recurvalis 50
Hypoglycemic effects 161
Hypogymnia physodes 94
Hypotensive activity 163

I

IBPGR 69
ICAR North Eastern Hill Complex, Shillong 47
Immunological activity 163
Important attributes of vegetable pigeonpea 262
In vitro conservation 8
In vitro lichen cultures 103
Indian mulberry 143
Indo-China 173
Indonesia 109, 114, 123
Insect pests 86, 129, 190, 214
Insect pests and diseases 190, 129
Insects 35
Inter-cropping 32, 189
Intercultural operations 85
Interculture 50
International prices of spearmint oil 211
Introduction 25, 118, 182
Iridaceae 203
Iron 44, 136
Irrigation 19, 31, 85, 205, 212, 271
Isolated compounds 147
Isoleucine 44
Isoprincepin 158
Isoscopoletin 156
Israel 109
Italian 215
Ixora nigricans R.Br. 223, 236

J

Jhuming/shifting cultivation 221

K

Kachri 62
Kaempferol 158
KAIR 55
Kalanchoe pinnata (Lamk.) Pers. 223, 236
Kathlitchi 109
Kay sweeney 114
Khadrawy 27
Khajoor 22
Khalas 24, 26
Kharek 22
Kheer 46
Khejri 62
Khuneizi 24, 26
Kohala 114
Kopyor 123
Kretch 256

L

Lack of infrastructural facilities 134
Lack of skilled or trained persons 134
Lactuca sativa L. 79
Laos 109, 114
Laphygma exigua Hub 214
Lasallia spp. 94
Lasiodiplodia theobromae 129
Lasiurus scindicus 64
Latin 82
Leaf 81
Leaf blight 50
Leaf eating caterpillar 192
Leaf Rollers (*Adoxophyes cyrtosema* Meyrick) 130
Leaves 112, 174
Lebakbooloos 184
Lepionurus sylvestris Blume 223, 236
Leptosphaeria capparidis 73
Lettuce 79
Lettuce big-vein 88
Lettuce drop 88
Lettuce mosaic virus 87

Leucine 44
Lichen as extremophile 100
Lichen diversity 101
Lichen products 104
Lichens 92
Lichens as medicines 96
Limonene: p-mentha-1,8-diene; cinene 215
Lindernia ruelloides (Colsm.) Penn. 224, 236
Lissa 249
Litchi fruit and shoot borer 191
Litchi fruit borer (*Conopomorpha sinensis* Bradley) 129
Lobaria 99
Lobaria amplissima 94
Lobaria pulmonaria 95, 96, 99
Lobaria scrobiculata 94
Lobelia angulata Forst. 224, 236
Lobelia nictianaefolia Heyne. 224, 236
Longan 109
Longan genetic resources 117
Longan witches' broom 131
Longhungyen 122
Longnha 122
Lotus stem 254
Lucidin 3-O-β-D-primeveroside 157
Lumbago 69
Lumut 123
Lygus bug 50
Lygus lineolarius 50
Lysine 44

M

Magnesium 136
Maize 44
Malaysia 109, 114
Mallow 246
Malva sylvestris 246
Malvaceae 246
Manganese 136
Manna lichen 93
Manure and fertilizer requirements 84
Manures and fertilizers 31
Marcotting or air-layering 186
Marketing of vegetable pigeonpea 274
Maturity indices 67
Mayanmar 109
Mealybugs 191
Medicinal applications 206
Medicinal uses 69, 116, 181
Medjool 26
Mentha spicata L. 209
Methionine 44
Micro-nutrient Analysis 144
Microflora 72
Micropropagation 66
Mint leaf roller 214
Mint oil 214
Mint rust 214
Mites 191
Mizoram 218
Molecular approaches 103
Monocotyledonous plant 25
Morinda citrifolia L. 143
Morindolin 158
Morindone 158
Morindone-6-O-β-D-primeveroside 157
Morphological 264
Morphological attributes 111, 174
Morphological characters 211
Morus australis Poir. 224, 237
MSS-1 213
MSS-5 213
Mulching 31, 126, 189
Musa glauca Roxb. 224, 237
Mutant clones 212
Myanmar 114, 173

N

Nadru 254
Narcissoside 157
National Academy of Sciences, USA 40
National guar scenario 14
Natural fats 136
Natural habitat 56
Natural variability 63
NBPGR 47
Neera 213
Nelumbo nucifera 254, 256
Nelumbonaceae 254
Nephelium lappaceum L. 172
Nephroma arcticum 94
Niacin 136
Nitric oxide 162
NO* 162
Non-climacteric 132
Non-determinate type 262
Noni 143
Noni juice 143
Nunar 248
Nursery 49
Nutrient management 19
Nutrition and irrigation 187
Nutritional composition 115
Nutritional importance 82
Nutritional value 3, 42
Nutritional value of Noni 144

Nutritive value 96
Nutritive value and uses 24

O

Oats 44
Ocimum tenuiflorum L. 224, 237
Odium nephelii 192
Off-season longan 126
Offshoots 29
Oil 69
Old varieties 19
Organic farming 206
Oriental fruit fly 191
Origin and distribution 4, 22
Origin and domestication 80
Origin and history 42
Origin and spread 110, 173
Origin and taxonomy 261
Origin of saffron 203
Other pests 130, 192
Other uses 117, 181

P

P. horizontalis 96
P. polydactyla 96
Packaging 194
Pajanelia longifolia (Willd.) K. Schum. 224, 237
Palmae 25
Pambh hawk 250
Parmelia 93, 99
Parmelia abessinica 95
Parmelia cirrhata 97
Parmelia quercina 97
Parmelia sulcata 99
Parmelia tinctorum 97
Parmotrema tinctorum 95
Pathology 213
Peltigera 99
Peltigera canina 95, 96, 97, 98, 99
Penicillium nigricans 72
Penicillium sp. 161
Pentapetes phoenicea L. 224, 237
Peri-urban commercial crop 271
Pericarp browning 128
Pertusaria 99
Pestalotiopsis sp. 129
Pests and diseases 6
Phenolics 155
Phenylalanine 44
Phetsakon 123
Phoenix dactylifera L. 22
Phoenix genus 25
Phoma capparidina 73
Phoma capparidis 73
Phosphorus 136
Phyllanthus airy-shawii Brunal & Roux. 224, 237
Phyllanthus fraternus Webs. 224, 238
Phyllosticta capparidis 73
Physcia 93, 99
Physiography of Mizoram 219
Physiological disorders 86, 128, 189
Phytophthora botryosa 192
Phytophthora Leaf Blight 130, 192
Pinju 62
Pink disease 192
Planococcus citri 191
Planococcus lilacinus Cockerell 191
Plant growth regulators 126, 189
Plant habit 5
Plant protection 19
Planting 30, 49, 125, 187, 204
Pod and seed size 263
Pod colour 263
Pollination 5, 33, 59
Pollination and fruit set 177
Polygonaceae 250, 252
Polynesian herbal medicinal plant 143
Poor filling of fruit 190
Portulaca 248
Portulaca oleracea L. 248
Portulacaceae 248
Postharvest maintenance 194
Postharvest maintenance of quality fruit 132
Postharvest management 36, 85
Postharvest technology 67
Potassium 136
Powdery mildew 192, 213, 214
PPO enzymatic activity 133
Pran 253
Processing 37
Production and maintenance of quality seed 272
Production and productivity 113, 179
Production in India 211
Production technology for higher yield 16
Productivity constraints 51
Progeny selection 118, 182, 267
Promising types 48
Propagating materials 204
Propagation 28, 124, 186
Propagation and Management 65
Prosopis cineraria 62
Protein 136

Proteus morgaii 160
Proximate composition 144
Pruning and training 127, 189
Pseudevernia furfuracea 94
Pseudomonas aeruginosa 160
Punjab spearmint 213
Pyralid 214
Pyxine 93

Q

Quality parameters 263
Quality traits 263
Queensland 110
Quercetin 156

R

Ramalina farinacea 95
Ramalina fastigiata 95
Ramalina fraxinea 95
Ramalina sinensis 97
Rambutan 172
Rambutan genetic resources 182
Rambutan stink bugs 190
Rapiah 184
Rauvolfia serpentina Benth. 224, 238
Reasons of low productivity 15
Red palm weevil 35
Released varieties 18
Research centres 47
Retarded fruits 190
Rhaphidophora hookeri Schott. 224, 238
Rheum rhabarbarum 250
Rheumatism 69, 70
Rhizopus spp. 161
Rhubarb 250
Rhynchophorus ferrugineus 35
Riboflavin 136
Rice 44
Ricinus communis Linn. 224, 238
Roccella 99
Roccella montagnei 97
Role of underutilized vegetables 244
Romaine 81, 82
Rong Rian 185
Rootstocks for budding and grafting 186
Rubia cordifolia L. 224, 238
Ruellia suffruticosa Roxb. 224, 238
Rumex 252
Rumex obtusifolius 252
Rust 213

S

Saffron 203, 206
Saffron production 204
Salmonella 160
Sapindaceae 110, 173
Saraca asoca (Roxb.) de Wilde 224, 238
Scale insects 35, 192
Sechium edule (Jacq.) Swartz. 1
Sectangkooweh 185
Section frantzia 5
Section sechium 4
Seed and sowing 16
Seed composition 159
Seed dispersal 63
Seed/sexual propagation 124, 186
Seeds 28, 57, 112
Seelengkeng 185
Seematjan 185
Seenjonja 185
Selection 26, 118, 182
Senecio scandens Buch-Ham. Ex D. Don 225, 238
Sepals 57
Septoria capparidis 73
Serine 44
Sewi 27
Shade or shelter 71
Shamran 27
Shigela 160
Shixia 121
SiChomphu 185
SiThong 185
Skin darkening 128
Skin splitting 189
Smilax glabra Roxb. 225, 238
SO_2 treatment 133
Soft dates 37
Soft rot 72
Soil 28, 204, 212
Soil and climatic requirements 16, 83
Soils 56
Solanum khasianum var. *Chatterjeeanum* Sen Gupta 239
Sonchal 246
Sorting 194
Sorting, grading and packaging 133
Southern and Central America and the Caribbean Regions 268
Southern China 110
Sowing 271
Sowing and transplanting requirement 84
Soybean 44
Spathe emergence and flowering 32
Spearmint 209

Spinach flea beetle 50
Spiny amaranth 249
Staphylococcus aureus 160
State released varieties 19
Status of MAPs in Mizoram 220
Stem 82
Stem canker 193
Stemona tuberosa Lour. 225, 239
Stephania japonica var *discolor* (Miq.) Forman. 225, 239
Sticta glomulerifera 94
Stigmasterol 156
Stink bugs [*Tessaratoma papillosa* (Drury)] 129
Stolon rot 213
Stomachache 70
Stomachic 116
Storage 134, 197
Suckers 29
Sweet porridge 46
Syngamia abruptalis Walk 214
Synonyms 1
System of pre-harvest 134

T

Taiwan Province of China 121
Talub Nak 123
Taraktegenos kurzii (King) Warb 225, 239
Taraxacum (Dandelion) 244
Taraxacum officinale 244, 245
Tarnished plant bug 50
Taxonomy 4
Termites 35
Tessaratoma javanica Thnb. 190
Tetranychus cinnabarius Boisd. 191
Thailand 109, 122, 173
Thiamin 136
Thinning of fruit panicles 127
Threatened medicinal 218
Threonine 44
Thunbergia grandiflora Roxb. 225, 239
Tieuhue 122
Timber 69
Tinospora cordifolia (DC.) Miers. Ex Hook. f. & Thoms. 225, 239
Tip Burn 86
Tissue culture plants 30
Tissue culture techniques 29
Toiletries 143
Tonic 69
Toothache 70
Top onion 253
Topset onions 253
Traditional uses of lichens 93
Training and pruning 31
Transportation and marketing 133, 196
Trapa natans var. *Bispinosa* (Roxb.) Makino 225, 239
Tree 111, 174
Tyrosine 44

U

Umbilicaria esculenta 94
Umbilicaria muehlenbergii 94
Umbilicaria spp. 94
Underutilized and future vegetables 243
Underutilized vegetables 253
Underutilized vegetables of Kashmir 243
Untapped resources 96
USA 109
USDA-NASS 79
Uses 68, 210, 246, 250, 251, 253, 256, 258
Uses and properties 1
Uses in classical codified literature 70
Usnea 99
Usnea florida 95
Usnea longissima 97
Usnea orientalis 97
Usnea plicata 96
Usnea stirtoniana 97

V

Valine 44
Varieties 120, 183
Varieties and hybrids 9
Vegetable pigeonpea 260, 269
Vegetable pigeonpea germplasm 266
Vegetative propagation 65, 186
Vermifuge 116
Vernacular name 244, 246, 249, 250, 252
Vietnam 109, 122
Vining 272
Vitamin B_1 136
Vitamin B_2 136
Vitamin B_3 136
Vitamin C 136
Vitamins 154
Vivekananda Parvatiya Krishi Anusandhanshala (VPKAS) 47

W

Weed control 19, 126, 188, 271
Weed management 32
Weeds and herbicides 212
West Malaysia 173

Wheat 44
White rust 50
WHO 43
Withholding period (WHP) 133
Woodfordia fructicosa Kurz. 225, 240
World production 211
Wound healing activity 162
Wulongling 121
Wuyuan 121

X

Xanthoria 99
Xanthoria parietina 96

Y

Yield 36, 85

Z

Zagloul 27
Zahidi 27
Zanonia indica L. 225, 240
Zanthoxylum armatum DC. 225, 240
Zingiber purpureum Rosc. 225, 240